AF581805

# Emerging Non-Thermal Food Processing Technologies

# Emerging Non-Thermal Food Processing Technologies

Editors

**Asgar Farahnaky**
**Mahsa Majzoobi**
**Mohsen Gavahian**

Basel • Beijing • Wuhan • Barcelona • Belgrade • Novi Sad • Cluj • Manchester

*Editors*
Asgar Farahnaky
RMIT University
Melbourne
Australia

Mahsa Majzoobi
RMIT University
Melbourne
Australia

Mohsen Gavahian
National Pingtung University of Science and Technology
Pingtung
Taiwan

*Editorial Office*
MDPI
St. Alban-Anlage 66
4052 Basel, Switzerland

This is a reprint of articles from the Special Issue published online in the open access journal *Foods* (ISSN 2304-8158) (available at: http://www.mdpi.com).

For citation purposes, cite each article independently as indicated on the article page online and as indicated below:

| Lastname, A.A.; Lastname, B.B. Article Title. *Journal Name* **Year**, *Volume Number*, Page Range. |
|---|

**ISBN 978-3-0365-8934-3 (Hbk)**
**ISBN 978-3-0365-8935-0 (PDF)**
**doi.org/10.3390/books978-3-0365-8935-0**

# Contents

# About the Editors

**Asgar Farahnaky**

Prof Asgar Farahnaky is a highly motivated and experienced university academic/researcher/mentor with demonstrated outcomes. His research philosophy is to carry out innovative fundamental and applied research to solve technological problems and create new windows for in-depth understanding of food systems. He received a PhD in 2002 in Food Physics from The University of Nottingham (UK) after completing a BSc and MSc. He was then employed by the University of Nottingham as a Research Fellow (2002–2005) and from 2006 to 2013 worked for Shiraz University as a senior academic, led the Department of Food Sci and Tech for 5 years and received several research productivity awards. In 2014, he moved to Australia and led the Food Science and Nutrition Discipline Group of Charles Sturt University with major contributions to teaching and research. From 2014 to 2018, he secured a number of external research grants with colleagues and was the Discipline Leader for Food Processing in the ARC Industrial Transformation Training Centre for Functional Grains supervising researchers and leading industrial projects on processing and value addition to food. Upon accepting a new position, he moved to RMIT University in 2019 and has led large research projects.

**Mahsa Majzoobi**

Dr. Mahsa Majzoobi is a Senior Research Fellow and Senior Lecturer in Food Sciences and Technology at RMIT University, Australia. Mahsa holds B.Sc., MS.c. and Ph.D. degrees in Food Science and Technology. After graduating from the University of Nottingham in England, her active academic journey has marked as an academic member of international universities and research institutes. She has undertaken key leader and research roles and has been Editorial Board Member, guest editor and reviewer of well-recognized journals. Mahsa's research addresses the major challenges within the food industry including food security and sustainability, waste reduction, environmental impact of food processing, promotion of healthy foods, and reinforcement of clean labelling and green food processing. She is acknowledged among the top 2% of global food scientists. Additionally, she has been the recipient of various research awards, notably the "RMIT Vice-Chancellor's Fellowship" and the "Women in STEM award". She is a prolific author with numerous publications in scientific journals and books.

**Mohsen Gavahian**

Dr. Mohsen Gavahian is a highly accomplished researcher with expertise in food science and technology. He holds B.Sc., M.Sc., and Ph.D. degrees in the field and has dedicated his career to exploring emerging food processing technologies. He has established a reputation for developing innovative, energy-saving systems to valorize agri-food waste and promote sustainability. Dr. Gavahian has successfully led academic and industrial projects and has received numerous accolades for his outstanding contributions to the field. He received recognition as "IUFoST Young Scientist" and "world's top 2% scientists". Additionally, he has contributed to the field through his editorial work, serving as Editorial Board Member for several international journals. In addition, Dr. Gavahian is a prolific author, with over one hundred well-cited scientific papers published in leading journals. He is also the editor of "Protocols in Emerging Food Processing Technologies," the first internationally recognized book in the field.

*Editorial*

# Emerging Non-Thermal Food Processing Technologies: Editorial Overview

**Asgar Farahnaky [1], Mahsa Majzoobi [1] and Mohsen Gavahian [2,*]**

[1] Biosciences and Food Technology, School of Science, RMIT University, Bundoora West Campus, Plenty Road, Melbourne, VIC 3083, Australia; asgar.farahnaky@rmit.edu.au (A.F.); mahsa.majzoobi@rmit.edu.au (M.M.)
[2] Department of Food Science, College of Agriculture, National Pingtung University of Science and Technology, 1, Shuefu Road, Neipu, Pingtung 91201, Taiwan
* Correspondence: mohsengavahian@yahoo.com or mg@mail.npust.edu.tw; Tel.: +886-8-7703202

According to the statistics, there is a strong consumer trend towards high-quality and healthy foods with "fresh-like" characteristics. On the other hand, thermal processing technologies, especially conventional ones, negatively affect the sensory and nutritional properties of foods. At the same time, the limited shelf-life and safety concerns of fresh foods necessitate food processing. Therefore, scientists are exploring the possibility of using nonthermal technologies for various purposes, such as shelf-life extension and the improvement of safety. These technologies include cold plasma, plasma-activated water, pulsed electric fields, moderate electric fields, ultrasound, high-pressure technologies, and other innovative approaches. However, their applicability and scalability are still under intensive investigation.

In this sense, a Special Issue entitled "Emerging Non-Thermal Food Processing Technologies" was launched in *Foods* (MDPI) to provide a forum for researchers to communicate some of their most recent findings on the applications of emerging thermal technologies for food processing. A total number of 17 manuscripts were submitted to this Special Issue from different regions of the world (Figure 1). According to the peer review process, 11 original research articles were included in this Special Issue of *Foods*.

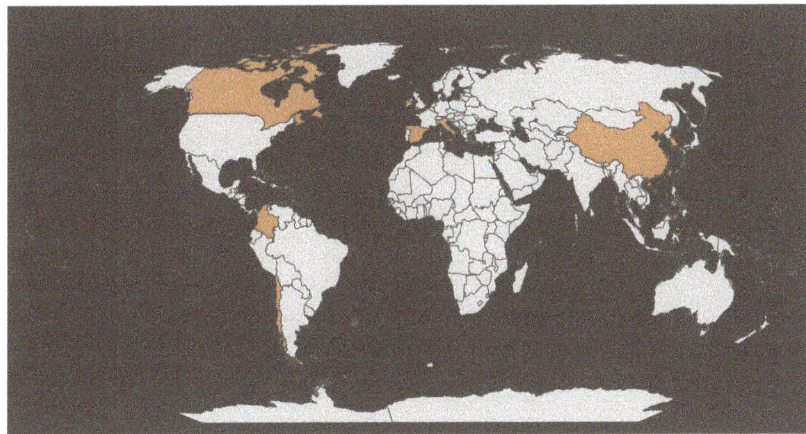

**Figure 1.** Origins of the submitted papers to this Special Issue based on the affiliations of the authors of the accepted papers. They include Canada, Chile, China, Spain, Taiwan, South Korea, Columbia, Ireland, and Italy.

Among the research papers published in this Special Issue, an attractive study investigated the effects of ultra-high-pressure homogenization technology treatment (UHPH) on

**Citation:** Farahnaky, A.; Majzoobi, M.; Gavahian, M. Emerging Non-Thermal Food Processing Technologies: Editorial Overview. *Foods* **2022**, *11*, 1003. https://doi.org/10.3390/foods11071003

Received: 24 March 2022
Accepted: 28 March 2022
Published: 29 March 2022

**Publisher's Note:** MDPI stays neutral with regard to jurisdictional claims in published maps and institutional affiliations.

the structural characteristics of egg yolk granules. The authors observed the high stability of yolk granules during UHPH and reported a restructuration of the granules via the generation of a protein network while the protein profile and proximate composition remained unchanged in a single pass of up to 300 MPa [1]. Another research team conducted an interesting study on the impacts of dual-frequency (28 and 40 kHz) ultrasound-assisted thawing on *Pseudosciaena crocea* (large yellow croaker) to elaborate on the effects of such a process on the microstructure, thawing rate, and quality properties. They reported that dual-frequency sonication can improve the thawing rate without compromising the water-holding capacity, color, texture, and water distribution, while inhibiting the disruption of the microstructure [2]. In another interesting study, the impact of cold plasma on the spores of naturally present fungi as well as the physical and chemical properties of sun-dried tomatoes was investigated. The germination of Aspergillus spores was found to be "species-dependent". According to the results of this paper, this process inactivated spores by percentages of 88 for *A. rugulovalvus* and 32 for *A. niger*, indicating them to be sensitive and resistant strains, respectively. This research also elaborated on the sporicidal effect of cold plasma on *A. rugulovalvus*. Cold plasma was found to be a promising tool for dried tomato decontamination (in terms of spores) without compromising the physicochemical aspects of the product (e.g., pH and water activity), while increasing the lycopene content significantly [3]. Another paper explained that "pre-crystallization of nougat by seeding with cocoa butter crystals enhances the bloom stability of nougat pralines". This informative study showed that the precrystallization of nougat via seeding with cocoa butter crystals, under controlled conditions, can enhance the physical storage stability of pralines and delay the onset of fat blooming [4]. A research team explored the "mathematical modelling of ultrasound-assisted extraction kinetics of bioactive compounds from artichoke by-products". This investigation showed that both sonication and temperature can significantly affect the yield of bioactive compounds during the bioactive extraction process from the stem of artichoke. It also proposed models that can be used for process prediction [5]. Another study explored the application of ultraviolet light-emitting diodes (UV-LEDs) to inactivate *Listeria monocytogenes*, *Escherichia coli*, *Bacillus subtilis*, and *Salmonella enterica* Serovar Typhimurium in powdered food ingredients. This innovative study proposed UV-LEDs as alternatives to UV technology, as they have the potential to enhance the decontamination efficiency [6]. Another group of researchers studied the effects of nitrogen-assisted high-pressure processing on the quality indices and microbial quality of bell peppers (fresh-cut). This food processing approach yielded a product with a greater microbial reduction and better microbiological stability during storage [7]. Furthermore, a manuscript discussed the results of a well-designed study on the impact of centrifugal block freeze crystallization (CBFC) on the quality properties of different types of berry juices. CBFC makes it feasible to obtain greater amounts of solutes, in comparison with the original sample, while intensifying the fresh juices' natural colors and increasing the phenolic compounds and antioxidant capacity in the final products [8]. Besides, a research team conducted an attractive study on high-hydrostatic-pressure technology. They investigated the antioxidant and anti-inflammatory effects of broccoli treated with high hydrostatic pressure in cell models, and reported that this emerging nonthermal technology can enhance the isothiocyanate content of broccoli with antioxidant anti-inflammatory effects, providing an innovative strategy with which to develop healthy food products in the future [9]. Moreover, an insightful manuscript reported a significant improvement in the antioxidant capacity of puffed turmeric by the application of high hydrostatic pressure extraction. This study elaborated on the effects of processing parameters and reported that 400 MPa for 20 min with 70% ethanol were the optimal extraction conditions for the highest antioxidant activity [10]. In another attractive study, researchers investigated the antibacterial efficacy and mechanisms of plasma-activated water (PAW) against *Salmonella enterica* serovar Enteritidis on eggshells, and reported that PAW is a promising decontamination technology due to the presence of ozone, nitrate, and other reactive species. They also provided images from scanning electron microscopes to elaborate on the effects of this technology on eggshells [11].

In summary, the findings published in this Special Issue clearly indicate both the breadth and depth of the recent studies on nonthermal processing technologies. They cover both a wide range of foods products as well as nonthermal technologies. Consumer demand for less-processed foods with a fresh flavor and delicate texture, plus better nutritional quality, are the main driving forces behind the continued research in this area, and it is expected that these technologies will be acquired at a higher pace by the food industry over the next decade.

**Author Contributions:** M.G. contributed to drafting the specific editorial note. A.F., M.M. and M.G. contributed to editing the manuscript. All authors have read and agreed to the published version of the manuscript.

**Funding:** This research received no external funding.

**Acknowledgments:** The Guest Editors would like to appreciate the journal and respective Editors for the invitation and for providing the opportunity. They also want to thank the authors and reviewers who contributed to this Special Issue.

**Conflicts of Interest:** The authors declare no conflict of interest.

## References

1. Gaillard, R.; Marciniak, A.; Brisson, G.; Perreault, V.; House, J.D.; Pouliot, Y.; Doyen, A. Impact of ultra-high pressure homogenization on the structural properties of egg yolk granule. *Foods* **2022**, *11*, 512. [CrossRef] [PubMed]
2. Cheng, H.; Bian, C.; Chu, Y.; Mei, J.; Xie, J. Effects of dual-frequency ultrasound-assisted thawing technology on thawing rate, quality properties, and microstructure of large yellow croaker (Pseudosciaena crocea). *Foods* **2022**, *11*, 226. [CrossRef] [PubMed]
3. Molina-Hernandez, J.B.; Laika, J.; Peralta-Ruiz, Y.; Palivala, V.K.; Tappi, S.; Cappelli, F.; Ricci, A.; Neri, L.; Chaves-López, C. Influence of Atmospheric Cold Plasma Exposure on Naturally Present Fungal Spores and Physicochemical Characteristics of Sundried Tomatoes (Solanum lycopersicum L.). *Foods* **2022**, *11*, 210. [CrossRef] [PubMed]
4. Böhme, B.; Bickhardt, A.; Rohm, H. Pre-crystallization of nougat by seeding with cocoa butter crystals enhances the bloom stability of nougat pralines. *Foods* **2021**, *10*, 1056. [CrossRef] [PubMed]
5. Reche, C.; Rosselló, C.; Umaña, M.M.; Eim, V.; Simal, S. Mathematical modelling of ultrasound-assisted extraction kinetics of bioactive compounds from artichoke by-products. *Foods* **2021**, *10*, 931. [CrossRef] [PubMed]
6. Nyhan, L.; Przyjalgowski, M.; Lewis, L.; Begley, M.; Callanan, M. Investigating the use of ultraviolet light emitting diodes (Uv-leds) for the inactivation of bacteria in powdered food ingredients. *Foods* **2021**, *10*, 797. [CrossRef] [PubMed]
7. Zhang, F.; Chai, J.; Zhao, L.; Wang, Y.; Liao, X. The impact of n2-assisted high-pressure processing on the microorganisms and quality indices of fresh-cut bell peppers. *Foods* **2021**, *10*, 508. [CrossRef] [PubMed]
8. Guerra-Valle, M.; Lillo-Perez, S.; Petzold, G.; Orellana-Palma, P. Effect of freeze crystallization on quality properties of two endemic patagonian berries juices: Murta (ugni molinae) and arrayan (luma apiculata). *Foods* **2021**, *10*, 466. [CrossRef] [PubMed]
9. Ke, Y.Y.; Shyu, Y.T.; Wu, S.J. Evaluating the anti-inflammatory and antioxidant effects of broccoli treated with high hydrostatic pressure in cell models. *Foods* **2021**, *10*, 167. [CrossRef] [PubMed]
10. Choi, Y.; Kim, W.; Lee, J.-S.; Youn, S.J.; Lee, H.; Baik, M.-Y. Enhanced antioxidant capacity of puffed turmeric (Curcuma longa L.) by high hydrostatic pressure extraction (HHPE) of bioactive compounds. *Foods* **2020**, *9*, 1690. [CrossRef] [PubMed]
11. Lin, C.M.; Hsiao, C.P.; Lin, H.S.; Liou, J.S.; Hsieh, C.W.; Wu, J.S.; Hou, C.Y. The antibacterial efficacy and mechanism of plasma-activated water against salmonella enteritidis (ATCC 13076) on shell eggs. *Foods* **2020**, *9*, 1491. [CrossRef] [PubMed]

*Article*

# Effects of Dual-Frequency Ultrasound-Assisted Thawing Technology on Thawing Rate, Quality Properties, and Microstructure of Large Yellow Croaker (*Pseudosciaena crocea*)

**Hao Cheng [1], Chuhan Bian [1], Yuanming Chu [1], Jun Mei [1,2,3,4] and Jing Xie [1,2,3,4,*]**

1 College of Food Science and Technology, Shanghai Ocean University, Shanghai 201306, China; m200300836@st.shou.edu.cn (H.C.); m200300826@st.shou.edu.cn (C.B.); m190300744@st.shou.edu.cn (Y.C.); jmei@shou.edu.cn (J.M.)
2 National Experimental Teaching Demonstration Center for Food Science and Engineering, Shanghai Ocean University, Shanghai 201306, China
3 Shanghai Engineering Research Center of Aquatic Product Processing and Preservation, Shanghai 201306, China
4 Shanghai Professional Technology Service Platform on Cold Chain Equipment Performance and Energy Saving Evaluation, Shanghai 201306, China
* Correspondence: jxie@shou.edu.cn; Tel.: +86-21-61900351

**Abstract:** This research evaluated the effects of dual-frequency ultrasound-assisted thawing (UAT) on the thawing time, physicochemical quality, water-holding capacity (WHC), microstructure, and moisture migration and distribution of large yellow croaker. Water thawing (WT), refrigerated thawing (RT), and UAT (single-frequency: 28 kHz (SUAT-28), single-frequency: 40 kHz (SUAT-40), dual-frequency: 28 kHz and 40 kHz (DUAT-28/40)) were used in the current research. Among them, the DUAT-28/40 treatment had the shortest thawing time, and ultrasound significantly improved the thawing rate. It also retained a better performance from the samples, such as color, texture, water-holding capacity and water distribution, and inhibited disruption of the microstructure. In addition, a quality property analysis showed that the pH, total volatile basic nitrogen (TVB-N), and K value were the most desirable under the DUAT-28/40 treatment, as well as this being best for the flavor of the samples. Therefore, DUAT-28/40 treatment could be a possible thawing method because it improves the thawing rate and maintains the quality properties of large yellow croaker.

**Keywords:** thawing rate; ultrasound-assisted thawing; quality properties; microstructure; large yellow croaker

**Citation:** Cheng, H.; Bian, C.; Chu, Y.; Mei, J.; Xie, J. Effects of Dual-Frequency Ultrasound-Assisted Thawing Technology on Thawing Rate, Quality Properties, and Microstructure of Large Yellow Croaker (*Pseudosciaena crocea*). *Foods* **2022**, *11*, 226. https://doi.org/10.3390/foods11020226

Academic Editors: Asgar Farahnaky, Mahsa Majzoobi and Mohsen Gavahian

Received: 24 November 2021
Accepted: 12 January 2022
Published: 14 January 2022



## 1. Introduction

Large yellow croaker (*Pseudosciaena crocea*) is rich in lipids and protein and widely cultured in China [1]. Due to the high moisture content, nutrient composition, and enzyme activity of large yellow croaker, it is highly vulnerable to contamination by pathogenic bacteria, which causes lipid oxidation, protein denaturation, and texture changes, resulting in a loss of commercial value [2,3]. To solve these problems, low-temperature frozen storage is usually applied. However, the quality of frozen products is influenced by both the freezing and thawing processes, especially the thawing process [4]. Thawing is an indispensable part of the production and processing of aquatic products. Unsuitable thawing methods may lead to quality loss [5]. Research on new thawing methods is currently in demand for the efficient utilization of aquatic products.

Ultrasound-assisted thawing (UAT) has been widely studied and applied in recent years. Studies have shown that UAT can reduce the thawing time and maintain the quality of frozen food [4,5]. UAT could also speed up the thawing rate and maintain the quality of fish, in aspects such as WHC (water-holding capacity) and texture [6]. Guo et al. reported that UAT at 400 W could better improve the thawing process efficiency and quality of frozen foods [7]. Moreover, under UAT (power ultrasound intensity: 2.39, 6.23, 11.32, and 20.96 W $cm^{-2}$; treatment time: 30, 60, 90, and 120 min), the gel properties and WHC of proteins were also improved [8]. The advantages of using ultrasound-assisted thawing with the right frequency and power were decreased thawing time, more uniform thawing temperature of the products, and better impact on the myofibrillar protein gel formation compared with other traditional thawing methods [9]. Li et al. [10] reported that ultrasound-assisted thawing (20 °C, 45 min) of frozen porcine longissimus muscle had reduced the damage to conformational change and gel formation of myofibrillar protein when compared with water thawing. In addition, Chen et al. [11] found that ultrasound-assisted thawing could save 13.5–40.4% in thawing time, and high ultrasonic frequency (80 kHz and 100 kHz) did not cause high thawing loss. Thawing at higher ultrasonic frequency could reduce the damage of the secondary and tertiary structure of myofibrillar proteins. Recently, Bian et al. [12] reported that multifrequency UAT significantly reduced the thawing time of frozen samples and maintained better quality attributes in comparison with SUAT-20 kHz, DUAT-20/28 kHz, and TUAT-20/28/40 kHz. It similarly focused on the multifrequency effect: what distinguished our study was that only two different single-frequency treatments were applied to explore the different effects of the two frequencies in combination and in isolation. Wang et al. [13] studied the effect of sweep frequency ultrasound thawing (SFUT) on gel, and the experiment was set with fluctuating frequency instead of constant frequency. It mainly paid attention to effects on gel, and SFUT improved gel properties, as shown in the study. In addition, it was found that DUAT-20/40 kHz had more desirable behavior than SUAT-20 kHz and TUAT-20/40/60 kHz [14], and it also focused on the gel. Based on this research, it was determined that DUAT performed better than SUAT and TUAT. It was considered worth exploring the different effects of two frequencies in combination and in isolation. The effects on quality under UAT were the focus of the study.

Frequency is an important parameter of ultrasound and poses no potential danger. The use of different frequencies can effectively improve the thawing process as described in the above research. There are few studies on UAT methods at different frequencies. Therefore, in this study, three types of UAT (SUAT-28, SUAT-40, and DUAT-28/40; ultrasonic power: 200 W) were used to study the effects of UAT on the quality of large yellow croaker, especially its water-holding capacity, water distribution and migration, K value, TVB-N, and the microstructure of the muscle. The study mainly focused on the effects of ultrasound-assisted thawing under different frequencies.

## 2. Materials and Methods

### 2.1. Preparation of Fish Samples

Large yellow croaker was acquired from Luchao Harbor Market and kept alive for prompt movement back to the laboratory. After being killed quickly, the fish were moved to a spiral freezer (Yantai Moon Co., Ltd., Yantai, Shandong, China) at a temperature of −35 °C and a wind velocity of 10 m/s for freezing for about half an hour. In order to facilitate further sample taking and measurement of experimental indicators, the heads and tips of the fish were all removed. Then the fish were quickly transferred to a −20 °C refrigerator and stored for three days. After that, the frozen samples were thawed using WT (water thawing), RT (refrigerator thawing), SUAT-28 (single-frequency ultrasound-assisted thawing at 28 kHz), SUAT-40 (single-frequency ultrasound-assisted thawing at 40 kHz), and DUAT-28/40 (dual-frequency ultrasound-assisted thawing at 28 kHz and 40 kHz). This was finished when the internal temperature of the fish reached 4 °C on the temperature monitor. WT and RT were used as control treatments, and SUAT-28, SUAT-40, and DUAT-28/40 were used as experiment treatments (Figure 1). The treatments were conducted at room temperature and RT was conducted in the refrigerator at 4 °C. They were considered finished when the internal temperature reached 4 °C. The power of all ultrasound-assisted thawing was 200 W.

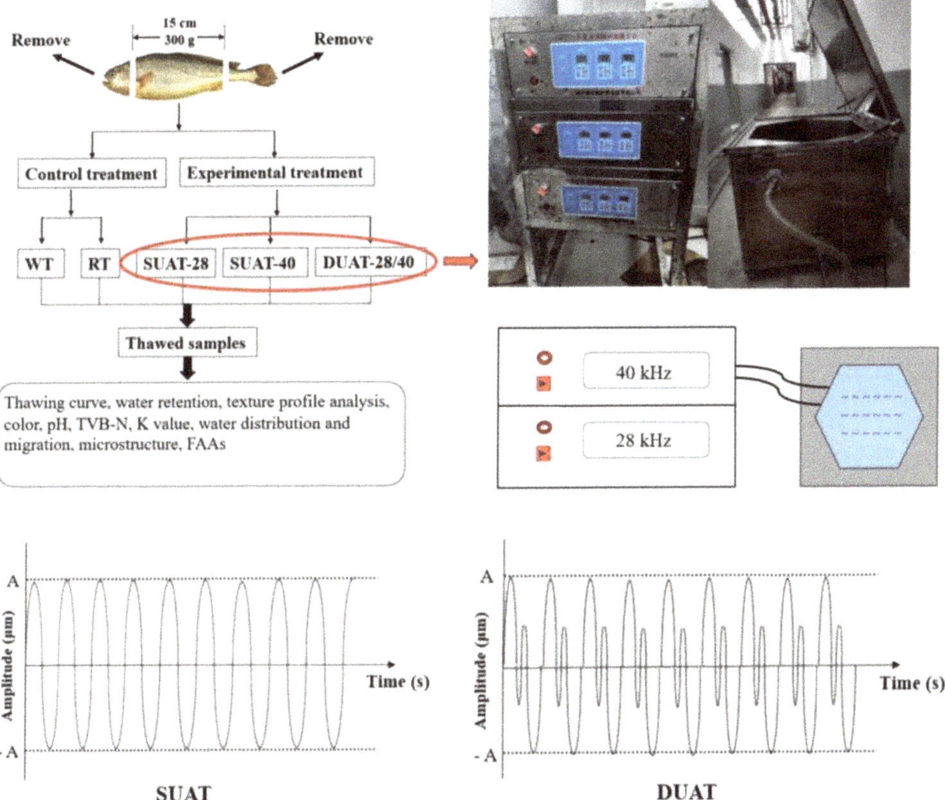


**Figure 1.** Diagram and possible mechanisms of the ultrasonic thawing system.

Figure 1 shows the main research content and ultrasonic-assisted thawing device used in the study. The device consisted of a control panel and sink. The control panel was used to monitor power and frequency, and fish were thawed in the sink. The possible mechanisms of ultrasound-assisted thawing under different frequencies were also proposed to account for the more uniform impact from ultrasound.

### 2.2. Thawing Curve

The temperature change was monitored by inserting the sensor of a multiple-point thermometer (Fluke 2640A, Everett, WA, USA) into the muscle center of the large yellow croaker. For each thawing method, three temperature-recording spots were set up, and the most appropriate thawing curve was selected, as shown in Figure 2. Each measurement was taken in triplicate.

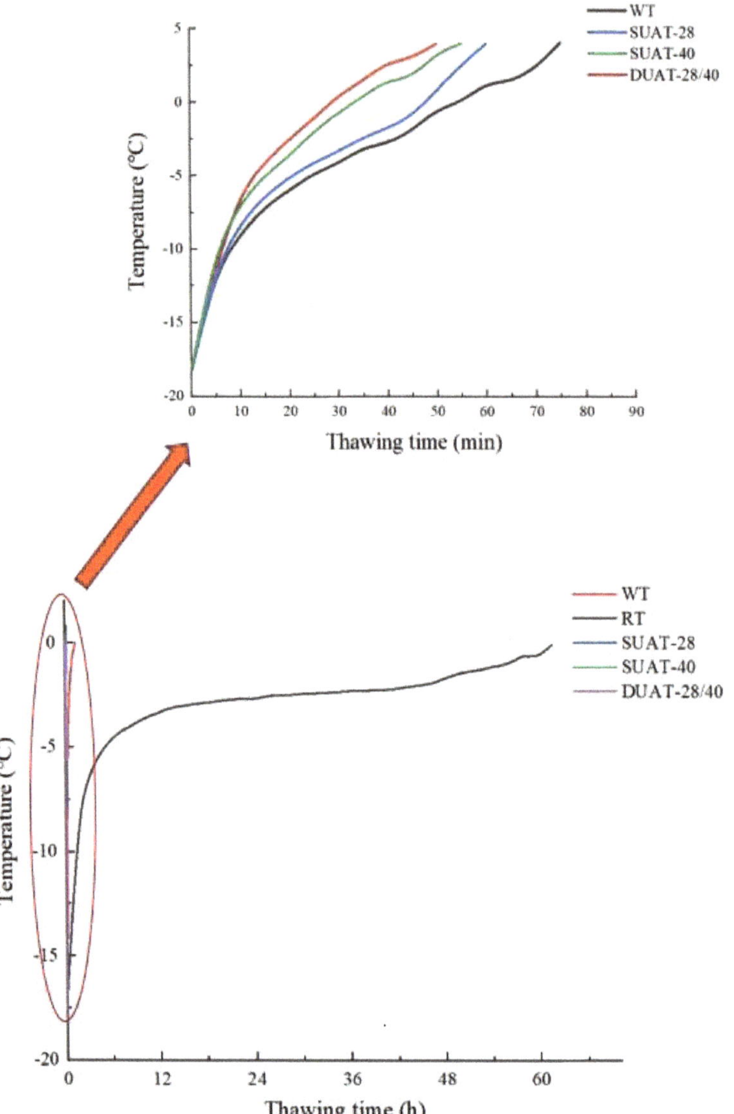


**Figure 2.** Differences in thawing curve of large yellow croaker under different thawing methods. (WT: water thawing; RT: refrigerated thawing; SUAT-28: single-frequency ultrasound-assisted thawing at 28 kHz; SUAT-40: single-frequency ultrasound-assisted thawing at 40 kHz; DUAT-28/40: dual-frequency ultrasound-assisted thawing at 28 kHz and 40 kHz).

### 2.3. *Determination of Water Retention*

#### 2.3.1. Thawing Loss

The water on the surface of the thawed fish was wiped, and the weight was measured as $W_1$ [15]. $W_0$ meant the weight of fish before thawing. This is shown in Equation (1). Only one trial was conducted per experiment.

$$\text{Thawing loss } \% = \frac{W_0 - W_1}{W_0} \times 100\% \quad (1)$$

where $W_0$ and $W_1$ refer to the weight before and after thawing.

2.3.2. Water-Holding Capacity

After defrosting the fish, the water on the surface of the fish was wiped, and 2.00 g of the sample was weighed. Centrifugation was conducted at 5000× *g* at 4 °C for 10 min [16]. The water-holding capacity was then calculated based on the formula below. The weight before centrifugation was called $M_1$, and the weight after centrifugation was called $M_2$, as shown in Equation (2). Each measurement was taken in triplicate.

$$\text{Water-holding capacity } (\%) = \frac{M_2}{M_1} \times 100\% \quad (2)$$

where $M_1$ and $M_2$ refer to the weight before and after centrifugation.

2.3.3. Cooking Loss

5.00 g of fish samples were weighed as $W_1$ and cooked in a water bath at 85 °C for 10 min. Then the weight was measured after cooking, denoted as $W_2$ [17]. Each measurement was taken in triplicate. It was calculated using Equation (3).

$$\text{Cooking loss } \% = \frac{W_1 - W_2}{W_1} \times 100\% \quad (3)$$

where $W_1$ and $W_2$ refer to the weight before and after cooking.

## *2.4. Determination of the Texture Profile Analysis (TPA)*

The samples were cut into 20 mm × 15 mm × 10 mm pieces, and the texture properties were analyzed using a texture analyzer [6]. The analysis software name was TA.XTplusC Texture Analyzer (Stable Micro Systems, Ltd., Godalming, Surrey, UK). The pre-test, test, and post-test speeds were 3, 1, and 1 mm/s, respectively, with a probe number (code) of Auto-5 g and a trigger force of 5 g. Each measurement was taken in triplicate.

## *2.5. Determination of Color*

The samples were placed on a flat plate, and the $\Delta L^*$ (brightness/darkness), $\Delta a^*$ (redness/greenness), and $\Delta b^*$ (yellowness/blueness) values were calculated using a colorimeter (CR-400, Konica Minolta, Tokyo, Japan). Changes in color were calculated as $\Delta E$ using Equation (4). Each measurement was taken in triplicate.

$$\triangle E = \sqrt{(\triangle L^*)^2 + (\triangle a^*)^2 + (\triangle b^*)^2} \quad (4)$$

where $\Delta L^*$, $\Delta a^*$, $\Delta b^*$ refer to the value of brightness/darkness, redness/greenness, and yellowness/blueness after calibration. In addition, $\Delta L^*$, $\Delta a^*$, and $\Delta b^*$ were the differences between $L^*$, $a^*$, and $b^*$ values before samples were UAT-treated and after samples were UAT-treated, respectively [18].

## *2.6. Determination of the pH Value*

According to the method of Lv et al. [19], 2.00 g of the chopped fish samples were mixed with 20 mL of physiological saline (0.9%, *m/v*) and centrifuged at 1000× *g* for 5 min at 4 °C. The pH value was measured using a pH meter (Mettler Toledo, Greifensee, Switzerland). Each measurement was taken in triplicate.

## *2.7. Determination of TVB-N*

The fish samples were minced and weighed to approximately 5 g for measurement. The TVB-N was calculated via microtitration and shown as mg N/100 g of the fish samples [20]. Each measurement was taken in triplicate.

### 2.8. Determination of the K Value

A 2.00 g sample was removed and homogenized with a 10% ($v/v$) perchloric acid solution. The sample was centrifuged at 8000× $g$ for 15 min at 4 °C, and the precipitate was mixed with a 5% ($v/v$) perchloric acid solution and centrifuged under the same conditions. The supernatants were combined, and the pH was adjusted to 6.5 with the phosphoric acid. The mixture was filtered through a 0.22 μm disposable filter, and 1 mL of the mixture was extracted [21]. The K value was then analyzed using the High-Performance Liquid Chromatography (HPLC) (Waters 2695, Milford, MA, USA) procedure suggested by Li and Zhan [22]. Each measurement was taken in triplicate.

The HPLC system with a Diamonsil-C18 column (250 × 4.6 mm) and a 254 nm UV detector was used to quantify adenosine triphosphate (ATP) and its catabolites including adenosine diphosphate (*ADP*), adenosine monophosphate (*AMP*), inosine monophosphate hypoxanthine (*IMP*), riboside (*HxR*), and hypoxanthine (*Hx*). Mobile phase A (phosphate) and mobile phase B (methanol) were subjected to gradient elution using the following procedure: 0 min to 20 min, 95% A and 5% B. The flow rate was 1 mL/min, the column temperature was 30 °C, and the injection volume was 10 μL. The K value was calculated using Equation (5):

$$K\ value\ (\%) = \frac{HxR + Hx}{ATP + ADP + AMP + IMP + HxR + Hx} \times 100 \quad (5)$$

where *ADP, AMP, ATP, IMP, Hx*, and *HxR* refer to adenosine diphosphate, adenosine monophosphate, adenosine triphosphate, inosine monophosphate hypoxanthine, hypoxanthine, and riboside, separately.

### 2.9. Water Distribution and Migration

The water distribution and migration within the fish were assessed using a low-field nuclear magnetic resonance (LF-NMR) analyzer (MesoMR23-060 H.I, NiuMeng, Shanghai, China). The specific settings were as follows: receiver bandwidth frequency (SW) = 100 kHz, SF = 21 MHz, RFD = 0.08 ms, RG1 = 20 db, P1 = 19 us, DRG1 = 6, TD = 400,100, DR = 1, TW = 2000 ms, NS = 4, P2 = 37 us, TE = 0.5 ms, and NECH = 8000 [23]. Each measurement was taken in triplicate.

### 2.10. Scanning Electron Microscope (SEM)

The microstructure was observed using an SEM (Hitachi HT 7700, Tokyo, Japan) as proposed by Cao et al. [24]. The samples were cut into 3 mm × 3 mm × 3 mm pieces, fixed with a 2.5% glutaraldehyde solution for 24 h, and then washed with distilled water to resist the fixer deposits. The samples were dewatered under an alcohol gradient (30%, 50%, 70%, 80%, 90%, 95%, and 100%) for 15 min for each elution. The samples were then lyophilized and coated with a palladium coater (Bal-TEC, Manchester, NH, USA). Each measurement was taken five times and the best image was selected.

### 2.11. Free Amino Acids (FAAs) Determination

A 2.00 g fish sample and 10 mL of 5% trichloroacetic acid were centrifuged at 10,000× $g$ for 10 min. The extraction and centrifugation were repeated and the combination was diluted to 25 mL. 1 mL of the extract was filtered through a 0.22 μm disposable filter and assessed using an amino acid analyzer (Hitachi L-8800, Tokyo, Japan). Only one trial was undertaken per experiment.

### 2.12. Statistical Analysis

The data were analyzed via one-way ANOVA using SPSS 23.0 followed by Duncan's test, and the results were expressed as the means ± standard deviation. The significance level was 0.05.

## 3. Results and Discussion

### 3.1. Thawing Rates and Curves

The thawing process of frozen foods is an essential factor in retaining their qualities, such as textural properties and color [25]. The RT treatment had the longest thawing time compared with the other treatments (Figure 2), and reached 64 h. The thawing time of the WT treatment was 70 min, and SUAT-28, SUAT-40, DUAT-28/40 took 45 min, 32 min, and 28 min, respectively. The UAT treatments were much faster, especially DUAT-28/40. The vibrational energy was attracted by the sample; thus, the frozen part of the sample quickly passed through the maximum ice crystal zone [26]. Among the three UAT treatments, DUAT was faster than SUAT, which was due to the cavitation effects produced by different ultrasound frequencies with mass transfer [1,27]. From −18 °C to −5 °C, the thawing rate was rapid and the central temperature rose sharply due to the difference between the sample temperature and the environmental temperature. However, from −5 °C to 0 °C, which was in the period of phase change of water in food, the thawing rate decreased, and the central temperature rose slowly [28]. The duration of this process is particularly important for the quality of frozen foods, and the UAT treatment significantly reduced the thawing time. Some studies have also indicated that UAT treatments shortened the thawing time of silver carp surimi [11].

### 3.2. Water Retention

Water retention is an important quality parameter that affects the texture, appearance, and storage stability of aquatic products [29]. Table 1 shows that the DUAT-28/40 treatment had the lowest thawing loss (1.70%) and the highest water-holding capacity (81.49%). However, there was no significant difference between the DUAT and WT samples in terms of the cooking loss ($p > 0.05$), which may be because the DUAT-28/40 treatment produced uniform force and internal stresses on the ice crystals of samples, resulting in less damage to cells from ice crystals. The increase in the thawing loss not only reduces the product weight, but also affects the product quality and eventually causes economic loss. Cooking loss includes large amounts of liquid and small amounts of soluble material from fish [30], which reflects structural damage to the muscle because of heat-induced denaturation of myofibrillar proteins. Therefore, the DUAT-28/40 treatment could better maintain water during the thawing process.

**Table 1.** Differences in thawing loss, water-holding capacity, and cooking loss of large yellow croaker under different thawing methods.

| Treatments | Thawing Loss (%) | Water-Holding Capacity (%) | Cooking Loss (%) |
|---|---|---|---|
| WT | 3.28 | 79.43 ± 3.21 $^{a}$ | 14.44 ± 2.48 $^{c}$ |
| RT | 1.97 | 75.08 ± 2.34 $^{c}$ | 26.88 ± 1.70 $^{a}$ |
| SUAT-28 | 1.97 | 79.45 ± 2.76 $^{a}$ | 20.52 ± 1.13 $^{b}$ |
| SUAT-40 | 3.22 | 77.58 ± 4.45 $^{b}$ | 21.42 ± 1.31 $^{b}$ |
| DUAT-28/40 | 1.70 | 81.49 ± 2.19 $^{a}$ | 16.65 ± 1.40 $^{bc}$ |

WT: water thawing; RT: refrigerated thawing; SUAT-28: single-frequency ultrasound-assisted thawing at 28 kHz; SUAT-40: single-frequency ultrasound-assisted thawing at 40 kHz; DUAT-28/40: dual-frequency ultrasound-assisted thawing at 28 kHz and 40 kHz. Different superscript letters indicate significant difference ($p < 0.05$).

### 3.3. Texture and Color

Texture is related to physical properties and affects the sensory and functional characteristics of fish [31]. The hardness, springiness, chewiness, and resilience values of the samples which underwent DUAT-28/40 treatment were all higher than those of the other samples (Table 2). There were no significant differences between the SUAT-28 treatment and the SUAT-40 treatment ($p > 0.05$). This result was similar to the results of another study [5]. It was also determined that the cavitation effect generated by ultrasonic action possibly modified the mechanical radiation, to some extent. Because different thawing methods had different effects on the changes in the water retention and protein denaturation in fish meat, the changes in the texture characteristics of fish meat were also different [32–34]. This could also be explained by the fact that the radius of the bubbles in the DUAT field was greater than the radius of the bubbles in the SUAT fields. Under the same UAT time, the DUAT treatment produced a more pronounced cavitation effect than SUAT [1]. In addition, the ice crystals also caused different degrees of damage to tissue. Irregular ice crystals in the WT and RT treatments caused more damage to meat than those in the UAT treatments. There are two main reasons for the rapid thawing assisted by ultrasonic technology: first of all, the attenuation of ultrasound in the medium produces high-frequency oscillations that are converted to thermal energy [33], which could accelerate melting of ice crystals; in addition, ultrasonic treatment could produce cavitation and microstreaming during the UAT process. The cavitation bubbles could be used as the primary ice nuclei and the mechanical force generated by the bursting of the cavitation bubbles breaks the ice crystals into smaller pieces [33], which affects the ice morphology and formation.

**Table 2.** Differences in hardness, springiness, chewiness, resilience, L*, a*, b*, and ΔE of large yellow croaker under different thawing methods.

| Treatments | Hardness (g) | Springiness | Chewiness | Resilience | L* | a* | b* | ΔE |
|---|---|---|---|---|---|---|---|---|
| WT | $2025.48 \pm 68.43^{b}$ | $0.41 \pm 0.03^{b}$ | $288.13 \pm 48.37^{ab}$ | $0.137 \pm 0.02^{ab}$ | $47.31 \pm 1.54^{ab}$ | $0.02 \pm 0.19^{c}$ | $-1.67 \pm 0.39^{c}$ | $47.35 \pm 1.53^{ab}$ |
| RT | $1891.09 \pm 61.10^{b}$ | $0.44 \pm 0.01^{ab}$ | $222.66 \pm 27.78^{bc}$ | $0.093 \pm 0.01^{c}$ | $43.39 \pm 0.96^{c}$ | $0.28 \pm 0.05^{bc}$ | $0.05 \pm 0.22^{a}$ | $43.39 \pm 0.96^{c}$ |
| SUAT-28 | $1067.77 \pm 43.34^{c}$ | $0.43 \pm 0.01^{ab}$ | $185.64 \pm 22.73^{c}$ | $0.114 \pm 0.01^{bc}$ | $45.02 \pm 1.23^{bc}$ | $0.57 \pm 0.01^{b}$ | $-1.66 \pm 0.45^{c}$ | $45.06 \pm 1.21^{bc}$ |
| SUAT-40 | $1427.11 \pm 24.13^{c}$ | $0.41 \pm 0.01^{b}$ | $262.31 \pm 12.69^{abc}$ | $0.103 \pm 0.01^{c}$ | $42.95 \pm 0.63^{c}$ | $0.72 \pm 0.43^{a}$ | $-1.68 \pm 0.56^{c}$ | $43.00 \pm 0.63^{c}$ |
| DUAT-28/40 | $2482.59 \pm 37.21^{a}$ | $0.47 \pm 0.01^{a}$ | $349.69 \pm 35.53^{a}$ | $0.152 \pm 0.01^{a}$ | $49.80 \pm 0.75^{a}$ | $0.72 \pm 0.64^{a}$ | $-0.78 \pm 0.89^{b}$ | $49.84 \pm 0.74^{a}$ |

WT: water thawing; RT: refrigerated thawing; SUAT-28: single-frequency ultrasound-assisted thawing at 28 kHz; SUAT-40: single-frequency ultrasound-assisted thawing at 40 kHz; DUAT-28/40: dual-frequency ultrasound-assisted thawing at 28 kHz and 40 kHz. Different superscript letters indicate significant difference ($p < 0.05$).

Consumers often subjectively judge the quality of fish by its color, which is the external expression of the physiological, biochemical, and microbiological changes of the fish [35] One of the most commonly used indicators is L* (brightness/darkness). The L* value of the DUAT-28/40 treatment was higher than that of the other treatments, and there were no significant differences between the WT treatment and the DUAT-28/40 treatment (Table 2). Li et al. [6] found that the L* values of UAT treatments were higher than those of the control, which was the same as the result of this research. There were two possible mechanisms mentioned in the article for changes in the color values caused by UAT treatment: (1) the production of free radicals to promote oxidation, resulting in destabilization of heme pigments, and (2) possible changes in the chemical structure of myoglobin and heme pigments due to the thermic and acoustic influence of ultrasound [36]. This theory could explain the phenomenon in the study. Moreover, the formation of extracellular space was also indicated as a main reason for the color change in a study [37].

### 3.4. pH Value

The pH value is one of the indicators used to evaluate the freshness of fish samples, and it also has a great influence on the protein properties of fish [38]. The pH range of all treatments was 6.5–6.8, where the SUAT-40 treatment was highest, and the RT treatment was lowest (Figure 3). Ultrasound treatment might have damaged the cells, leading to endolysis. Nitrogenous substances decompose to alkaline substances, resulting in higher pH. The DUAT-28/40 treatment was not significantly different from the WT treatment ($p > 0.05$). Studies have shown that the decrease in the pH value is due to an increase in the solute concentration caused by the loss of water and the release of hydrogen ions [39]; this affected pH the most. A decrease in pH might have led to changes in the muscle structure which reduced its water retention. However, it was determined that pH was not the main factor leading to changes in quality [40].

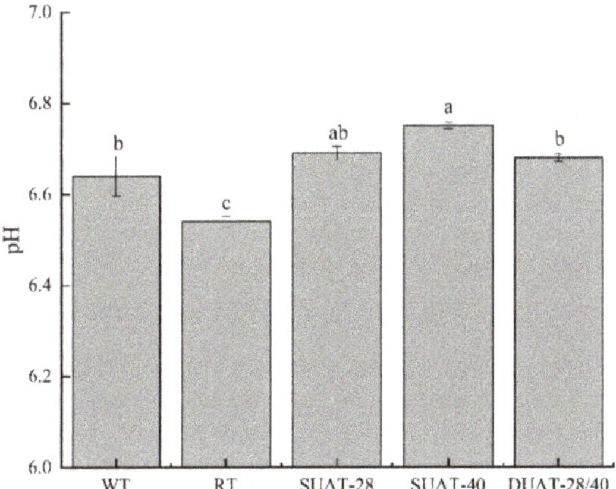


**Figure 3.** Differences in pH value of large yellow croaker under different thawing methods. (WT: water thawing; RT: refrigerated thawing; SUAT-28: single-frequency ultrasound-assisted thawing at 28 kHz; SUAT-40: single-frequency ultrasound-assisted thawing at 40 kHz; DUAT-28/40: dual-frequency ultrasound-assisted thawing at 28 kHz and 40 kHz). Different superscript letters indicate significant difference ($p < 0.05$).

### 3.5. Analysis of TVB-N and the K Value

TVB-N generally increases with a decrease in freshness, so it is widely used as an important indicator to reflect the freshness of fish [41,42]. The DUAT-28/40 treatment had the lowest value at 10.46 mg/100 g (Figure 4A), which was still at the first-grade freshness standard [23]. This result indicated that fish spoilage was minimal with this thawing method [1]. In addition, there were no significant differences between the DUAT-28/40 treatment group and two controls (WT and RT) ($p > 0.05$). The results of TVB-N were better under the DUAT-28/40 treatment.

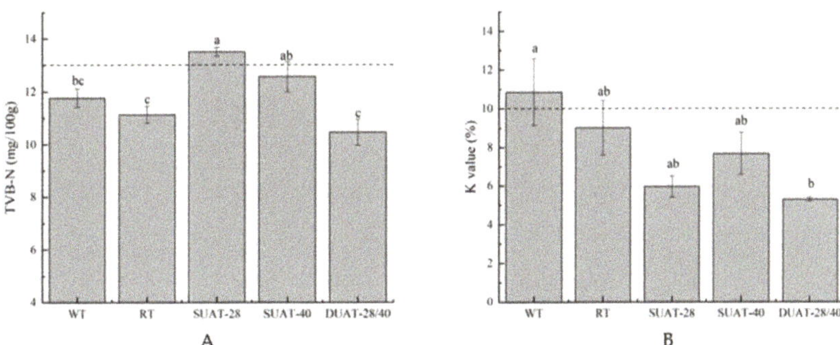


**Figure 4.** Differences in TVB-N value (mg/100 g) (**A**) and K value (%) (**B**) of large yellow croaker under different thawing methods. (WT: water thawing; RT: refrigerated thawing; SUAT-28: single-frequency ultrasound-assisted thawing at 28 kHz; SUAT-40: single-frequency ultrasound-assisted thawing at 40 kHz; DUAT-28/40: dual-frequency ultrasound-assisted thawing at 28 kHz and 40 kHz). Different superscript letters indicate significant difference ($p < 0.05$).

The K value is a relative value based on the quantification of ATP and its decomposition products, reflecting the degree of ATP degradation reaction after death. Moreover, it is one of the indices used to evaluate the freshness of fish [43]. The K value of the DUAT-28/40 treatment was 5.29% (Figure 4B), which was still at the first-grade freshness standard and similar to the result for the TVB-N. The same trend was found for the K value and TVB-N in another study, which was consistent with the results of this research [1]. From both of these sets of results it could be concluded that the DUAT-28/40 treatment showed a better freshness. Moreover, there were no significant differences between the sample which underwent the DUAT-28/40 treatment and the other samples, except the sample which underwent WT ($p > 0.05$). There was no direct explanation for the change in the K value under the ultrasound treatment. It was hypothesized that the reason for the lower K value was due to lower damage to muscle tissue. The K value was related to ATP degradation. The regular and small ice crystals under ultrasound treatment were less destructive to the tissue and delayed the degradation of ATP.

### 3.6. *Analysis via LF-NMR*

LF-NMR and magnetic resonance imaging (MRI) can be used to evaluate the distribution of water and the mobility of the muscle to learn about water migration due to different freezing treatments [44]. There are three peaks corresponding to the three relaxation components, named $T_{21}$ (<10 ms), $T_{22}$ (20–400 ms), and $T_{23}$ (>1000 ms), which represent bound water, fixed water, and free water, respectively [24]. The peak of the samples changed under different thawing conditions, indicating that the thawing methods influenced the distribution of water in the muscle of fish. The DUAT-28/40 treatment had the highest $T_{21}$ and $T_{22}$ values (Figure 5). The high peak of bound water showed that a high level of bound water could be maintained and the damage to the fish was minimized. It also indicated that its combining ability was the strongest and had the lowest transfer of fixed water to free water, possessing a better water-migration effect and quality [1]. Studies have shown that the redistribution of ice crystals and mechanical damage are the root causes of water transfer and loss [45]. Irregular and large ice crystals disrupted the cell structure, leading to the release of cell contents and facilitating protein denaturation. Under UAT treatment, the ice crystals were small and uniform to avoid this problem. Harnkarnsujarit et al. [46] also indicated that the ice formation also modified protein conformation and denaturation of protein in fish muscles. Thus, the bound water content of the samples that underwent WT and RT was lower than that in those that underwent the UAT treatment.

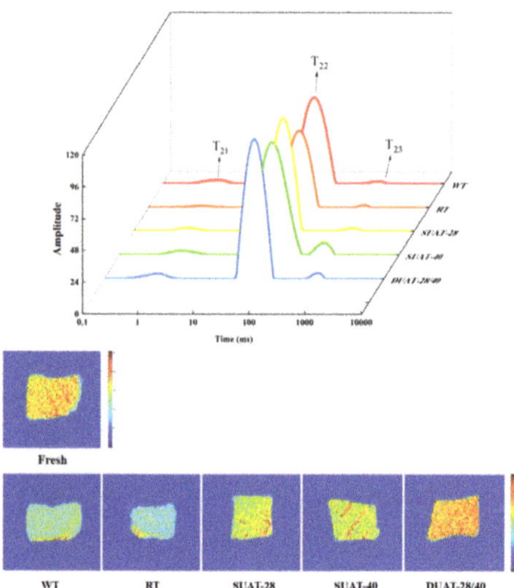


**Figure 5.** Differences in water distribution and magnetic resonance imaging of large yellow croaker under different thawing methods and in fresh sample. (WT: water thawing; RT: refrigerated thawing; SUAT-28: single-frequency ultrasound-assisted thawing at 28 kHz; SUAT-40: single-frequency ultrasound-assisted thawing at 40 kHz; DUAT-28/40: dual-frequency ultrasound-assisted thawing at 28 kHz and 40 kHz).

A brighter red color indicates a higher $^1$H proton density and a higher moisture mass fraction, and a blue color indicates a low moisture content. The change in image color can reflect the water distribution of fish samples [47]. The color of the samples that underwent DUAT-28/40 treatment was redder than that of those that underwent the other treatments (Figure 5), indicating that the DUAT-28/40 treatment retained higher moisture levels. The control treatment samples all showed as blue in MRI, reflecting the higher free water content, severe moisture loss, and quality damage.

### 3.7. Microstructure of the Fish Muscle

Muscle microstructure is an important indicator of direct changes in muscle structural integrity. Studies have shown that muscle microstructure is related to water-holding capacity, texture, and myofibrillar protein properties [48]. As seen in Figure 6, the microstructure of the samples that underwent the WT and RT treatments was irregular and not flattened, and they were surrounded by adhesion on the surface. This result indicated that the bundles of myofibrils were severely damaged, resulting in protein denaturation, cell disruption, and damage to muscle structure [49]. The microstructure of the samples that underwent UAT was flatter, more regular, and smoother, especially after DUAT-28/40 treatment. Although the outer membrane exposed to the myofibrils was inevitably damaged, the overall integrity of the shape was essentially maintained. Irregular ice crystals damaged the structure of the muscles [50]. Under ultrasonic treatment, the ice crystals first melt into uniform and regular small ice crystals and then disappear slowly. The melting of ice crystals is an orderly process, which is consistent with the microstructure. Sun et al. [33] found that muscles treated with UAT-100 (UAT power at 100 W) and UAT-300 (UAT power at 300 W) had a more complete microstructure than the WT samples, with a slightly larger gap between the fascicles than the controls. Although the outer membrane exposed to the myofibrils was injured in the muscle, the inner membrane attached to myofibrils was relatively intact. This might be because the ultrasound decayed more rapidly in the unthawed region than in the thawed region, and the absorption of energy was mainly focused on the freeze–thaw border and the frozen part. This avoided the development of local overheating and thus decreased damage to the muscle bundle membrane. However, Li et al. [6] found that ultrasound treatment had little effect on the muscle tissue of fish.

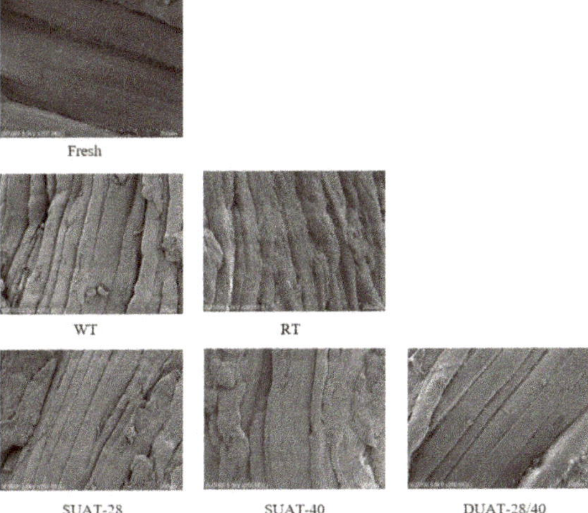


**Figure 6.** Differences in scanning electron microscope image of large yellow croaker under different thawing methods and fresh sample. (WT: water thawing; RT: refrigerated thawing; SUAT-28: single-frequency ultrasound-assisted thawing at 28 kHz; SUAT-40: single-frequency ultrasound-assisted thawing at 40 kHz; DUAT-28/40: dual-frequency ultrasound-assisted thawing at 28 kHz and 40 kHz).

### 3.8. Analysis of Free Amino Acids (FAAs)

The content of free amino acids is closely related to the flavor of fish [51]. Usually, there are three types of amino acids, including umami amino acids, sweet amino acids, and bitter amino acids. Among them, phenylalanine, tyrosine, lysine, and arginine play key roles, as they are the main biogenic amine precursors in fish [52]. As Table 3 shows, the highest contents of all FAAs were alanine, glycine, and lysine, and they played important roles in the taste and flavor of the fish. Kimbuathong et al. [53] also found that formation of amine compounds including trimethyl amine influenced the sensory characteristics of seafoods. Rich flavor amino acids contributed to improving the taste of fish. Alanine and glycine appeared to be umami and sweet, whereas lysine was bitter [54]. In terms of the total amino acids, the DUAT-28/40 treatment had the highest content, reaching 916.44 mg/100 g. This was obviously higher than that of the control, and DUAT increased the content of FAAs. The study by Kang [55] indicated that the cavitation caused by ultrasound could produce free radicals, which enhanced the level of protein degradation. The degradation of protein led to an increase in the FAA contents. In addition, it was discussed that high levels of ultrasound led to $OH^-$ production from water molecules, which resulted in protein degradation and an increase in the FAA contents [56]. Among all the treatments, the sweet amino acids had the highest content, and the bitter amino acids were lower (Figure 7), which was related to the taste of the large yellow croaker [23]. The change in the taste presentation might be due to the degradation of ATP and associated compounds after fish spoilage, a loss of taste-presenting substances during thawing, and microbial metabolism. In summary, the cavitation effect of ultrasound in water could affect the degradation of proteins, which led to changes in the FAA contents. Moreover, Laorenza et al. [57] found that the enzymatic activity in seafoods caused the changes of protein conformation and degradation of protein into amino acids as well as degradation of lipid substances that influenced organoleptic qualities. The effect of the interaction between free amino acids and nucleotides on taste presentation should also be considered. Moreover, the Gly content was much higher in fish treated with DUAT-28/40 than other treatments. Because Gly is both an umami amino acid and a sweet amino acid, this phenomenon showed that the DUAT treatment contributed better to the flavor of fish.

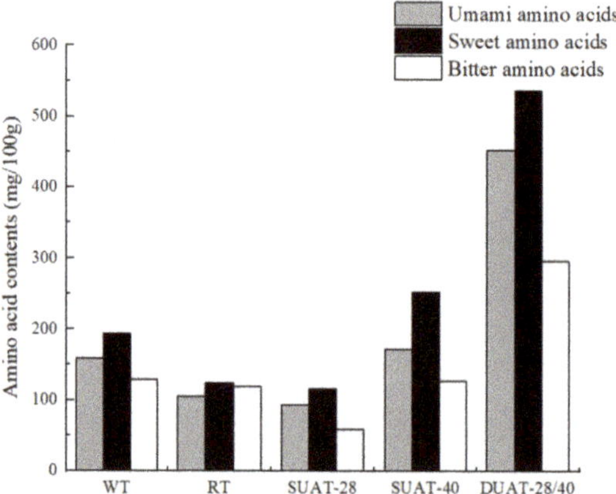


**Figure 7.** Differences in free amino acid contents (mg/100 g) of large yellow croaker under different thawing methods. (WT: water thawing; RT: refrigerated thawing; SUAT-28: single-frequency ultrasound-assisted thawing at 28 kHz; SUAT-40: single-frequency ultrasound-assisted thawing at 40 kHz; DUAT-28/40: dual-frequency ultrasound-assisted thawing at 28 kHz and 40 kHz).

**Table 3.** Differences in free amino acids contents (mg/100 g) of large yellow croaker under different thawing methods.

| Free Amino Acids | Presentation of Taste | Treatments | | | | |
|---|---|---|---|---|---|---|
| | | WT | RT | SUAT-28 | SUAT-40 | DUAT-28/40 |
| Aspartic acid (Asp) | umami | 10.56 ± 0.55 | 4.19 ± 0.10 | 5.27 ± 0.01 | 9.18 ± 0.32 | 13.12 ± 0.42 |
| Threonine (Thr) | sweet | 28.59 ± 0.16 | 12.59 ± 0.25 | 14.18 ± 0.17 | 26.31 ± 1.18 | 37.05 ± 0.42 |
| Serine (Ser) | sweet | 30.39 ± 0.30 | 19.96 ± 0.30 | 16.63 ± 0.12 | 55.25 ± 2.58 | 88.15 ± 2.65 |
| Glutamic acid (Glu) | umami | 29.95 ± 0.24 | 15.95 ± 0.29 | 7.48 ± 0.09 | ND | 64.67 ± 1.27 |
| Glycine (Gly) | umami, sweet | 38.17 ± 0.51 | 25.64 ± 0.46 | 22.50 ± 0.08 | 58.23 ± 2.29 | 237.45 ± 6.58 |
| Alanine (Ala) | umami, sweet | 80.07 ± 0.90 | 59.37 ± 1.11 | 57.93 ± 0.40 | 103.98 ± 4.59 | 137.14 ± 3.92 |
| Cysteine (Cys) | - | 2.00 ± 0.08 | 1.96 ± 0.09 | 1.43 ± 0.04 | 3.26 ± 0.17 | 6.08 ± 0.18 |
| Valine (Val) | bitter | 21.95 ± 0.29 | 10.81 ± 0.24 | 8.00 ± 0.10 | 8.86 ± 0.45 | 12.24 ± 0.27 |
| Methionine (Met) | bitter | 10.17 ± 0.10 | 7.64 ± 0.19 | 6.68 ± 0.19 | 6.08 ± 0.39 | 9.33 ± 0.16 |
| Isoleucine (Ile) | bitter | 12.76 ± 0.15 | 6.29 ± 0.11 | 4.92 ± 0.06 | 4.76 ± 0.21 | 6.72 ± 0.19 |
| Leucine (Leu) | bitter | 23.35 ± 0.25 | 11.71 ± 0.27 | 9.19 ± 0.06 | 9.56 ± 0.42 | 12.54 ± 0.25 |
| Tyrosine (Tyr) | bitter | 9.64 ± 0.04 | 6.26 ± 0.22 | 6.90 ± 0.09 | 6.21 ± 0.32 | 8.67 ± 0.24 |
| Phenylalanine (Phe) | bitter | 9.06 ± 0.16 | 6.55 ± 0.20 | 5.53 ± 0.10 | 5.34 ± 0.29 | 7.72 ± 0.25 |
| Lysine (Lys) | bitter | 29.71 ± 0.35 | 55.77 ± 1.35 | 8.46 ± 0.16 | 52.37 ± 2.82 | 171.82 ± 3.74 |
| Histidine (His) | bitter | 12.56 ± 0.19 | 13.40 ± 0.33 | 9.41 ± 0.08 | 34.10 ± 1.81 | 67.18 ± 1.57 |
| Arginine (Arg) | sweet | 0.66 ± 0.02 | 1.59 ± 0.07 | 0.04 ± 0.04 | 2.38 ± 0.18 | 2.63 ± 0.00 |
| Proline (Pro) | sweet | 15.82 ± 5.45 | 4.42 ± 1.07 | 5.13 ± 0.01 | 6.25 ± 1.04 | 33.93 ± 11.13 |
| Total | | 365.40 ± 9.73 | 264.07 ± 6.66 | 189.66 ± 1.50 | 392.12 ± 19.04 | 916.44 ± 33.23 |

WT: water thawing; RT: refrigerated thawing; SUAT-28: single-frequency ultrasound-assisted thawing at 28 kHz; SUAT-40: single-frequency ultrasound-assisted thawing at 40 kHz; DUAT-28/40: dual-frequency ultrasound-assisted thawing at 28 kHz and 40 kHz.

## 4. Conclusions

In this experiment, we investigated the effects of single-frequency and dual-frequency ultrasonic treatment on the quality of large yellow croaker, including the thawing rate, water-holding capacity, and quality characteristics. Among all the thawing methods applied in the study, the DUAT-28/40 treatment was considered to be the best thawing method. The results showed that the UAT treatment significantly accelerated the thawing rate, and that the DUAT-28/40 treatment had the fastest thawing rate. It also had a better performance in terms of water-holding capacity, texture, color, and water distribution. Furthermore, the DUAT-28/40 treatment prevented the quality from being seriously damaged compared with other thawing methods in terms of the pH, TVB-N, and K values. It still stayed at the first-grade freshness standard under the DUAT-28/40 treatment. The microstructure analysis revealed less damage to the structure of the muscle under DUAT-28/40 treatment as the flatter, more regular, and smoother myofibrils showed. At the same time, the samples retained an impressive flavor, which was showed by the free amino acid results. Overall, the DUAT-28/40 treatment was a desirable thawing method for accelerating the thawing rate and improving the quality of large yellow croaker. Further research could explore the synergy between different ultrasound frequencies and power and how these apply in the thawing process. In addition, the mechanism is also expected to be studied.

**Author Contributions:** Conceptualization, H.C.; data curation, Y.C.; formal analysis, H.C., C.B. and J.M.; funding acquisition, J.X.; investigation, Y.C.; methodology, H.C., C.B. and Y.C.; project administration, J.M. and J.X.; resources, C.B.; software, H.C. and C.B.; validation, J.X.; writing—original draft, H.C.; writing—review & editing, J.M. All authors have read and agreed to the published version of the manuscript.

**Funding:** The authors gratefully acknowledge the funding from China Agriculture Research System (CARS-47-G26), Shanghai Municipal Science and Technology Project to Enhance the Ca-pabilities of the PlatForm (20DZ2292200, 19DZ2284000).

**Data Availability Statement:** Not applicable.

**Conflicts of Interest:** The authors declare no conflict of interest.

## References

1. Wang, Y.-Y.; Yan, J.-K.; Rashid, M.T.; Ding, Y.; Chikari, F.; Huang, S.; Ma, H. Dual-frequency sequential ultrasound thawing for improving the quality of quick-frozen small yellow croaker and its possible mechanisms. *Innov. Food Sci. Emerg. Technol.* **2021**, *68*, 102614. [CrossRef]
2. Li, D.; Qin, N.; Zhang, L.; Li, Q.; Prinyawiwatkul, W.; Luo, Y. Degradation of adenosine triphosphate, water loss and textural changes in frozen common carp (*Cyprinus carpio*) fillets during storage at different temperatures. *Int. J. Refrig.* **2019**, *98*, 294–301. [CrossRef]
3. Jun, M.; Xuan, M.; Jing, X. Review on natural preservatives for extending fish shelf life. *Foods* **2019**, *8*, 490. [CrossRef]
4. Liu, Y.; Chen, S.; Pu, Y.; Muhammad, A.I.; Hang, M.; Liu, D.; Ye, T. Ultrasound-assisted thawing of mango pulp: Effect on thawing rate, sensory, and nutritional properties. *Food Chem.* **2019**, *286*, 576–583. [CrossRef] [PubMed]
5. Li, X.-X.; Sun, P.; Jia, J.-Z.; Cai, L.-Y.; Li, J.-R.; Lv, Y.-F. Effect of low frequency ultrasound thawing method on the quality characteristics of Peru squid (*Dosidicus gigas*). *Food Sci. Technol. Int.* **2019**, *25*, 171–181. [CrossRef]
6. Li, D.; Zhao, H.; Muhammad, A.I.; Song, L.; Guo, M.; Liu, D. The comparison of ultrasound-assisted thawing, air thawing and water immersion thawing on the quality of slow/fast freezing bighead carp (*Aristichthys nobilis*) fillets. *Food Chem.* **2020**, *320*, 126614. [CrossRef]
7. Guo, Z.; Ge, X.; Yang, L.; Ma, G.; Ma, J.; Yu, Q.L.; Han, L. Ultrasound-assisted thawing of frozen white yak meat: Effects on thawing rate, meat quality, nutrients, and microstructure. *Ultrason. Sonochem.* **2021**, *70*, 105345. [CrossRef]
8. Kang, D.C.; Zou, Y.H.; Cheng, Y.P.; Xing, L.J.; Zhou, G.H.; Zhang, W.G. Effects of power ultrasound on oxidation and structure of beef proteins during curing processing. *Ultrason. Sonochem.* **2016**, *33*, 47–53. [CrossRef] [PubMed]
9. Li, X.-X.; Sun, P.; Ma, Y.; Cai, L.; Li, J.-R. Effect of ultrasonic thawing on the water-holding capacity, physicochemical properties and structure of frozen tuna (*Thunnus tonggol*) myofibrillar proteins. *J. Sci. Food Agric.* **2019**, *99*, 5083–5091. [CrossRef]
10. Li, F.; Wang, B.; Liu, Q.; Chen, Q.; Zhang, H.; Xia, X.; Kong, B. Changes in myofibrillar protein gel quality of porcine longissimus muscle induced by its stuctural modification under different thawing methods. *Meat Sci.* **2019**, *147*, 108–115. [CrossRef] [PubMed]
11. Chen, H.-Z.; Zhang, M.; Rao, Z. Effect of ultrasound-assisted thawing on gelling and 3D printing properties of silver carp surimi. *Food Res. Int.* **2021**, *145*, 110405. [CrossRef]
12. Bian, C.; Cheng, H.; Yu, H.; Mei, J.; Xie, J. Effect of multi-frequency ultrasound assisted thawing on the quality of large yellow croaker (*Larimichthys crocea*). *Ultrason. Sonochem.* **2022**, *82*, 105907. [CrossRef] [PubMed]
13. Wang, Y.-Y.; Yan, J.-K.; Ding, Y.; Rashid, M.T.; Ma, H. Effect of sweep frequency ultrasound and fixed frequency ultrasound thawing on gelling properties of myofibrillar protein from quick-frozen small yellow croaker and its possible mechanisms. *LWT* **2021**, *150*, 111922. [CrossRef]
14. Wang, Y.-Y.; Tayyab Rashid, M.; Yan, J.-K.; Ma, H. Effect of multi-frequency ultrasound thawing on the structure and rheological properties of myofibrillar proteins from small yellow croaker. *Ultrason. Sonochem.* **2021**, *70*, 105352. [CrossRef] [PubMed]
15. Sun, Q.; Sun, F.; Xia, X.; Xu, H.; Kong, B. The comparison of ultrasound-assisted immersion freezing, air freezing and immersion freezing on the muscle quality and physicochemical properties of common carp (*Cyprinus carpio*) during freezing storage. *Ultrason. Sonochem.* **2019**, *51*, 281–291. [CrossRef]
16. Tan, M.; Ye, J.; Chu, Y.; Xie, J. The effects of ice crystal on water properties and protein stability of large yellow croaker (*Pseudosciaena crocea*). *Int. J. Refrig.* **2021**, *130*, 242–252. [CrossRef]
17. Jinfeng, W.; Wenhui, Y.; Jing, X. Effect of glazing with different materials on the quality of tuna during frozen storage. *Foods* **2020**, *9*, 231. [CrossRef]
18. Cartagena, L.; Puértolas, E.; Martínez de Marañón, I. Impact of different air blast freezing conditions on the physicochemical quality of albacore (*Thunnus alalunga*) pretreated by high pressure processing. *LWT* **2021**, *145*, 111538. [CrossRef]
19. Lv, Y.; Xie, J. Effects of Freeze–Thaw Cycles on Water Migration, Microstructure and Protein Oxidation in Cuttlefish. *Foods* **2021**, *10*, 2576. [CrossRef] [PubMed]
20. Chu, Y.; Tan, M.; Bian, C.; Xie, J. Effect of ultrasonic thawing on the physicochemical properties, freshness, and protein-related properties of frozen large yellow croaker (*Pseudosciaena crocea*). *J. Food Sci.* **2021**, *79*, 105787. [CrossRef] [PubMed]
21. Chu, Y.; Tan, M.; Yi, Z.; Ding, Z.; Yang, D.; Xie, J. Shelf-Life Prediction of Glazed Large Yellow Croaker (*Pseudosciaena crocea*) during Frozen Storage Based on Arrhenius Model and Long-Short-Term Memory Neural Networks Model. *Fishes* **2021**, *6*, 39. [CrossRef]
22. Li, N.; Zhan, X. Identification of clinical trait-related lncRNA and mRNA biomarkers with weighted gene co-expression network analysis as useful tool for personalized medicine in ovarian cancer. *EPMA J.* **2019**, *10*, 273–290. [CrossRef] [PubMed]
23. Chu, Y.; Cheng, H.; Yu, H.; Mei, J.; Xie, J. Quality enhancement of large yellow croaker (*Pseudosciaena crocea*) during frozen (−18 °C) storage by spiral freezing. *CyTA J. Food* **2021**, *19*, 710–720. [CrossRef]

24. Cao, M.; Cao, A.; Wang, J.; Cai, L.; Regenstein, J.; Ruan, Y.; Li, X. Effect of magnetic nanoparticles plus microwave or far-infrared thawing on protein conformation changes and moisture migration of red seabream (*Pagrus major*) fillets. *Food Chem.* **2018**, *266*, 498–507. [CrossRef]
25. Jia, G.; Liu, H.; Nirasawa, S.; Liu, H. Effects of high-voltage electrostatic field treatment on the thawing rate and post-thawing quality of frozen rabbit meat. *Innov. Food Sci. Emerg. Technol.* **2017**, *41*, 348–356. [CrossRef]
26. Jiang, Q.; Nakazawa, N.; Hu, Y.; Osako, K.; Okazaki, E. Changes in quality properties and tissue histology of lightly salted tuna meat subjected to multiple freeze-thaw cycles. *Food Chem.* **2019**, *293*, 178–186. [CrossRef] [PubMed]
27. Yang, X.; Li, Y.; Li, S.; Oladejo, A.O.; Wang, Y.; Huang, S.; Zhou, C.; Ye, X.; Ma, H.; Duan, Y. Effects of ultrasound-assisted alpha-amylase degradation treatment with multiple modes on the extraction of rice protein. *Ultrason. Sonochem.* **2018**, *40*, 890–899. [CrossRef]
28. Amiri, A.; Mousakhani-Ganjeh, A.; Shafiekhani, S.; Mandal, R.; Singh, A.P.; Kenari, R.E. Effect of high voltage electrostatic field thawing on the functional and physicochemical properties of myofibrillar proteins. *Innov. Food Sci. Emerg. Technol.* **2019**, *56*, 102191. [CrossRef]
29. Peng-cheng, Z.; Jing, X. Effect of different thawing methods on the quality of mackerel (*Pneumatophorus japonicus*). *Food Sci. Biotechnol.* **2021**, *30*, 1213–1223. [CrossRef]
30. Dalvi-Isfahan, M.; Hamdami, N.; Le-Bail, A. Effect of freezing under electrostatic field on the quality of lamb meat. *Innov. Food Sci. Emerg. Technol.* **2016**, *37*, 68–73. [CrossRef]
31. Mousakhani-Ganjeh, A.; Hamdami, N.; Soltanizadeh, N. Impact of high voltage electric field thawing on the quality of frozen tuna fish (*Thunnus albacares*). *J. Food Eng.* **2015**, *156*, 39–44. [CrossRef]
32. Wang, Y.Y.; Yan, J.K.; Ding, Y.; Ma, H. Effects of ultrasound on the thawing of quick-frozen small yellow croaker (*Larimichthys polyactis*) based on TMT-labeled quantitative proteomic. *Food Chem.* **2022**, *366*, 130600. [CrossRef]
33. Sun, Q.; Kong, B.; Liu, S.; Zheng, O.; Zhang, C. Ultrasound-assisted thawing accelerates the thawing of common carp (*Cyprinus carpio*) and improves its muscle quality. *LWT* **2021**, *141*, 111080. [CrossRef]
34. Liu, D.; Liang, L.; Xia, W.; Regenstein, J.M.; Zhou, P. Biochemical and physical changes of grass carp (*Ctenopharyngodon idella*) fillets stored at −3 and 0 degrees C. *Food Chem.* **2013**, *140*, 105–114. [CrossRef]
35. Cropotova, J.; Tappi, S.; Genovese, J.; Rocculi, P.; Dalla Rosa, M.; Rustad, T. The combined effect of pulsed electric field treatment and brine salting on changes in the oxidative stability of lipids and proteins and color characteristics of sea bass (*Dicentrarchus labrax*). *Heliyon* **2021**, *7*, e05947. [CrossRef]
36. Stadnik, J.; Dolatowski, Z.J. Influence of sonication on Warner-Bratzler shear force, colour and myoglobin of beef (*m. semimembranosus*). *Eur. Food Res. Technol.* **2011**, *233*, 553–559. [CrossRef]
37. Sun, Q.; Zhao, X.; Zhang, C.; Xia, X.; Sun, F.; Kong, B. Ultrasound-assisted immersion freezing accelerates the freezing process and improves the quality of common carp (*Cyprinus carpio*) at different power levels. *LWT* **2019**, *108*, 106–112. [CrossRef]
38. Diao, Y.; Cheng, X.; Wang, L.; Xia, W. Effects of immersion freezing methods on water holding capacity, ice crystals and water migration in grass carp during frozen storage. *Int. J. Refrig.* **2021**, *131*, 581–591. [CrossRef]
39. Zhang, M.; Haili, N.; Chen, Q.; Xia, X.; Kong, B. Influence of ultrasound-assisted immersion freezing on the freezing rate and quality of porcine longissimus muscles. *Meat Sci* **2018**, *136*, 1–8. [CrossRef] [PubMed]
40. Bekhit, A.E.-D.A.; Holman, B.W.B.; Giteru, S.G.; Hopkins, D.L. Total volatile basic nitrogen (TVB-N) and its role in meat spoilage: A review. *Trends Food Sci. Technol.* **2021**, *109*, 280–302. [CrossRef]
41. Chunlin, H.; Jing, X. The effect of multiple freeze-thaw cycles on the microstructure and quality of *Trachurus murphyi*. *Foods* **2021**, *10*, 1350. [CrossRef]
42. Agyekum, A.A.; Kutsanedzie, F.Y.H.; Annavaram, V.; Mintah, B.K.; Asare, E.K.; Wang, B. FT-NIR coupled chemometric methods rapid prediction of K-value in fish. *Vib. Spectrosc.* **2020**, *108*, 103044. [CrossRef]
43. Zhang, M.; Li, F.; Diao, X.; Kong, B.; Xia, X. Moisture migration, microstructure damage and protein structure changes in porcine longissimus muscle as influenced by multiple freeze-thaw cycles. *Meat Sci* **2017**, *133*, 10–18. [CrossRef]
44. Mingtang, T.; Jing, X. Exploring the effect of dehydration on water migrating property and protein changes of large yellow croaker (*Pseudosciaena crocea*) during frozen storage. *Foods* **2021**, *10*, 784. [CrossRef]
45. Li, F.; Wang, B.; Kong, B.; Shi, S.; Xia, X. Decreased gelling properties of protein in mirror carp (*Cyprinus carpio*) are due to protein aggregation and structure deterioration when subjected to freeze-thaw cycles. *Food Hydrocoll.* **2019**, *97*, 105223. [CrossRef]
46. Harnkarnsujarit, N.; Kawai, K.; Suzuki, T. Effects of Freezing Temperature and Water Activity on Microstructure, Color, and Protein Conformation of Freeze-Dried Bluefin Tuna (*Thunnus orientalis*). *Food Bioprocess Technol.* **2015**, *8*, 916–925. [CrossRef]
47. Wang, X.-Y.; Xie, J. Evaluation of water dynamics and protein changes in bigeye tuna (*Thunnus obesus*) during cold storage. *LWT* **2019**, *108*, 289–296. [CrossRef]
48. Jiang, Q.; Jia, R.; Nakazawa, N.; Hu, Y.; Osako, K.; Okazaki, E. Changes in protein properties and tissue histology of tuna meat as affected by salting and subsequent freezing. *Food Chem.* **2019**, *271*, 550–560. [CrossRef] [PubMed]
49. Xiao-Fei, W.; Min, Z.; Benu, A.; Jincai, S. Recent developments in novel freezing and thawing technologies applied to foods. *Crit. Rev. Food Sci. Nutr.* **2017**, *57*, 3620–3631. [CrossRef]
50. Anggo, A.D.; Ma'ruf, W.F.; Swastawati, F.; Rianingsih, L. Changes of Amino and Fatty Acids in Anchovy (*Stolephorus Sp*) Fermented Fish Paste with Different Fermentation Periods. *Procedia Environ. Sci.* **2015**, *23*, 58–63. [CrossRef]

51. Xicai, Z.; Wenbo, H.; Jing, X. Effect of different packaging methods on protein oxidation and degradation of grouper (*Epinephelus coioides*) during refrigerated storage. *Foods* **2019**, *8*, 325. [CrossRef]
52. Dabade, D.S.; Jacxsens, L.; Miclotte, L.; Abatih, E.; Devlieghere, F.; De Meulenaer, B. Survey of multiple biogenic amines and correlation to microbiological quality and free amino acids in foods. *Food Control* **2021**, *120*, 107497. [CrossRef]
53. Kimbuathong, N.; Leelaphiwat, P.; Harnkarnsujarit, N. Inhibition of melanosis and microbial growth in Pacific white shrimp (*Litopenaeus vannamei*) using high $CO_2$ modified atmosphere packaging. *Food Chem.* **2020**, *312*, 126114. [CrossRef] [PubMed]
54. Takahashi, M.; Hirose, N.; Ohno, S.; Arakaki, M.; Wada, K. Flavor characteristics and antioxidant capacities of hihatsumodoki (*Piper retrofractum* Vahl) fresh fruit at three edible maturity stages. *J. Food Sci. Technol.* **2018**, *55*, 1295–1305. [CrossRef]
55. Kang, D.C.; Gao, X.Q.; Ge, Q.F.; Zhou, G.H.; Zhang, W.G. Effects of ultrasound on the beef structure and water distribution during curing through protein degradation and modification. *Ultrason. Sonochem.* **2017**, *38*, 317–325. [CrossRef] [PubMed]
56. Zou, Y.; Kang, D.; Liu, R.; Qi, J.; Zhou, G.; Zhang, W. Effects of ultrasonic assisted cooking on the chemical profiles of taste and flavor of spiced beef. *Ultrason. Sonochem.* **2018**, *46*, 36–45. [CrossRef]
57. Laorenza, Y.; Harnkarnsujarit, N. Carvacrol, citral and alpha-terpineol essential oil incorporated biodegradable films for functional active packaging of Pacific white shrimp. *Food Chem.* **2021**, *363*, 130252. [CrossRef]

Article

# Mathematical Modelling of Ultrasound-Assisted Extraction Kinetics of Bioactive Compounds from Artichoke By-Products

**Cristina Reche, Carmen Rosselló, Mónica M. Umaña, Valeria Eim and Susana Simal ***

Department of Chemistry, University of the Balearic Islands, Ctra. Valldemossa km 7.5, 07122 Palma de Mallorca, Spain; cristina.reche@uib.es (C.R.); carmen.rossello@uib.es (C.R.); monica.umana@uib.es (M.M.U.); valeria.eim@uib.es (V.E.)

* Correspondence: susana.simal@uib.es; Tel.: +34-971172757

**Abstract:** Valorization of an artichoke by-product, rich in bioactive compounds, by ultrasound-assisted extraction, is proposed. The extraction yield curves of total phenolic content (TPC) and chlorogenic acid content (CAC) in 20% ethanol (*v/v*) with agitation (100 rpm) and ultrasound (200 and 335 W/L) were determined at 25, 40, and 60 °C. A mathematical model considering simultaneous diffusion and convection is proposed to simulate the extraction curves and to quantify both temperature and ultrasound power density effects in terms of the model parameters variation. The effective diffusion coefficient exhibited temperature dependence (72% increase for TPC from 25 °C to 60 °C), whereas the external mass transfer coefficient and the equilibrium extraction yield depended on both temperature (72% and 90% increases for TPC from 25 to 60 °C) and ultrasound power density (26 and 51% increases for TPC from 0 (agitation) to 335 W/L). The model allowed the accurate curves simulation, the average mean relative error being 5.3 ± 2.6%. Thus, the need of considering two resistances in series to satisfactorily simulate the extraction yield curves could be related to the diffusion of the bioactive compound from inside the vegetable cells toward the intercellular volume and from there, to the liquid phase.

**Keywords:** artichoke; by-products; ultrasound-assisted extraction; phenolic compounds; chlorogenic acid; diffusion model

**Citation:** Reche, C.; Rosselló, C.; Umaña, M.M.; Eim, V.; Simal, S. Mathematical Modelling of Ultrasound-Assisted Extraction Kinetics of Bioactive Compounds from Artichoke By-Products. *Foods* **2021**, *10*, 931. https://doi.org/10.3390/foods10050931

Academic Editor: Javier Raso

Received: 19 March 2021
Accepted: 21 April 2021
Published: 23 April 2021



## 1. Introduction

Large quantities of by-products are produced through the processing of plant-based foods whose re-introduction into a circular sustainable economy is important for reducing environmental problems generated by agro-industrial processes [1,2]. Moreover, in recent decades there has been considerable interest in the food industry in finding new sources of bioactive compounds with health and nutritional benefits. One of these sources is vegetable by-products because they are usually rich in antioxidants and food fiber [3].

Artichokes (*Cynara scolymus* L.) are widely grown and consumed in the Mediterranean region, and as such are considered a fundamental food of the Mediterranean diet [4]. During the industrial processing of artichoke, about 80–85% of the total plant biomass is rejected, being the heart of the artichoke the edible part [5]. This waste consists mainly of the artichoke bracts, stems, and leaves. Artichokes are a rich source of phenolic compounds, mainly chlorogenic acid, cynarine, caffeoylquinic acid, and flavonoids [6,7]. These compounds are responsible for the high antioxidant activity (16.76–180 mg of Trolox equivalents/g dry matter, dm) and other properties such as hepatoprotective, choleretics, anticarcinogenic, anti-HIV, diuretic, antifungal, and antibacterial properties [8,9]. Chlorogenic acid is a major artichoke phenolic compound, displaying a strong scavenging activity against reactive oxygen species and free radicals, and showing protective activity against oxidative damage [10].

For the recovery of bioactive compounds retained in the vegetable matrix, solid-liquid extraction is one of the most commonly used unit operations in the food industry [11].

One of the important variables that can be optimized to minimize the energy cost of the process is the temperature [12]. This variable affects not only the equilibrium and the mass transfer rates during the extraction but also, the yield in thermally labile compounds, so, a sufficiently high temperature can contribute to their degradation [13]. Standard extraction techniques usually use large amounts of organic solvents and imply high operating costs and long extraction times [14]. Thus, intensification of the extraction process is one of the areas with the greatest potential for development in the food industry. Among the techniques proposed to do this, power ultrasound is one of the most promising, allowing higher yields in shorter times and with lower energy costs [15].

The intensification of the extraction process by power ultrasound is related to several phenomena occurring simultaneously. One of the most important and studied is the cavitation phenomenon, which consists of the formation, growth and implosion of gas nano-microbubbles in the liquid, causing the molecules to oscillate around their equilibrium position, producing intermolecular distances to be continuously modified following alternate cycles of compression and decompression [16]. The implosion of cavitation bubbles leads to macro-turbulences and micromixing in the liquid media and microjets on the solid surface, which provokes surface peeling, erosion, and particle breakdown [17]. Furthermore, sonoporation, the cell walls' permeabilization, and the sonocapillary effect, both due to ultrasound, can increase the release of cellular content into the extractive medium.

Overall, the effects of the ultrasound on the solid/liquid system can alter the cell walls microstructure [18]. Vallespir et al. [19] observed the formation of micro-channels in mushroom cell walls after ultrasound-assisted convective drying, which were wider when the temperature increased. Llavata et al. [16] described in a review that according to a number of different authors, ultrasound treatment causes the weakening of the cellular structure and the creation of microcracks, which facilitate the mass transfer.

Mathematical modelling allows us to estimate, control and predict the behavior of a process under different extraction conditions; at the same time, it can contribute to the optimization of the process [20].

When two phases with different compositions, as in an extraction process, come into contact, the diffusing substance leaves its place in an area of high concentration and moves to an area of low concentration. To understand the relationship between the parameters involved in ultrasound-assisted extraction and the efficiency of the antioxidant compounds' recovery from natural sources, different modeling strategies have been proposed, most of them, by using empirical or kinetic models [21]. One of the proposed approaches considers that the extraction of active compounds is controlled by two-phase boundaries. The first explains the washing step where the compounds are dissolved into the solvent, and the second considers a diffusion step where the solutes from the internal wall cells diffuse into the solvent, this step being slower than the first extraction step due to mass transfer limitations. Different mathematical expressions are used to empirically represent this situation such as the parabolic diffusion model and the double exponential or two-site kinetic model. Mustapa et al. [22] and Natolino and Da Porto [23] used a double exponential model to satisfactorily simulate the extraction curves of phenolic compounds assisted by microwave from the Lindau herb, and assisted by ultrasound from grape marc, respectively.

Other authors have used models with a phenomenological approach. For the extraction of vegetable oils, the broken plus intact cells model has been proposed. This model considers the presence of broken cell walls, usually as a result of a pre-treatment, and intact cells that add an additional mass transfer resistance. In the broken cells, the main mechanism of mass transfer is convection, whereas the intact cell mass transfer is governed by diffusion [24].

To the best of our knowledge, the extraction process of bioactive compounds using power ultrasound from artichokes stems has not been modelled and simulated with a phenomenological model. The main objective of this work was to evaluate the application of acoustic energy as a technology to intensify the bioactive compounds extraction process (phenolic compounds and chlorogenic acid) from artichoke by-products and to propose a mathematical model to simulate the extraction kinetics at different temperatures (25, 40, and 60 °C) and ultrasound power densities (at 200 W/L and 335 W/L), comparing them with conventional mechanical agitation.

## 2. Materials and Methods

### 2.1. Chemicals

The following reagents were used for the different determinations: methanol HPLC grade, ethanol 96% (*v/v*) extra pure, acetonitrile HPLC grade, and Folin-Ciocalteu reagent, which were all from Scharlau (Barcelona, Spain); and orto phosphoric acid 85% (*v/v*) purchased from Panreac (Barcelona, Spain). The standard of chlorogenic acid > 95% was purchased from Sigma Aldrich Company Ltd. (Gillingham, UK).

### 2.2. Sample and Preparation

Artichokes (*Cynara scolymus*) grown in Majorca (Spain), selected with a stem diameter of 1.9 ± 0.2 cm, were purchased from a local supermarket, and stored at 4 °C for a maximum of one week.

To determine the total phenolic compounds content (TPC) and the chlorogenic acid content (CAC) of the artichoke parts, the artichokes were divided into three fractions: stem, bracts, and heart. Each part was frozen at −80 °C in an ultra-freezer (CVF 525/86 Ing. Climas, Spain), subsequently subjected to lyophilization (LyoQuest, Telstar, Spain) at 0.3 mbar and −50 °C, ground (A10, Ika Werke, Germany), sieved (FIT-0200, Filtra, Spain) to a particle size of less than 0.5 mm, vacuum packed (EVT-10, Tecnotrip, Spain) in 20/70 polyamide/polyethylene bags with permeability to $O_2$ of $2.58 \times 10^{-7}$ mol/m$^2$·s·Pa (Guerrero Coves SL, Valencia, Spain) and stored at 4 °C protected from light until processed (for a maximum of one month).

Methanolic extracts were prepared to determine the TPC and CAC. For this, samples were accurately weighed (~1.0 g), 40 mL of methanol were added, and the mixture was homogenized with the help of an Ultra-Turrax T25 Digital (Ika, Germany) for 60 s, at 13,000 rpm and 4 °C protected from light. The homogenate was refrigerated for 24 h. Subsequently, it was centrifuged at 4000 rpm for 10 min (ALC 4218, Thermo Scientific, Milano, Italy). The supernatant was filtered through Whatman No. 4 paper and stored at 4 °C in the dark until analysis.

The moisture content of the raw and freeze-dried samples was determined according to the official AOAC method 934.06 and the results were expressed as g of water per 1 g of artichoke (on a dry matter basis, dm).

### 2.3. Extraction Experiments

The experimental equipment used to carry out the extraction treatments in liquid medium was a jacketed glass container with a capacity of 600 mL connected to a thermostatic bath (Frigedor, J.P. Selecta, Barcelona, Spain) for temperature control. The selected extraction solvent was water-ethanol (80–20 *v/v*) due to the solubility of phenolic compounds and chlorogenic acid [6,25] and low toxicity and cost, used in a total volume of 200 mL and a solid/solvent ratio of 1:5 (*w/v*, g/mL).

Conventional extraction (A) was conducted with mechanical agitation (100 rpm) by using a conventional stirrer (RZR 2021, Heidolph, Germany) equipped with a four-blade propeller that describes a 50 mm diameter circle. For the extractions with ultrasound assistance (U), the mechanical stirring was replaced by a titanium sonotrode of 100 mm length (Hielscher Ultrasonics, Schwabach, Germany) working in cycles of 0.5 s and immersed 0.5 cm into the extraction solution. The sonotrode was connected to an ultrasonic generator (UP400S, Hielscher Ultrasound Technology, Teltow, Germany). To carry out the extraction experiments, two different sonotrodes of 22 mm (Ua experiments) and 14 mm (Ub experiments) diameter were used, obtaining two different ultrasound power densities which were characterized.

For the extraction experiments, the stem of the artichoke was selected and cut into 0.5 and 1 cm thick slabs. Extractions from 0.5 cm thick artichoke stem slabs were carried out with mechanical agitation and assisted by ultrasound at three different temperatures ($25 \pm 2$ °C, $40 \pm 2$ °C, and $60 \pm 1$ °C), because higher temperatures could produce bio-compounds' degradation. The extraction times were 4, 6, 8, 10, 12, 15, 20, 25, 30, and 35 min, since high extraction yields were observed at 35 min. Two additional experiments were carried out with 1 cm thick artichoke stem slabs at 40 °C, the first one with mechanical agitation (100 rpm), and the second one with ultrasound assistance (22 mm diameter sonotrode). In all the experiments, after the corresponding time, 2 mL of extract were taken by syringe, filtered with a 0.45 μm polytetrafluoroethylene (PTFE) filter and stored at 4 °C to subsequently determine both the chlorogenic acid and the total phenolic compounds contents.

### 2.4. Characterization of the Acoustic Field

To measure the acoustic power (W) supplied to the immersion fluid, a calorimetric technique was used. This method consists of measuring the temperature increase during the first 300 s of ultrasound application by using two K-type thermocouple probes connected to a data acquisition equipment HP 34970A Data Logger (COMARK, London, United Kingdom) [26]. The calculation of the ultrasound power was performed using Equation (1), based on the experimentally tripled temperature-time curve.

$$P = MC_P \frac{dT}{dt} \quad (1)$$

where P is the ultrasound power (W), M the mass of solvent (kg) T the temperature (K), t the exposure time (s) and $C_P$ the heat capacity of the solvent (J/kg K). The acoustic power densities (W/L) were calculated as the ratio between the ultrasound power (P) and the total extraction volume.

### 2.5. Determination of Total Phenolic Content

The TPC was spectrophotometrically determined, from both the methanolic extracts and those taken at the different points of the extraction process, according to the method of quantification of Folin-Ciocalteu phenols adapted to 96-well microplates described by Eim et al. [27]. The determinations were performed in triplicate and the results were expressed as mg of gallic acid equivalent (GAE) per 1 g of artichoke (on a dry matter basis, dm).

### 2.6. Determination of Chlorogenic Acid Content

The CAC was determined using the method described by Wang et al. [28], using a Thermo-Finnigan Surveyor HPLC analytical system (Thermo Scientific, Paris, France) equipped with a Waters 600E pump, a Waters 2966 photodiode matrix detector and a software-controlled Waters 717 plus autosampler (Empower). For this, 10 μL of extract were injected into a Prodigy 5 μm ODS-3 100 Å (5 μm; 3.2 × 150 mm) (Phenomenex, Torrace, CA, USA). The mobile phases were formed by 0.2% $H_3PO_4/H_2O$ (*v/v*, solvent A) and CH3CN (solvent B). A gradient flow was used starting with 94% A and 6% B and changing to 70% A and 30% B after 20 min. The flow rate was set at 1 mL/min, the temperature at 25 °C and the UV/vis detection wavelength at 330 nm.

The identification and quantification of the main peak were carried out by calibration with external standards (y = 26,837.1x + 13,177.4, $R^2$ = 0.9996, LOD = 0.067 mg/L, LOQ = 0.223 mg/L). The experiments were performed in triplicate and the results were expressed in mg chlorogenic acid per 1 g of artichoke (on a dry matter basis, dm).

### 2.7. Mathematical Modelling

The transport of bioactive compounds during the extraction process could take place through different and simultaneous mechanisms that contribute to the total flux. A mathematical model has been proposed to explain the mass transport by combining Fick's second law with the microscopic mass transfer. A constant and effective diffusion coefficient ($D_{eff}$, $m^2/s$) was considered, only temperature-dependent [29], although other mechanisms could coexist. Due to the presence of the external stem skin, it was also assumed that the mass transfer took place mainly through the axial direction; therefore, the solid was considered as an infinite slab. The governing equation obtained is as follows Equation (2):

$$\frac{\partial C_{sol}(x,t)}{\partial t} = D_{eff} \frac{\partial^2 C_{sol}(x,t)}{\partial x^2} \tag{2}$$

where $C_{sol}$ is the local concentration of the bioactive compound in the solid phase (mg/g dm), x the axis distance (m) and t is the extraction time (s). To solve this partial differential equation, the initial bioactive compounds content ($C_{o\ sol}$) was considered uniform throughout the slab and both the shape and the size of the slab considered constant during the process. Both solid symmetry and surface convection were the boundary conditions considered.

By using the separation of variables method [30], the following expression (Equation (3)) was obtained for the bioactive compounds content profile in an infinite slab [31], where L (m) is the half-thickness of the slab.

$$\frac{C_{sol} - C_{e\ sol}}{C_{o\ sol} - C_{e\ sol}} = \sum_{\nu=1}^{\infty} \frac{2 \sin(\gamma_\nu)}{\sin(\gamma_\nu)\cos(\gamma_\nu) + \gamma_\nu} e^{\left[-\frac{D_{eff}\, t}{L^2} \gamma_\nu^2\right] \cos\left[\gamma_\nu \frac{x}{L}\right]} \tag{3}$$

$$\gamma_\nu \tan \gamma_\nu = Bi \tag{4}$$

$$Bi = \frac{h_m L}{D_{eff}} \tag{5}$$

where $C_{e\ sol}$ is the equilibrium bioactive compounds content in the solid phase, and the $\gamma_\nu$ coefficients in Equation (3) are the roots of Equation (4). The Biot number (Bi) (Equation (5)) expresses the relative importance of the internal and the external resistances to mass transfer ($h_m$, m/s). High values of the Biot number ($Bi > 1$) mean that the internal transfer is limiting and, therefore, the extraction process is mainly controlled by diffusion. Conversely, low values of the Biot number ($Bi < 1$) indicate that the external transfer resistance is important. The Biot number calculation allows us to determine if the limiting factor is internal or external. If the internal transfer is limiting, the intensification of surface exchanges will not be affected [32].

The average TPC and CAC in the slab were estimated as the average of the local content ($C_{sol}$) calculated using Equations (3)–(5) in 30 equally distributed positions along L and using enough terms of Equation (3) to achieve a change lower than 1%. The TPC and CAC in the liquid phase at time t were assumed to be the difference between the initial slab contents ($C_{o\ sol}$) and the estimated average slab contents at that time. The TPC and CAC extraction yields (%) were estimated in percentage as the average contents in the liquid phase divided by the initial contents in the solid phase.

The 'nlinfit' function of the optimization toolbox of Matlab R2020b (The MathWorks Inc., Natick, MA, USA), which estimates the coefficients of a nonlinear regression function and the residuals using least squares, was used to identify the model parameters. The mean relative error (MRE, Equation (6)), calculated by comparison of the experimental ($C_{exp}$) and calculated ($C_{cal}$) concentrations, was used as the objective function to minimize. N is the number of experimental data. The MRE has been used in the literature to assess the quality of fit for different mathematical models and it is generally accepted that MRE values below 10% provide a good fit [33].

$$MRE = \frac{100}{N} \sum_{i=1}^{N} \frac{|C_{exp\ i} - C_{cal\ i}|}{C_{exp\ i}} \tag{6}$$

### 2.8. Statistical Analysis

For the statistical analysis of the experimental results, R 3.1.0 software [34] together with RStudio IDE [35] were used. Shapiro-Wilk and Levene tests were used to evaluate the normality of data distribution and the homogeneity of variances, respectively. When data were normally distributed and exhibited homogeneity of variances, the parametric ANOVA and Tukey's tests were used to evaluate the existence and extent of significant differences among samples, respectively. These methods were replaced by robust statistical methods when data were not normally distributed and/or exhibited heterogeneity of variances. In this case, the 't1way' and 'lincon' R functions (WRS2 package), a heteroscedastic one-way ANOVA which uses a generalization of Welch's method, and pairwise comparisons at $\alpha = 0.05$, respectively [36], were used with trimmed means.

To perform a thorough and objective evaluation of the accuracy of the simulation provided by the proposed mathematical model, the MRE (Equation (6)) and the percentage of explained variance (%var, Equation (7) were determined by a comparison between the experimental results and those provided by the model. $S_{yx}$ and $S_y$ are the standard deviations of the estimation and the sample, respectively.

$$\%var = \left[1 - \frac{S_{yx}}{S_y}\right]100 \tag{7}$$

## 3. Results and Discussion

### 3.1. Phenolic Compounds of the Artichoke

The TPC and CAC of the three different parts of the artichoke (heart, stem, and bracts) were determined and compared in order to select the best by-product to work with in the subsequent extractions (stem or bracts). The results are shown in Table 1 together with the moisture contents.

**Table 1.** Moisture content, total phenolic content (TPC) and chlorogenic acid content (CAC) of the artichoke heart, stem, and bracts.

| | Moisture (g Water/g dm) | TPC (mg/g dm) | CAC (mg/g dm) |
|---|---|---|---|
| Heart | 4.38 ± 0.20 a | 48.9 ± 1.4 a | 24.2 ± 1.0 a |
| Stem | 7.00 ± 0.33 c | 45.7 ± 2.1 a | 21.4 ± 1.1 a |
| Bracts | 5.10 ± 0.29 b | 27.4 ± 1.6 b | 5.8 ± 1.3 b |

Means with different letter for the same column showed significant differences according to the Tukey's test ($p < 0.05$).

According to the FDA, the average moisture content of artichoke is of 5.28 g/g dm [37]. In this study, the moisture content of the three parts exhibited significant ($p < 0.05$) differences. The TPC was lower in the bracts than in both the heart and the stem, with no significant differences ($p < 0.05$) observed between the latter two. The CAC showed a similar trend to that observed in the phenolic compounds, being significantly lower ($p < 0.05$) in the bracts than in the other two parts. Lavecchia et al. [38] analyzed artichoke residues and determined the phenolic compounds in the stem and bracts obtaining values of 51.10 ± 0.74 and 24.58 ± 0.57 mg GAE/g dm, respectively. Song et al. [39] analyzed the chlorogenic acid content of the heart, obtaining a value of 23.81 ± 0.01 mg/g dm.

The heart is the consumed part and therefore is not a by-product. The TPC and CAC of the stem were similar to those in the artichoke heart and significantly ($p < 0.05$) higher than those in the bracts. Therefore, the stem was selected as the raw material for the extraction experiments.

### 3.2. Ultrasonic Power Characterization

The acoustic power density (Pot, W/L) was calculated, as described previously, as the ratio between the ultrasound power applied (P) and the total extraction volume. The measured ultrasound power densities were 200 ± 5 W/L for the 22 mm diameter sonotrode (Ua experiments) and 335 ± 31 W/L for the 14 mm diameter sonotrode (Ub experiments).

### 3.3. TPC and CAC Extraction Yields

The experiments of bioactive compounds extraction from the artichoke stem slabs (0.5 cm thickness) were carried out at different temperatures (25, 40, and 60 °C) with mechanical stirring (A) and with acoustic assistance (Ua at 200 W/L and Ub at 335 W/L) for 35 min. The TPC and CAC were measured in the extracts at different times, and the yields of extraction were estimated as the percentages of the initial contents in the artichoke stem that had moved from the solid matrix to the solvent. These results are shown in Figure 1 as average values and deviations (dots).

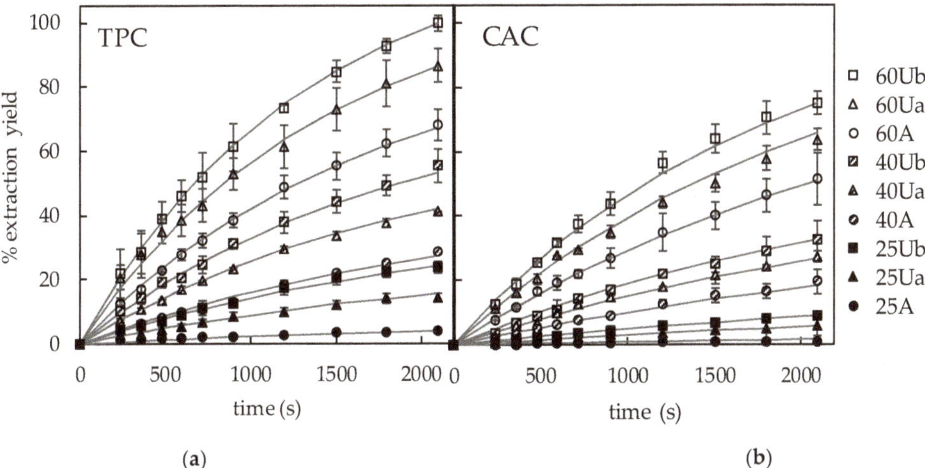


**Figure 1.** Experimental and simulated extraction yield curves of total phenolic content (TPC) (**a**) and chlorogenic acid content (CAC) (**b**) from artichoke stem slabs (0.5 cm thickness) at different temperatures (25, 40 and 60 °C) with agitation (A, 100 rpm) and acoustic assistance (Ua, 200W/L and Ub, 335 W/L).

A total of three groups of experiments can be observed in these figures depending on the temperature. The extraction yield curves at 25 °C were located at the bottom of each figure, those at 60 °C at the top, and the curves of experiments at 40 °C were located in the middle areas. Thus, the influence of temperature on the extraction yield was remarkable. Moreover, in each of these groups, the extraction yield increased when ultrasound was applied, the increase being higher when the ultrasound power density was higher.

The TPC yield increased, after 35 min of extraction with agitation, from 4–68% when the temperature increased from 25–60 °C. In the case of the ultrasound-assisted extraction, the yield increased from 13–82% in Ua experiments (200 W/L) and from 24–99% in Ub experiments (335 W/L) for the same time and temperature interval. With regard to the CAC extraction yield after 35 min, it increased from 1% at 25 °C to 46% at 60 °C for the experiments with agitation, while the increases in Ua experiments were from 6–60%, and from 8–75% in Ub experiments, for the same temperature increase.

Roselló-Soto et al. [40] observed a 10% increase in phenolic compounds extraction from tiger nut by-products in 50% ethanol at 50 °C with mechanical agitation, in comparison with the same extraction carried out at 25 °C. Irakli et al. [41] carried out the phenolic compounds extraction from olive leaves in an ultrasound bath (frequency 37 kHz) with different solvents and for different times and temperatures (25, 45, and 60 °C). These authors observed that when optimal extraction conditions were applied (50% acetone as solvent and 10 min of extraction time), TPC increased by 9% at 60 °C in comparison with the TPC at 25 °C. Angelov et al. [42] carried out the extraction of phenolic compounds from the stems and the leaves of an artichoke under agitation in thermostatic shaker and observed a 105% increase from the extraction carried out at 20 °C to the extraction carried out at 100 °C.

According to the literature, extracting bioactive compounds has a temperature-labile limit [43], and using temperature above 75 °C may promote significant degradation losses [44]. Moreover, mechanical agitation can be inefficient in extracting the parts of phenolic compounds that can be ester bound or trapped within proteins and polysaccharides, on cell walls [45]. After 35 min of extraction at 25 °C, the TPC and CAC yields increased from 4% and 1%, respectively, in the experiment with agitation, up to 24% and 8%, respectively, when 335 W/L of acoustic energy was applied. The application of ultrasound at 335 W/L caused increases in the TPC and CAC yields, in comparison to the experiments with agitation, of 93% and 67% at 40 °C, and 43% and 62% at 60 °C, respectively. Similar extraction yields of TPC were achieved at 25 °C with the application of power ultrasound (335 W/L) to those obtained at 40 °C by mechanical stirring (100 rpm).

Umaña et al. [33] observed a 193% increase in the ergosterol extraction yield from mushroom stalks in 70% ethanol at 25 °C and 321 W/L of ultrasound power density in comparison to the extraction with mechanical agitation assistance (130 rpm). Arruda et al. [46] carried out the extraction of phenolic compounds and chlorogenic acid from dried araticum peel with 50% ethanol-water (*v/v*), at a maximum temperature of 40.4 °C and applying ultrasound with a sonotrode of 13 mm of diameter, reaching ultrasound power from 160 W to 640 W. These authors reported extraction yields about 41% higher for TPC and 82% higher for CAC when extracting with the highest ultrasound power (640 W) compared with the lowest (160 W).

### 3.4. Mathematical Modelling

The proposed mathematical model parameters were identified for the TPC and CAC extraction curves. In a preliminary identification step, $D_{eff}$, Bi and $C_e$ for TPC and CAC extraction kinetics were identified for each tested extraction condition. Once the figures for these parameters were obtained, it was observed that the effective diffusion coefficient was temperature-dependent, according to an Arrhenius type equation (Equation (8)), where $D_0$ is the pre-exponential factor ($m^2/s$) and $E_a$ is the activation energy (J/mol K), R is the gas constant (8.31 J/mol K), and T is the extraction temperature (°C). Moreover, by assigning a zero for the ultrasound power density variable in the experiments carried out with mechanical agitation, the Biot number exhibited a linear dependence with the ultrasound power density (Equation (9)), and the equilibrium content in the liquid phase (extract) was linearly dependent on both the ultrasound power density and the extraction temperature (Equation (10)).

$$D_{eff} = D_0\, e^{\left[\frac{E_a}{R(T+273)}\right]} \tag{8}$$

$$Bi = Bi_0 + Bi_1\, Pot \tag{9}$$

$$C_e = C_{e0} + C_{e1}\, Pot + C_{e2}\, T \tag{10}$$

In a second identification step, all the experimental data obtained at the different temperatures, ultrasound power densities, and extraction times were simultaneously used to identify the seven model parameters to simulate the TPC and CAC extraction yields. The identified figures for the parameters of Equations (8)–(10) are shown in Table 2, for both the TPC and CAC extraction kinetics, together with the confidence interval ($p < 0.05$) and standard error for each parameter. As it can be seen, the standard errors were relatively high since a large number of parameters were simultaneously estimated.

According to Equation (8) and figures in Table 2, the effective diffusion coefficient varied from $1.42 \times 10^{-7}$ $m^2/s$ at 25 °C to $2.43 \times 10^{-7}$ $m^2/s$ at 60 °C for TPC extraction and from $9.44 \times 10^{-7}$ $m^2/s$ at 25 °C to $1.79 \times 10^{-6}$ $m^2/s$ at 60 °C for CAC extraction, thus, increases of 71.6% and 90.1%, respectively.

**Table 2.** Identified parameters for the proposed model and confidence interval (CI) and standard error (SE) for each parameter. Extraction of total phenolic content (TPC) and chlorogenic acid content (CAC) from artichoke stem at different temperatures under agitation and acoustic assistance.

| | TPC | | | CAC | | |
|---|---|---|---|---|---|---|
| | Value | CI | SE | Value | CI | SE |
| $D_o$ ($m^2/s$) | $2.41 \times 10^{-5}$ | $[0.93 \times 10^{-5}, 3.89 \times 10^{-5}]$ | $6.34 \times 10^{-6}$ | $4.25 \times 10^{-4}$ | $[2.73 \times 10^{-4}, 5.77 \times 10^{-4}]$ | $2.37 \times 10^{-1}$ |
| $E_a$ (kJ/mol K) | 12.7 | [12.0, 13.5] | $4.44 \times 10^{-5}$ | 15.1 | [12.7, 17.5] | $1.35 \times 10^{-4}$ |
| $Bi_0$ | $1.55 \times 10^{-2}$ | $[0.54 \times 10^{-2}, 2.56 \times 10^{-2}]$ | $3.36 \times 10^{-1}$ | $1.31 \times 10^{-3}$ | $[0.46 \times 10^{-3}, 2.16 \times 10^{-3}]$ | $7.33 \times 10^{-1}$ |
| $Bi_1$ (L/W) | $1.18 \times 10^{-5}$ | $[0.26 \times 10^{-5}, 2.15 \times 10^{-5}]$ | $2.85 \times 10^{-4}$ | $2.00 \times 10^{-6}$ | $[0.23 \times 10^{-6}, 3.77 \times 10^{-6}]$ | $1.12 \times 10^{-3}$ |
| $C_{e0}$ (mg/g dm) | −24.4 | [−25.2, −23.6] | 0.39 | −12.3 | [−12.9, −11.8] | 0.27 |
| $C_{e1}$ (mg L/g dm W) | $4.41 \times 10^{-2}$ | $[4.26 \times 10^{-2}, 4.57 \times 10^{-2}]$ | $7.8 \times 10^{-4}$ | $8.98 \times 10^{-3}$ | $[6.47 \times 10^{-3}, 11.49 \times 10^{-3}]$ | $5.3 \times 10^{-4}$ |
| $C_{e2}$ (mg/g dm °C) | 1.12 | [1.10, 1.13] | $7.4 \times 10^{-3}$ | 0.54 | [0.53, 0.55] | $5.0 \times 10^{-3}$ |

It can be assumed that the temperature increase favored extraction by improving the solubility of the phenolic compounds. This trend could be observed in the study conducted by Türker and Erdoğdu [47] who concluded that the diffusion coefficient increased by ca 106% in the extraction of pigments from black carrot (*Daucus carota* L.) when the temperature increased from 25 °C ($1.8 \times 10^{-11}$ $m^2/s$) to 50 °C ($3.7 \times 10^{-11}$ $m^2/s$).

The estimated activation energy ($E_a$) for the TPC and CAC extraction processes was 12.7 kJ/mol and 15.1 kJ/mol, respectively. The parameter $E_a$ may depend on different factors related to the extraction process, such as the structures of both the solid matrix and the bioactive compounds, the treatment of the sample before the extraction, the solvent used, among others [48]. According to the literature, when the $E_a$ is lower than 20 kJ/mol, the extraction rate is mainly controlled by the diffusion process [49]. Tao et al. [50] estimated 7.0 kJ/mol of activation energy for the extraction of phenolic compounds from grape pomace in an ultrasonic bath system at 25 kHz.

The low figures obtained for the Biot number (from $1.55 \times 10^{-2}$ at 0 W to $1.94 \times 10^{-2}$ at 335 W/L for the TPC extraction, and from $1.31 \times 10^{-3}$ at 0 W to $1.98 \times 10^{-3}$ at 335 W/L for the CAC extraction) are indicative of the relative importance of the external mass transfer resistance. The increase of Biot number was also observed in the extraction of paclitaxel from Taxus Chinensis in an ultrasound bath at 25 °C, by Yoo and Kim [51], from 4.77 at 180 W to 6.50 at 380 W of ultrasound power.

According to Equation (5), the external mass transfer coefficient ($h_m$) for the extraction of TPC varied with both the temperature and the ultrasound power density, from $8.79 \times 10^{-7}$ m/s for the extraction by mechanical agitation (0 W/L of ultrasound) at 25 °C to $1.89 \times 10^{-6}$ m/s at 335 W/L and 60 °C, thus a 71.6% increase from 25 °C to 60 °C and a 25.5% increase from 0 to 335 W/L. For the CAC extraction, $h_m$ varied from $4.95 \times 10^{-7}$ m/s for the extraction by mechanical agitation (0 W/L of ultrasound) at 25 °C to $1.42 \times 10^{-6}$ m/s at 335 W/L and 60 °C, thus a 90.1% increase from 25–60 °C and a 51.1% increase from 0–335 W/L. In both cases, it can be observed that the effect of the temperature was higher, although the ultrasound assistance also caused significant $h_m$ increases.

The equilibrium TPC and CAC in the liquid phase depended on both the ultrasonic power density and the temperature. When the ultrasonic power density increased from 0–335 W/L, $C_e$ increased 417%, 73% and 35% for the TPC extraction, and 262%, 33% and 15% for the CAC extraction, at 25, 40, and 60 °C, respectively. Thus, the effect of the ultrasound assistance on $C_e$ was higher at lower temperatures. On the other hand, when the temperature increased from 25–60 °C, $C_e$ increased by 1103%, 316%, and 214% for the TPC extraction, and 1642%, 641%, and 454% for the CAC extraction, at 0, 200, and 335 W/L of ultrasound power density, respectively.

Using the parameter figures shown in Table 2 together with the model equations (Equations (3)–(5) and (8)–(10)), the extraction curves were simulated for the different extraction conditions and represented in Figure 1 as continuous lines. The accuracy of the simulation obtained with the proposed model and, therefore, its ability to represent the experimental results was also statistically evaluated by using the mean relative error (MRE, Equation (6)) and the percentage of explained variance (%var, Equation (7)). Thus, Table 3 shows the MRE and %var obtained by comparison of the experimental and calculated TPC and CAC in the liquid phase (extracts). It can be seen in Figure 1 that the model provided a satisfactory simulation of the extraction curves. However, as the extraction yield at 25 °C with mechanical agitation was very low even after 35 min (<4% of TPC and <1.1% of CAC), therefore, the accuracy of the simulation was not satisfactory, with MRE>15% and %var < 95%. For the rest of the experiments, the average MRE and %var were 3.7 ± 0.5% and 99.4 ± 0.5% for the simulation of the TPC extraction curves and 6.4 ± 3.0% and 99.3 ± 0.4% for the simulation of the CAC extraction curves.

**Table 3.** Mean relative error (MRE) and percentage of explained variance (%var) estimated by comparison of experimental and simulated extraction kinetics of total phenolic content (TPC) and chlorogenic acid content (CAC) from artichoke stem slabs (0.5 cm thickness) at different temperatures with agitation (A) and ultrasound assistance (Ua, 200W/L and Ub, 335 W/L).

| | | TPC | | CAC | |
|---|---|---|---|---|---|
| | **T (°C)** | **MRE (%)** | **%var** | **MRE (%)** | **%var** |
| A | 25 | >15 | <95 | >15 | <95 |
| | 40 | 6.2 | 98.5 | 8.3 | 98.5 |
| | 60 | 1.9 | 99.9 | 1.8 | 99.9 |
| Ua | 25 | 5.4 | 98.9 | 9.7 | 99.2 |
| | 40 | 1.9 | 99.8 | 8.3 | 99.1 |
| | 60 | 4.5 | 99.5 | 3.7 | 99.3 |
| Ub | 25 | 4.7 | 99.4 | 7.8 | 99.2 |
| | 40 | 3.0 | 99.6 | 8.5 | 99.3 |
| | 60 | 1.9 | 99.9 | 3.3 | 99.6 |
| | Average: | 3.7 ± 0.5 | 99.4 ± 0.5 | 6.4 ± 3.0 | 99.3 ± 0.4 |

To further evaluate the robustness of the proposed model, two additional experiments, not used in the model parameters identification, carried out with artichoke slabs of double thickness (1 cm) at 40 °C and agitation (100 rpm) or 200 W/L of ultrasound assistance, were simulated. Figure 2 shows the experimental (dots) and simulated (lines) extraction TPC and CAC yield curves for these experiments. As can be seen in this figure, the simulation provided by the model was satisfactory, even when a different particle size was used. The average MRE and %var were 6.7 ± 2.5% and 99.3 ± 0.5%, respectively, for the TPC extraction, and 5.5 ± 0.2% and 99.3 ± 0.1%, respectively, for the CAC extraction.

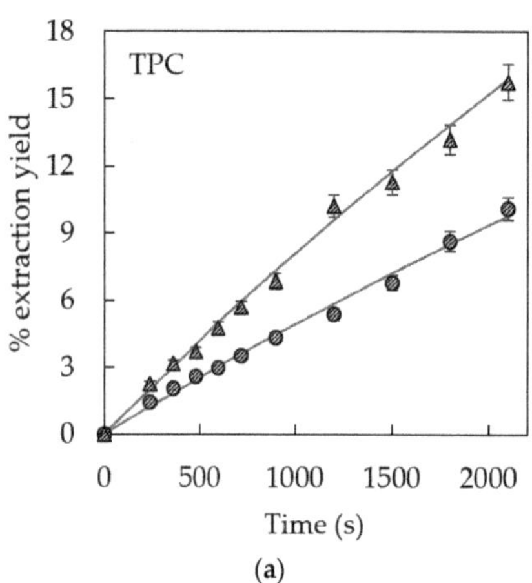


(a)

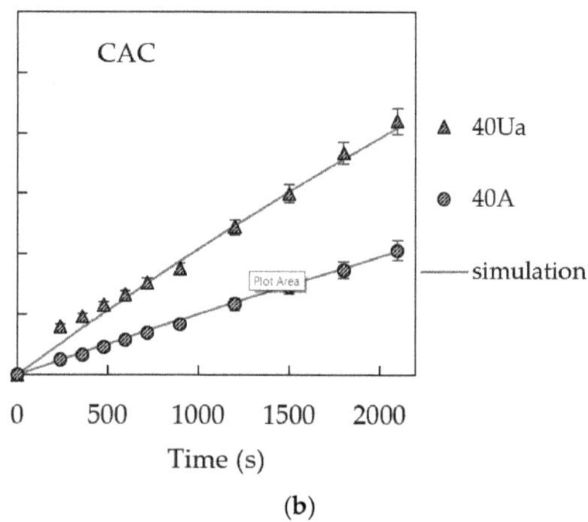


(b)

**Figure 2.** Experimental and simulated extraction yield curves of (**a**) total polyphenol content (TPC) and (**b**) chlorogenic acid content (CAC) from artichoke stem slabs (1 cm thickness) at 40 °C with agitation (A, 100 rpm) and acoustic assistance (Ua, 200 W/L).

Figure 3 shows the simulated vs. experimental TPC and CAC extraction yields for all the experiments carried out with both slab thicknesses, the linear regression of these data (slope and y-intercept) and predicted bounds at 95% of confidence. The coefficients of determination were >0.99, the slopes were not significantly different to 1 ($p > 0.05$) and the y-intercepts were not significantly different to zero ($p > 0.05$), indicating a good agreement between both groups of data, experimental and simulated.

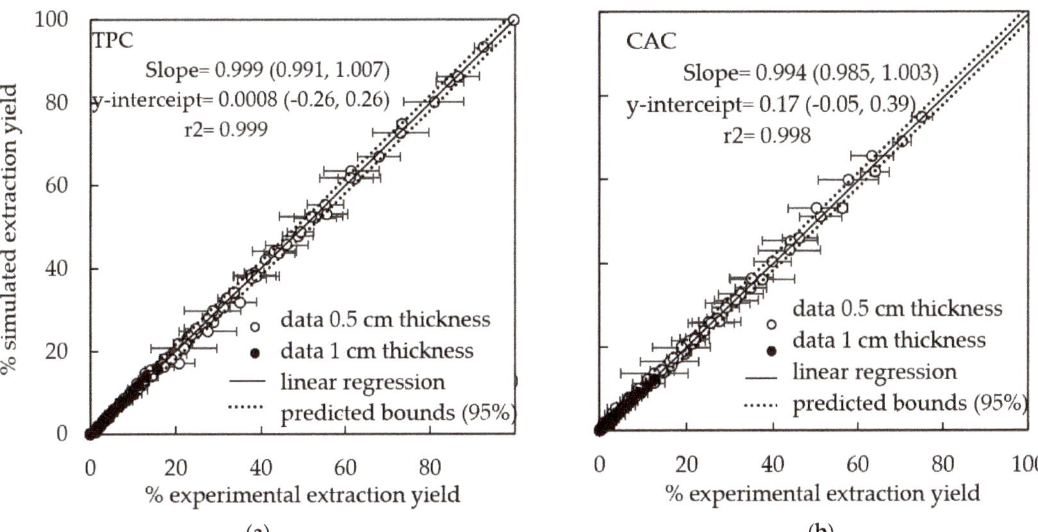


**Figure 3.** Simulated vs. experimental (**a**) total polyphenol content (TPC) and (**b**) chlorogenic acid content (CAC) extraction yields from artichoke stem, linear regression (slope and y-intercept), and predicted bounds at 95% of confidence. Experiments carried out with slabs of 0.5 cm thickness at different temperatures (25, 40 and 60 °C) with agitation (100 rpm) and acoustic assistance (200 W/L and 335 W/L) and with slabs of 1 cm thickness at 40 °C with agitation (100 rpm) and acoustic assistance at 200 W/L.

The proposed model was developed by assuming that two resistances in series simultaneously controlled the mass transfer rate during the extraction process, one associated with diffusion and a second one, to convection. To evaluate the necessity of these two resistances, the mathematical model was also solved by considering the external resistance to mass transfer to be negligible. In this case, the model simulation was very unsatisfactory, with MRE > 20% and %var < 80% (results not shown). The artichoke stem is made up of vegetable cells, containing the bioactive compounds mainly in their interior [52]. The major resistance of molecular diffusion in materials of plant origin always comes from the adhering membranes to the intercellular area and cell walls to the liquid phase [53]. Thus, as it is considered in the broken plus intact cells model, two mechanisms of mass transfer can contribute to the extraction kinetics: convection and diffusion. The contribution of each one is different depending on the extraction conditions, agitation or ultrasound assistance, and temperature. It can also change during the extraction time, as the effects of the ultrasound alter the microstructure of the vegetable matrix.

## 4. Conclusions

According to the results obtained in this study, it can be concluded that not only temperature causes a considerable increase in the bioactive compounds extraction yield from artichoke stem, but the ultrasound assistance as well, in comparison with conventional solid-liquid extraction with mechanical agitation assistance. The effects of the ultrasound assistance were greater at lower temperatures, i.e., after 35 min of extraction at 25 °C, the TPC and CAC yields increased by 525% and 700%, respectively, when 335 W/L of ultrasound power density were used, in comparison with the experiments with 100 rpm of mechanical agitation. Taking into account environmental aspects, it could be interesting for the food industry to consider ultrasound as a technology of extraction process intensification since ultrasound application could allow obtaining similar or higher extraction yields of phenolic compounds using lower temperatures and ecofriendly solvents. A phenomenological model has been proposed to simulate the extraction yield curves. This model allowed the quantification of both temperature and ultrasound power density effects in terms of the variation of the model parameters with both variables. The need to consider two series resistance to satisfactorily simulate the extraction yield curves was observed, which could be related to the diffusion of the bioactive compound from inside the vegetable cells toward the intercellular volume and from there, to the liquid phase. The effective diffusion coefficient exhibited temperature dependence according to an Arrhenius equation, whereas the external mass transfer coefficient and the equilibrium extraction yield depended on both the temperature and ultrasound power density. The model allowed the accurate simulation of the extraction yield curves even when different solid particle sizes were used, the average mean relative error of the simulation being of 5.3 ± 2.6%. Finally, the model proposed not only provides insights into the extraction process but is also a useful tool for prediction and optimization.

**Author Contributions:** Conceptualization, C.R. (Carmen Rosselló), V.E. and S.S.; Data curation, C.R. (Cristina Reche); Funding acquisition, S.S.; Investigation, C.R. (Cristina Reche); Methodology, C.R. (Cristina Reche) and M.M.U.; Project administration, C.R. (Carmen Rosselló) and S.S.; Resources, C.R. (Carmen Rosselló); Software, S.S.; Supervision, C.R. (Carmen Rosselló), V.E. and S.S.; Validation, C.R. (Cristina Reche) and M.M.U.; Visualization, S.S.; Writing—original draft, C.R. (Cristina Reche); Writing—review and editing, C.R. (Carmen Rosselló), M.M.U., V.E. and S.S. All authors have read and agreed to the published version of the manuscript.

**Funding:** This research was funded by the National Institute of Research and Agro-Food Technology (INIA) and co-financed with ERDF funds (RTA2015-00060-C04-03), the AEI (Spanish research agency), the MICI (Ministry of Science and Innovation), ERDF funds EU, and AEI (PID2019-106148RR-C43 / AEI / 10.13039/501100011033), and by the Balearic Government for the research fellowship (FPI/2211/2019) co-financed with ERDF funds. The APC was funded by PID2019-106148RR-C43 / AEI / 10.13039/501100011033.

**Institutional Review Board Statement:** Not applicable.

**Informed Consent Statement:** Not applicable.

**Data Availability Statement:** Data is contained within the article.

**Acknowledgments:** We thank Josep Agustí Pablo, from the "Serveis Cientificotècnics" of the University of the Balearic Islands for his assistance with HPLC.

**Conflicts of Interest:** The authors declare no conflict of interest.

## References

1. Bas-Bellver, C.; Barrera, C.; Betoret, N.; Seguí, L. Turning Agri-Food Cooperative Vegetable Residues into Functional Powdered Ingredients for the Food Industry. *Sustainability* **2020**, *12*, 1284. [CrossRef]
2. Ramos, M.; Dominici, F.; Luzi, F.; Jiménez, A.; Garrigós, M.C.; Torre, L.; Puglia, D. Effect of Almond Shell Waste on Physicochemical Properties of Polyester-Based Biocomposites. *Polymers* **2020**, *12*, 835. [CrossRef]
3. Kumar, K.; Srivastav, S.; Sharanagat, V.S. Ultrasound assisted extraction (UAE) of bioactive compounds from fruit and vegetable processing by-products: A review. *Ultrason. Sonochem.* **2021**, *70*, 105325. [CrossRef]
4. Sahebkar, A.; Pirro, M.; Banach, M.; Mikhailidis, D.P.; Atkin, S.L.; Cicero, A.F.G. Lipid-lowering activity of artichoke extracts: A systematic review and meta-analysis. *Crit. Rev. Food Sci. Nutr.* **2018**, *58*, 2549–2556. [CrossRef] [PubMed]
5. Boubaker, M.; El Omri, A.; Blecker, C.; Bouzouita, N. Fibre concentrate from artichoke (*Cynara scolymus* L.) stem by-products: Characterization and application as a bakery product ingredient. *Food Sci. Technol. Int.* **2016**, *22*, 759–768. [CrossRef] [PubMed]
6. Rabelo, R.S.; Machado, M.T.; Martínez, J.; Hubinger, M.D. Ultrasound assisted extraction and nanofiltration of phenolic compounds from artichoke solid wastes. *J. Food Eng.* **2016**, *178*, 170–180. [CrossRef]

7. El Sayed, A.M.; Hussein, R.; Motaal, A.A.; Fouad, M.A.; Aziz, M.A.; El-Sayed, A. Artichoke edible parts are hepatoprotective as commercial leaf preparation. *Rev. Bras. Farm.* **2018**, *28*, 165–178. [CrossRef]
8. Lucera, A.; Costa, C.; Marinelli, V.; Saccotelli, M.A.; Del Nobile, M.A.; Conte, A. Fruit and Vegetable by-Products to Fortify Spreadable Cheese. *Antioxidants* **2018**, *7*, 61. [CrossRef]
9. Yang, M.; Ma, Y.; Wang, Z.; Khan, A.; Zhou, W.; Zhao, T.; Cao, J.; Cheng, G.; Cai, S. Phenolic constituents, antioxidant and cytoprotective activities of crude extract and fractions from cultivated artichoke inflorescence. *Ind. Crops Prod.* **2020**, *143*, 111433. [CrossRef]
10. Avio, L.; Maggini, R.; Ujvári, G.; Incrocci, L.; Giovannetti, M.; Turrini, A. Phenolics content and antioxidant activity in the leaves of two artichoke cultivars are differentially affected by six mycorrhizal symbionts. *Sci. Hortic.* **2020**, *264*, 109153. [CrossRef]
11. Panzella, L.; Moccia, F.; Nasti, R.; Marzorati, S.; Verotta, L.; Napolitano, A. Bioactive Phenolic Compounds from Agri-Food Wastes: An Update on Green and Sustainable Extraction Methodologies. *Front. Nutr.* **2020**, *7*, 60. [CrossRef] [PubMed]
12. Rajha, H.N.; El Darra, N.; Hobaika, Z.; Boussetta, N.; Vorobiev, E.; Maroun, R.G.; Louka, N. Extraction of Total Phenolic Compounds, Flavonoids, Anthocyanins and Tannins from Grape Byproducts by Response Surface Methodology. Influence of Solid-Liquid Ratio, Particle Size, Time, Temperature and Solvent Mixtures on the Optimization Process. *Food Nutr. Sci.* **2014**, *5*, 397–409. [CrossRef]
13. López-Vidaña, E.C.; Figueroa, I.P.; Cortés, F.B.; Rojano, B.A.; Ocaña, A.N. Effect of temperature on antioxidant capacity during drying process of mortiño (*Vaccinium meridionale* Swartz). *Int. J. Food Prop.* **2017**, *20*, 294–305. [CrossRef]
14. Ivanović, M.; Razboršek, M.I.; Kolar, M. Innovative Extraction Techniques Using Deep Eutectic Solvents and Analytical Methods for the Isolation and Characterization of Natural Bioactive Compounds from Plant Material. *Plants* **2020**, *9*, 1428. [CrossRef] [PubMed]
15. Garofalo, S.F.; Tommasi, T.; Fino, D. A short review of green extraction technologies for rice bran oil. *Biomass Convers. Biorefinery* **2021**, *11*, 569–587. [CrossRef]
16. Llavata, B.; García-Pérez, J.V.; Simal, S.; Cárcel, J.A. Innovative pre-treatments to enhance food drying: A current review. *Curr. Opin. Food Sci.* **2020**, *35*, 20–26. [CrossRef]
17. Chemat, F.; Rombaut, N.; Sicaire, A.-G.; Meullemiestre, A.; Fabiano-Tixier, A.-S.; Abert-Vian, M. Ultrasound assisted extraction of food and natural products. Mechanisms, techniques, combinations, protocols and applications. A review. *Ultrason. Sonochem.* **2017**, *34*, 540–560. [CrossRef]
18. Khemakhem, I.; Ahmad-Qasem, M.H.; Catalán, E.B.; Micol, V.; García-Pérez, J.V.; Ayadi, M.A.; Bouaziz, M. Kinetic improvement of olive leaves' bioactive compounds extraction by using power ultrasound in a wide temperature range. *Ultrason. Sonochem.* **2017**, *34*, 466–473. [CrossRef]
19. Vallespir, F.; Crescenzo, L.; Rodríguez, Ó.; Marra, F.; Simal, S. Intensification of Low-Temperature Drying of Mushroom by Means of Power Ultrasound: Effects on Drying Kinetics and Quality Parameters. *Food Bioprocess Technol.* **2019**, *12*, 839–851. [CrossRef]
20. Giametta, F.; Fucci, F.; Catalano, P.; La Fianza, G. Thermo-fluid-dynamic modelling of a cold store for cheese maturation. *J. Agric. Eng.* **2013**, *43*, e26. [CrossRef]
21. Gullón, P.; Gullón, B.; Astray, G.; Costa, P.; Lorenzo, J.M. Modeling approaches to optimize the recovery of polyphenols using ultrasound-assisted extraction. In *Design and Optimization of Innovative Food Processing Techniques Assisted by Ultrasound*; Elsevier BV: Amsterdam, The Netherlands, 2021; pp. 15–38.
22. Mustapa, A.; Martin, A.; Gallego, J.; Mato, R.; Cocero, M. Microwave-assisted extraction of polyphenols from Clinacanthus nutans Lindau medicinal plant: Energy perspective and kinetics modeling. *Chem. Eng. Process. Process. Intensif.* **2015**, *97*, 66–74. [CrossRef]
23. Natolino, A.; Da Porto, C. Kinetic models for conventional and ultrasound assistant extraction of polyphenols from defatted fresh and distilled grape marc and its main components skins and seeds. *Chem. Eng. Res. Des.* **2020**, *156*, 1–12. [CrossRef]
24. Comerlatto, A.; Voll, F.A.; Daga, A.L.; Fontana, É. Mass transfer in soybean oil extraction using ethanol/isopropyl alcohol mixtures. *Int. J. Heat Mass Transf.* **2021**, *165*, 120630. [CrossRef]
25. Zuorro, A.; Maffei, G.; Lavecchia, R. Effect of solvent type and extraction conditions on the recovery of Phenolic compounds from artichoke waste. *Chem. Eng. Trans.* **2014**, *39*, 463–468. [CrossRef]
26. Umaña, M.M.; Dalmau, M.E.; Eim, V.S.; Femenia, A.; Rosselló, C. Effects of acoustic power and pH on pectin-enriched extracts obtained from citrus by-products. Modelling of the extraction process. *J. Sci. Food Agric.* **2019**, *99*, 6893–6902. [CrossRef] [PubMed]
27. Eim, V.S.; Urrea, D.; Rosselló, C.; García-Pérez, J.V.; Femenia, A.; Simal, S. Optimization of the Drying Process of Carrot (Daucus carotav. Nantes) on the Basis of Quality Criteria. *Dry. Technol.* **2013**, *31*, 951–962. [CrossRef]
28. Wang, M.; Simon, J.E.; Aviles, I.F.; He, K.; Zheng, Q.-Y.; Tadmor, Y. Analysis of Antioxidative Phenolic Compounds in Artichoke (*Cynara scolymus* L.). *J. Agric. Food Chem.* **2003**, *51*, 601–608. [CrossRef] [PubMed]
29. Castell-Palou, Á.; Rosselló, C.; Femenia, A.; Bon, J.; Simal, S. Moisture profiles in cheese drying determined by TD-NMR: Mathematical modeling of mass transfer. *J. Food Eng.* **2011**, *104*, 525–531. [CrossRef]
30. Souraki, B.A.; Tondro, H.; Ghavami, M. Modeling of Mass Transfer during Osmotic Dehydration of Apple Using an Enhanced Lumped Model. *Dry. Technol.* **2013**, *31*, 595–604. [CrossRef]
31. Coulson, C.A.; Crank, J. The Mathematics of Diffusion. *Math. Gaz.* **1958**, *42*, 165. [CrossRef]
32. Tao, Y.; Zhang, Z.; Sun, D.-W. Experimental and modeling studies of ultrasound-assisted release of phenolics from oak chips into model wine. *Ultrason. Sonochem.* **2014**, *21*, 1839–1848. [CrossRef] [PubMed]

33. Umaña, M.; Eim, V.; Garau, C.; Rosselló, C.; Simal, S. Ultrasound-assisted extraction of ergosterol and antioxidant components from mushroom by-products and the attainment of a β-glucan rich residue. *Food Chem.* **2020**, *332*, 127390. [CrossRef]
34. R Core Team. *R: A Language and Environment for Statistical Computing*; R Foundation for Statistical Computing: Vienna, Austria, 2019; Available online: https://www.R-project.org/ (accessed on 12 December 2020).
35. Rstudio, T. *RStudio: Integrated Development for R. Rstudio Team*; PBC: Boston, MA, USA, 2020; Available online: https://www.rstudio.com/ (accessed on 12 December 2020).
36. Mair, P.; Wilcox, R. Robust statistical methods in R using the WRS2 package. *Behav. Res. Methods* **2019**, *52*, 464–488. [CrossRef]
37. USDA U.S. Department of Agriculture, Agricultural Research Service. FoodData Central. Available online: https://fdc.nal.usda.gov/ (accessed on 24 February 2021).
38. Lavecchia, R.; Maffei, G.; Paccassoni, F.; Piga, L.; Zuorro, A. Artichoke Waste as a Source of Phenolic Antioxidants and Bioenergy. *Waste Biomass Valorization* **2018**, *10*, 2975–2984. [CrossRef]
39. Song, S.; He, H.; Tang, X.; Wang, W. Determination of polyphenols and chlorogenic acid in artichoke (*Cynara scolymus* L.). *Acta Hortic.* **2010**, 167–172. [CrossRef]
40. Roselló-Soto, E.; Martí-Quijal, F.J.; Cilla, A.; Munekata, P.E.S.; Lorenzo, J.M.; Remize, F.; Barba, F.J. Influence of Temperature, Solvent and pH on the Selective Extraction of Phenolic Compounds from Tiger Nuts by-Products: Triple-TOF-LC-MS-MS Characterization. *Molecules* **2019**, *24*, 797. [CrossRef]
41. Irakli, M.; Chatzopoulou, P.; Ekateriniadou, L. Optimization of ultrasound-assisted extraction of phenolic compounds: Oleuropein, phenolic acids, phenolic alcohols and flavonoids from olive leaves and evaluation of its antioxidant activities. *Ind. Crops Prod.* **2018**, *124*, 382–388. [CrossRef]
42. Angelov, G.; Georgieva, S.; Boyadzhieva, S.; Boyadzhiev, L. Optimizing the extraction of globe artichoke wastes. *C. R. Acad. Bulg. Sci.* **2015**, *68*, 1235–1240.
43. Rudić, S.; Dimitrijević-Branković, S.; Dimitrijević, S.; Milić, M. Valorization of unexploited artichoke leaves dust for obtaining of extracts rich in natural antioxidants. *Sep. Purif. Technol.* **2021**, *256*, 117714. [CrossRef]
44. Medina-Torres, N.; Ayora-Talavera, T.; Espinosa-Andrews, H.; Sánchez-Contreras, A.; Pacheco, N. Ultrasound Assisted Extraction for the Recovery of Phenolic Compounds from Vegetable Sources. *Agronomy* **2017**, *7*, 47. [CrossRef]
45. Drevelegka, I.; Goula, A.M. Recovery of grape pomace phenolic compounds through optimized extraction and adsorption processes. *Chem. Eng. Process. Process. Intensif.* **2020**, *149*, 107845. [CrossRef]
46. Arruda, H.S.; Silva, E.K.; Pereira, G.A.; Angolini, C.F.F.; Eberlin, M.N.; Meireles, M.A.A.; Pastore, G.M. Effects of high-intensity ultrasound process parameters on the phenolic compounds recovery from araticum peel. *Ultrason. Sonochem.* **2019**, *50*, 82–95. [CrossRef]
47. Türker, N.; Erdoğdu, F. Effects of pH and temperature of extraction medium on effective diffusion coefficient of anthocynanin pigments of black carrot (*Daucus carota* var. L.). *J. Food Eng.* **2006**, *76*, 579–583. [CrossRef]
48. Sant'Anna, V.; Brandelli, A.; Marczak, L.D.F.; Tessaro, I.C. Kinetic modeling of total polyphenol extraction from grape marc and characterization of the extracts. *Sep. Purif. Technol.* **2012**, *100*, 82–87. [CrossRef]
49. Wang, Y.G.; Yue, S.T.; Li, D.Q.; Jin, M.J.; Li, C.Z. Kinetics and Mechanism of Y(III) Extraction with Ca-100 Using a Constant Interfacial Cell with Laminar Flow. *Solvent Extr. Ion Exch.* **2002**, *20*, 345–358. [CrossRef]
50. Tao, Y.; Zhang, Z.; Sun, D.-W. Kinetic modeling of ultrasound-assisted extraction of phenolic compounds from grape marc: Influence of acoustic energy density and temperature. *Ultrason. Sonochem.* **2014**, *21*, 1461–1469. [CrossRef]
51. Yoo, K.-W.; Kim, J.-H. Kinetics and Mechanism of Ultrasound-assisted Extraction of Paclitaxel from Taxus chinensis. *Biotechnol. Bioprocess Eng.* **2018**, *23*, 532–540. [CrossRef]
52. Gil-Chávez, G.J.; Villa, J.A.; Ayala-Zavala, J.F.; Heredia, J.B.; Sepulveda, D.; Yahia, E.M.; González-Aguilar, G.A. Technologies for Extraction and Production of Bioactive Compounds to be Used as Nutraceuticals and Food Ingredients: An Overview. *Compr. Rev. Food Sci. Food Saf.* **2013**, *12*, 5–23. [CrossRef]
53. Zhao, S.; Baik, O.-D.; Choi, Y.J.; Kim, S.-M. Pretreatments for the Efficient Extraction of Bioactive Compounds from Plant-Based Biomaterials. *Crit. Rev. Food Sci. Nutr.* **2014**, *54*, 1283–1297. [CrossRef]

*Article*

# Influence of Atmospheric Cold Plasma Exposure on Naturally Present Fungal Spores and Physicochemical Characteristics of Sundried Tomatoes (*Solanum lycopersicum* L.)

**Junior Bernardo Molina-Hernandez [1], Jessica Laika [1], Yeimmy Peralta-Ruiz [1,2], Vinay Kumar Palivala [1], Silvia Tappi [3,4], Filippo Cappelli [5], Antonella Ricci [1], Lilia Neri [1] and Clemencia Chaves-López [1,*]**

1 Faculty of Bioscience and Technology for Food, Agriculture and Environment, University of Teramo, Via R. Balzarini 1, 64100 Teramo, Italy; jbmolinahernandez@unite.it (J.B.M.-H.); jlaika@unite.it (J.L.); yyperaltaruiz@unite.it (Y.P.-R.); vinaykumar.palivala@studenti.unite.it (V.K.P.); aricci@unite.it (A.R.); lneri@unite.it (L.N.)
2 Programa de Ingeniería Agroindustrial, Facultad de Ingeniería, Universidad del Atlántico, Carrera 30 Número 8-49, Puerto Colombia 081008, Colombia
3 Department of Agricultural and Food Sciences, University of Bologna, 47521 Cesena, Italy; silvia.tappi2@unibo.it
4 Inter-Departmental Centre for Agri-Food Industrial Research, University of Bologna, Via Quinto Bucci 336, 47521 Cesena, Italy
5 Alma Plasma s.r.l., 40136 Bologna, Italy; filippo.capelli@almaplasma.com
* Correspondence: cchaveslopez@unite.it

**Citation:** Molina-Hernandez, J.B.; Laika, J.; Peralta-Ruiz, Y.; Palivala, V.K.; Tappi, S.; Cappelli, F.; Ricci, A.; Neri, L.; Chaves-López, C. Influence of Atmospheric Cold Plasma Exposure on Naturally Present Fungal Spores and Physicochemical Characteristics of Sundried Tomatoes (*Solanum lycopersicum* L.). *Foods* **2022**, *11*, 210. https://doi.org/10.3390/foods11020210

Academic Editors: Mohsen Gavahian, Asgar Farahnaky and Mahsa Majzoobi

Received: 2 December 2021
Accepted: 2 January 2022
Published: 13 January 2022



**Abstract:** This research aimed to evaluate the impact of atmospheric cold plasma (ACP) treatment on the fungal spores naturally present in sundried tomatoes, as well as their influence on the physicochemical properties and antioxidant activity. ACP was performed with a Surface Dielectric Barrier Discharge (SDBD), applying 6 kV at 23 kHz and exposure times up to 30 min. The results showed a significant reduction of mesophilic aerobic bacteria population and of filamentous fungi after the longer ACP exposure. In particular, the effect of the treatment was assessed on *Aspergillus rugulovalvus* (as sensible strain) and *Aspergillus niger* (as resistant strain). The germination of the spores was observed to be reliant on the species, with nearly 88% and 32% of non-germinated spores for *A. rugulovalvus* and *A. niger*, respectively. Fluorescence probes revealed that ACP affects spore viability promoting strong damage to the wall and cellular membrane. For the first time, the sporicidal effect of ACP against *A. rugulovalvus* is reported. Physicochemical parameters of sundried tomatoes such as pH and water activity ($a_w$) were not affected by the ACP treatment; on the contrary, the antioxidant activity was not affected while the lycopene content was significantly increased with the increase in ACP exposure time ($p \leq 0.05$) probably due to increased extractability.

**Keywords:** atmospheric cold plasma (ACP); sporicidal activity; *Aspergillus niger*; *Aspergillus rugulovalvus*; lycopene; antioxidant activity

## 1. Introduction

The consumption of dried food products and the related risk to human health have increased globally, and concerns have been raised about the microbial quality of the products [1]. In fact, food security is considered a very important problem worldwide, and the potential effects of climate change on yields and quality of food crops, including mycotoxins, are of particular relevance. It is estimated that around 20–25% of harvested fruits and vegetables decompose during the post-harvest stage, even in developed countries [2]. In this context, tomato (*Lycopersicon esculentum* L. var. Excell and Aranca) is a very delicate and fragile vegetable, highly susceptible to microbial contamination, mechanical damage during transport, processing, and storage [3]. The most important dry-tomatoes producer in the world is China, followed by India, the United States, Turkey, Egypt, Iran, and Italy [4].

In 2018, 182 million tons were produced worldwide, of which Italy produced 4% of the total production [5]. Tomatoes are usually consumed fresh, but they are also processed into a diversity of foodstuffs, such as pulp, sauces, paste, juices, and dried tomatoes [6]. In particular, dried tomatoes production presents different advantages: It increases aroma, flavor, and shelf life, and reduces transportation costs thanks to the decrease in volume and weight [7]. However, drying is considered one of the critical points during the processing of dry tomatoes, because of the high microbial contamination found in the environment [6]. Contamination mainly comes from pathogenic or saprophytic fungi that are found both in soil and air and that have colonized food in the form of spores, probably carried by plant materials and packaging [8]. In this context, fungal spores are found ubiquitously and can tolerate many harsh conditions that are deleterious to many other life-forms such as austere temperatures, dryness, and high levels of radiation and radical exposure [9]. Thus, like other dry fruits, dry tomatoes could be susceptible to contamination with filamentous fungi such as *Aspergillus* spp., *Fusarium* spp., *Alternaria* spp., *Penicillium* spp., and their associated toxins during processing and storage [10]. The importance of fungal contamination in foodstuffs does not only refer to the possible degradative activity but also to the capability of many of them to produce mycotoxins to which humans are susceptible [11].

In recent years, novel food processing technologies have developed, each with peculiar strengths, with specific applications of the greatest significance for increasing foods shelf life, such as ohmic heating [12], high hydrostatic pressure, supercritical carbon dioxide [13], irradiation [14,15], and gaseous chlorine dioxide, alone and in combination with biological controls [16], ultraviolet treatment, and a pulsed electric field (PEF) [17]. Among these, cold plasma technology is gaining importance due to its effect in minimizing food-borne pathogens. Plasma can be generated under different conditions, both at low or at atmospheric pressure; the most exploited architectures use radiofrequency waves, microwaves, inductively coupled plasma (ICP), Dielectric Barrier Discharges (DBDs), and plasma jets. Among these, DBD and plasma jets are prominently used in food applications. A DBD consists of two parallel electrodes separated by a gap, and at least one of the electrodes must be covered by a dielectric layer. In this configuration, one of the electrodes is connected to a high-voltage generator while the other acts as the ground. As soon as the potential difference in the gap exceeds the breakdown voltage of the gas, the ionization process begins and micro-discharges are formed (plasma). However, to date, little research has addressed the cold plasma decontamination of dry fruit. In addition, since cold plasma contains various reactive species that can interact with food components, it is important to investigate the effects of such treatments on food quality [13,18,19]. Therefore, the aims of the present study were to (i) investigate the sporicidal effects of ACP treatments on the spores naturally present in sundried tomatoes and (ii) to evaluate the ACP impact on the physicochemical characteristics and antioxidant properties of the sundried tomatoes.

## 2. Materials and Methods

Sundried tomatoes (10 kg) were purchased from a marketplace at Teramo, Italy, in spring 2021, and stored in polyethylene bags at 4 °C until use.

### 2.1. Plasma Treatments

Atmospheric cold plasma (ACP) was generated by a Surface Dielectric Barrier Discharge (SDBD), where the setup included 4 rectangular high-voltage electrodes (115 $cm^2$ each), and a mica dielectric layer over 2 mm thick. The ground electrode had the shape of a mesh, and it was in contact with the dielectric layer, whereas the plasma was formed in the holes of the mesh, producing an indirect treatment. The dry tomatoes were located under the ground electrode. A sinusoidal waveform was applied to the high-voltage electrode with a 6 kV of amplitude at 23 kHz.

ACP SDBD treatments were performed at room temperature (26 ± 1 °C). The sundried tomatoes samples were exposed to ACP treatment for 5, 10, 20 and 30 min and successively stored in polyethylene bags at 5 °C, until analysis. Each treatment was replicated three times in two different experiments. From each replicate, three sundried tomatoes samples were randomly selected ($n$ = 9). In order to avoid introducing a new study variable (different batch contamination), all the experiment was repeated using the same sundried tomatoes batch.

### 2.2. Changes in the Natural Microbiota Associate to Sundried Tomatoes after ACP

In order to determine the effect of ACP on the microbiota naturally associated with sundried tomatoes, treated and untreated samples were analyzed after treatment ($t_0$) and after 22 days ($t_{22}$) of storage (20 °C).

Random aliquots of 10 g sundried tomatoes, untreated and treated with cold plasma, were aseptically placed in sterile stomacher bags containing 90 mL of sterile physiological solution (0.9% *w*/*v*), and then successively homogenized in a Stomacher-400 Circulator (Seward, West Sussex, UK). Ten-fold dilutions from the homogenate were prepared, and 100 µL was inoculated on Plate Count Agar (PCA) for total mesophilic aerobic bacteria, Violet Red Bile Glucose Agar plates (VRBGA) for *Enterobacteriaceae* count, Violet Red Bile Agar plates (VRBA) for *coliforms* counts, Potato Dextrose Agar (PDA) for filamentous fungi and yeast, and DG18 for xerophilic filamentous fungi. Duplicate plates were made at each dilution. All the culture media were purchased from Liofilchem (Liofilchem, Roseto degli Abruzzi, Italy). PCA Plates were incubated at 30 °C for 48 h, PDA and DG18 agar plates for 5 days, and VRBGA and MRS plates at 37 °C for 48 h, with the last ones in anaerobic conditions. Coliforms were searched at 42 °C under microaerophilic conditions. The microbial counts were reported as logarithm colony-forming units per gram of sample (Log CFU/g).

#### Phenotypical and Molecular Identification of Filamentous Fungi

A random number of each colony type was recovered from the different petri dishes of the untreated and ACP-treated sundried tomatoes. In order to identify the isolates on the basis of the morphological and growth characteristics, colonies were inoculated in two different culture media, malt extract agar (MEA) and Czapeck yeast extract agar (Liofilchem, Roseto degli Abruzzi, Italy). The isolation incidence of genus and species were calculated according to [20].

Subsequently, the molecular identification was determined according to the methodology reported by [21]. A PCR assay was performed using primers reported in Table 1. All genes were amplified using universal primers purchased from Sigma Aldrich (Saint Louis, MO, USA).

**Table 1.** Primers used for PCR assay.

| Gene Target | Length bp | Primer | Sequences (5′ → 3′) | Reference |
|---|---|---|---|---|
| *Internal transcribed Spacer (ITS)* | 420–825 | ITS1 (F)<br>ITS4 (R) | 5′TCCGTAGGTGAACCTGCGG3′<br>5′TCCTCCGCTTATTGATATGC3′ | [22] |
| β-tubulin (*BenA*) | 1125 | β-tub 2a (F)<br>β-tub 2b (R) | 5′GGTAACCAAATCGGTGCTTTC3′<br>5′ACCCTCAGTGTAGTGACCCTTGGC3′ | [22]<br>[23] |
| Calmodulin (*CaM*) | 543 | Cmd5 (F)<br>Cmd6 (R) | 5′-CCGAGTACAAGGAGGCCTTC-3′<br>5′-CCGATAGAGGTCATAACGTGG-3′ | [23] |

Abbreviation: F: Forward, R: Reverse. BenA means B-tubulin and CaM mean calmodulin.

### 2.3. Preparation of Spore Suspensions

In order to evaluate the resistance or sensitivity of the fungal spores to the ACP exposure, spores from *A. niger* and *A. rugulovalvus* were collected with sterilized physiological solution for five- (*A. niger*) and eight-day (*A. rugulovalvus* due to their low growth rate) cultures in MEA. Spores were optically standardized at a wavelength of 620 nm to obtain an optical density (OD) of 0.1 AU, corresponding to $3.0 \times 10^4$ CFU/mL. The spore solution was diluted in a relation 1:2 with a sterilized physiological solution, and an aliquot (10 µL) was placed on sterile glass slides and dried with forced convection in a biosafety cabinet in sterility for 3 h. Afterward, glass slides were treated with cold ACP for 30 min. After treatment, 20 µL of malt extract broth (MEB) was added to the treated spore inoculation point successively, and the inoculated glasses slides were incubated in a humidity chamber for 24 h at 30 °C. Non-treated spores were considered the control. After that, spore germination was observed using a light microscope. Spores were considered germinated when their germ tube was longer than the same conidia [24]. All the experiments were realized in triplicate, and a total of 200 spores were counted for each sample.

Spore Viability after ACP Treatment

In order to analyze the spore viability immediately after ACP treatment, treated and untreated spores from *A. rugulovalvus* (as sensible strain) and *A. niger* (as resistant strain) were stained with a mixture of CFDA (carboxyfluorescein diacetate) and propidium iodide (PI). While the green fluorescent dye CFDA is able to permeate both intact and damaged cell membranes, the red fluorescent dye propidium iodide (PI) can enter only the cells with significant membrane damage [25]. A $10^4$ spore solution in PBS was treated using 30 min of ACP. Subsequently, spores were stained as reported by [25]. A Nikon A1R confocal imaging system (Nikon Corp., Tokyo, Japan) was used to observe the spore viability.

### 2.4. Moisture, Water Activity and pH

Moisture content was measured according to the gravimetric method [26]. Water activity ($a_w$) was determined using a hygrometer AquaLab CX 4-TE (Decagon Devices Inc., Pullman, WA, USA). The pH values were determined with a laboratory pH-meter (MP220, Mettler Toledo International, Polaris Parkway, OH, USA). All measurements were performed in triplicate.

### 2.5. Color Analysis

The tomato color was evaluated in the epicarp area before and immediately after the ACP treatments. The analysis was conducted using a Konica Minolta Chroma Meter CR-5 spectrophotocolorimeter (Konica Minolta, Osaka, Japan) equipped with a D65 illuminant. Measurements were performed directly on a target mask with a measurement area of 8 mm using standard 10° observers. For each sample, 9 different halves were analyzed, and for each half, the colorimetric CIELab coordinates were collected in two different areas. L*, b*, and a* coordinates were thus used in order to calculate the redness (a*/b*), the hue angle (h°), and Chroma (C*), according to the following equations:

$$h^\circ = \arctan^{-1}(b^*/a^*); \quad (1)$$

$$C^* = (a^2 + b^2)^{0.5}; \quad (2)$$

### 2.6. Lycopene Determination

Lycopene content was measured according to the methodology described by [27] with some modifications. Briefly, 0.5 g of sundried tomato were added to 5 mL of distilled water and homogenized for 5 min, using a dispersing homogenizer (Ultra-Turrax, Yellowline DI25basic, IKA, Staufen, Germany). After mixing, 10 mL of n-Hexane was added to the solution vortexed and centrifuged for 10 min at 4000 rpm (4 °C) (NEYA 16R centrifuge, Remi, Italy). The supernatant was recovered, and the extraction step was repeated until complete discoloration of the pellet, adding 5 mL of hexane. The absorbance of the lycopene dissolved in hexane was detected at 503 nm using a spectrophotometer (Lambda 25 UV/VIS, Perkin Elmer, MA, USA) and the results were expressed in µg lycopene/g of the sample, using the formula reported by [28]:

$$\text{Lycopene}(\mu g/g) = ((A_{503}/\varepsilon \times b) \times MW \times V/M) \times 103 \quad (3)$$

where:

$A_{503}$ = absorbance of the hexane phase at 503 nm;
$\varepsilon$ = molar extinction coefficient (L $mol^{-1}$ $cm^{-1}$) in the appropriate solvent;
b = cell optical path (cm);
MW = lycopene molecular weight = 536.9 g $mol^{-1}$;
V = volume of hexane (mL);
M = sample mass (g).

### 2.7. Determination of the Antioxidant Capacity

Antioxidant capacity was evaluated on sundried tomatoes before and after ACP treatments using two different methods i.e., (i) the Folin–Ciocalteu (FC) assay, which is based on a redox reaction (single electron transfer—SET) and measures the capability of an extract to reduce a mixture of phosphomolybdic/phosphotungstic acid complexes (FC reagent) in alkaline medium, and (ii) the TEAC (Trolox Equivalent Antioxidant Capacity) assay that presents a mixed mechanism of action towards antioxidants since ABTS radicals can undergo reduction by both single electron transfer (SET) and hydrogen atom transfer (HAT).

Sample extraction was performed by adding an aliquot (2.0 ± 0.2 g) of sundried tomatoes to 10 mL of MeOH:$H_2O$ (80:20; *v*/*v*). Homogenization was performed for 5 min at 13,500 rpm using a dispersing homogenizer (Ultra-Turrax, Yellowline DI25basic, IKA, Staufen, Germany). The sample obtained was thus centrifuged at 4000 rpm for 20 min at 4 °C (NEYA 16R centrifuge, Remi, Italy) and the recovered supernatant was immediately used for analyses.

The FC assay was carried out using the method reported by [29] with modifications. In brief, 120 µL of extracts was added to 600 µL of diluted Folin–Ciocalteu reagent (1:10) and vortexed. After 2 min, 960 µL of $Na_2CO^3$ (7.5%) was added and incubated for 5 min in a water bath at 50 °C. The absorbance was measured with a spectrophotometer at 760 nm (Lambda 25 UV/VIS, Perkin Elmer, Waltham, MA, USA). The final values were calculated in equivalents of gallic acid, according to the calibration curve of the method. Analyses were performed in triplicate.

The TEAC assay was performed according to [30], with slight modifications. $ABTS^{\bullet+}$ (2,2′-azino-bis-(3-ethylbenzothiazoline-6-sulfonic-acid) was dissolved in water to a 7 mM concentration; the ABTS radical was formed by reacting ABTS stock solution with 2.45 mM potassium persulphate ($K_2S_2O_8$) and allowing the mixture to stand in the dark at room temperature for 12–16 h. Before use, this solution was diluted with distilled water to obtain an absorbance of 0.700 ± 0.020 at 734 nm. The reaction was performed by mixing in cuvettes 30 µL of tomato extracts with 2970 µL of the $ABTS^{\bullet+}$ solution. The reaction mixture was incubated in the dark for 7 min at room temperature and the percentage of discoloration was used as the measure of antioxidant activity. The percent scavenging of $ABTS^{\bullet+}$ (cation radical) was calculated as:

$$(\text{Absorbance Control} - \text{Absorbance sample})/\text{Absorbance control} \times 100 \quad (4)$$

The antioxidant activity was expressed as µmol TEAC (Trolox Equivalent Antioxidant Capacity)/g tomato (dry matter). The TEAC value was calculated by the ratio of the regression coefficient of the dose–response curve of the sample and the regression coefficient of the dose–response curve of Trolox and was expressed as µmoles of Trolox equivalents per g of dry matter. The analysis was carried out on each extract in triplicate.

### 2.8. Statistical Analysis

Data were expressed as mean and standard deviation calculated on three replicated treatments and additionally analyzed by one-way ANOVA analysis. Significant differences between means were computed by Tukey's multiple comparison test at a significance level of $p \leq 0.05$. Data were processed using STATISTICA 12 for Windows (StatSoftTM, Tulsa, OK, USA) software.

## 3. Results

### 3.1. Plasma Gas and Reactive Species Analysis

The literature regarding ACP is accurate about the chemistry that governs the atmosphere inside a plasma reactor, and depending on the surface power density (SPD) absorbed by the plasma, two different regimes may occur [31,32]. For small values of SPD, the chemistry is governed by the ozone formation reactions, and in this configuration, the concentration of ozone increases with time, eventually reaching a plateau. This regime is known as the ozone regime, and under these circumstances, the main reactive species that can be found in an ACP are ozone and atomic oxygen, both with a relevant antimicrobial effect [33,34].

Increasing the SPD leads to a greater formation of ozone up to a certain threshold, and once this value of SPD is exceeded, the $NO_x$ formation reactions begin [34]. This regime is known as the transition regime. In fact, ozone has a transitory behavior: At first. the ozone concentration increases, reaches a maximum, and then decreases until it disappears. Many reactive species characterize this transition-regime: NO, N, $NO_2$, $NO_3$, $N_2O_5$, O, $O_3$.

To address the regime of the plasma process under consideration, two long-living species were measured: Ozone and nitrogen dioxide. The ozone concentration increased during the whole treatment and no $NO_2$ was detected; both results led to the conclusion that the performed process occurs in an ozone regime.

The plasma source used in the present study presents the peculiarity that the plasma is confined in the holes of the mesh, therefore it is safe to assume that the greater part of ions and electrons are confined in these holes too. Furthermore, the short-living species (such as O and OH) do not have the time to reach the substrate to be treated due to their slow diffusion velocity.

These considerations together with the fact that the plasma process occurs in the ozone regime led us to the conclusion that the main reactive species responsible for the biological effect is ozone.

The concentration of ozone was measured using the optical absorption spectroscopy technique that exploits the Lambert–Beer law. The behavior of $O_3$ showed a strong increase, which reached a concentration of 350 ppm in 100 s; after this first phase, the concentration growth continued with a less steep trend, and the concentration reached after 600 s was about 600 ppm; at the end of the treatment, the concentration was about 900 ppm. The high-voltage generator (AlmaPlasma s.r.l., Bologna, Italy) was specifically designed for this application.

### 3.2. Effect of Atmospheric Cold Plasma on the Total Mesophilic Aerobic Bacteria and Filamentous Fungi Count Naturally Present in Sundried Tomatoes

Natural Mesophilic Aerobic Bacteria (MAB) and filamentous fungi counts in sundried tomatoes at 5–30 min of ACP treatment periods compared to the untreated ones (control) are presented in Table 2. In the untreated samples, MAB and filamentous fungi were detected in low levels (3.06 ± 0.78 log CFU/g and 1.7 ± 0.82 log CFU/g, respectively) that were considered acceptable. Lactic acid bacteria, Coliforms, and *Enterobacteriaceae* were not detected in the analyzed samples. The absence of coliforms and *Enterobacteriaceae* could indicate the good hygienic and handling practices used to produce sundried tomatoes.

The application of ACP for 10 min allowed us to obtain a slight (0.76 ± 0.32 log CFU/g) but significant reduction ($p \leq 0.05$) of MAB, whereas further exposure at 20 and 30 min, did not show additional improvement. It is known that the conditions of availability of water activity found in a product such as sundried tomatoes do not allow the development of bacteria; however. foodborne pathogens and spores from bacteria and filamentous fungi have the potential to survive for an extended period on dried fruits [35]. In our study, we found that the *Bacillus* genus was the predominant bacterial population in the sundried tomato samples, as indicated by the presence of endospores in the isolated colonies grown in PCA that were evidenced by microscopically analysis. Studies on the efficacy of ACP against bacteria have been conducted overall in their vegetative states; however, there is little information on its potential for bacterial spore inactivation [36]. For example, a study conducted on *Bacillus cereus* and *Bacillus anthracis* spores revealed that 6 log CFU/g spores treated in liquid or air-dried on a solid surface were efficiently inactivated within 1 min of DBD plasma treatment at a discharge power of 0.3 $W/cm^2$ [37]. On the other hand, other researchers obtained an inactivation of 3.5 to 4.8 log CFU/g of *B. subtilis* wild-type spores after 7 min of exposure to ACP; however, the inactivation depended on the process gas used [38]. In the present experiment, we used ambient air, which during the process generated high quantities of $O_3$ (at 10 min is near 600 ppm), which were, however, not enough to kill spores. The small spore reduction observed in the present experiment might be explained by a different qualitative and quantitative composition of the plasma discharge and by the different sensitivities of the specific spore types [37]. Some authors suggested that the primary mechanism of spore inactivation is the diffusion of ROS (e.g., $H_2O_2$) into spores, followed by the damage of internal macromolecules or molecular systems. On the other hand, it has been reported that Alpha/beta-type small, acid-soluble proteins (SASPs) contribute to the resistance of spores to ACP and that the spores' coat represents a protective layer against many oxidizing agents [38,39]. In addition, Patil et al. [40] reported that detoxifying enzymes present in the bacteria coat of the spore play an important role in detoxifying chemicals.

**Table 2.** Total mesophilic aerobic bacteria and filamentous fungi count naturally present in sundried tomatoes immediately ($t_0$) and at 22 ($t_{22}$) days after ACP treatment.

| Time of Treatment (min.) | Mesophilic Bacteria (log UFC/g) | | Filamentous Fungi (log CFU/g) | |
|---|---|---|---|---|
| | $t_0$ | $t_{22}$ | $t_0$ | $t_{22}$ |
| Control | 3.06 ± 0.78 [a] | 3.12 ±0.54 [a] | 1.79 ± 0.82 [a] | 1.56 ± 0.75 [a] |
| 5 | 2.95 ± 0.45 [a] | 3.06 ±0.34 [a] | 1.68 ± 0.4 [a] | 1.54 ± 0.56 [a] |
| 10 | 2.30 ± 0.50 [b] | 2.40 ±0.10 [b] | 0.77 ± 0.45 [a] | 0.47 ± 0.52 [a] |
| 20 | 2.15 ± 0.28 [b] | 2.30 ±0.19 [b] | 0.62 ± 0.6 [a] | 0.67 ± 0.58 [a] |
| 30 | 2.20 ± 0.56 [b] | 2.10 ±0.64 [b] | 0.44 ± 0.52 [b] | 0.51 ± 0.61 [a] |

Mean and standard deviation of three repetitions in two different experiments. Different letters in the same line mean significant differences between the treatments ($p < 0.05$; Tukey HSD post-hoc test).

With regard to the filamentous fungi counts, inactivation due to ACP was slightly higher compared to *Bacillus* spores. In fact, immediately after the 30 min treatment, a reduction of 1.35 log CFU/g was observed. As mentioned before, the comparison of the efficacy of the ACP treatment with previous studies is very difficult because it depends on the voltage, exposure time, initial microbial density, process gas, working distance, and plasma exposure [41]. However, in other studies performed on blueberries, cherries, tomatoes, and meat, the authors also reported a small reduction (from 0.8 to 1.5 log CFU/g) of yeast and filamentous fungi after ACP treatment [42–44]. In general, information on the sensitivity of fungal spores to plasma is scarce and focused on the decontamination of seeds, nuts, and powdered food [45–47].

As expected, 22 days after the ACP treatment ($t_{22}$), there were no significant differences ($p \geq 0.05$) between the control and the treated samples. In fact, as observed in Table 2, the water activity in the samples did not increase during storage, therefore the spores present on the sundried tomatoes surface were not able to germinate or form mycelia.

### 3.3. *Effect of the CAP on the Fungal Species*

In order to verify if ACP treatments could have a species-dependent effect, the filamentous fungi associated with treated and untreated samples were identified. From 58 fungal isolates, a total of 23 different colonies of fungal morphotypes were found in the analyzed samples, but only 4 different genera were identified. In particular, *Aspergillus* genera represented the vast majority (78.2%) of the fungal colonies, followed by *Rhizopus* (13.04%), *Corynascus* (4.34%), and *Cladosporium* (4.34%). All the isolated specimens were identified at the molecular level and the gene sequences obtained were aligned with the accession numbers of the NCBI BLAST database and reported with their respective ones (Supplementary Table S1).

In our study, the ITS gene was able to discriminate only four species, and for this reason we used other genes (*BenA, CaM,*) that gave us more information (Table S1). As evidenced, untreated sundried tomatoes harbored eight different species, such as *Aspergillus rugulovalvus* formerly *Aspergillus rugulosus* (syn *Emericella rugulosa* var. lazuline), *A. niger*, *A. amstelodami*, *A. tubingensis*, *A. cristatus*, *R. oryzae*, *Cladosporium cladosporioides*, and *Corynascus sepedonium*. As observed in Table S1, all *A. tubingensis* and *A. niger* strains, which belong to the *Nigri* section, were identified by means of amplification and sequencing of the *CamA* and *BenA* genes, respectively. These genes have been considered important for the identification of *Aspergillus* species, and in particular for the *Nigri* section [48]. It is important to underline that only *Aspergillus cristatus* was identified with the primers ITS1 and ITS4.

A great number of the species isolated here are of particular interest to food as they can grow rapidly over wide temperature and $a_w$ ranges (minimum ~0.70–0.72 $a_w$). Moreover, the presence of *Aspergillus* ssp. is of special concern since several species of this genus are capable of producing mycotoxins, for example *A. rugulovalvus* potentially produce Sterigmatoxin; *A. niger* and *A. tubingensis* have the potential to secrete Ochratoxin A (OTA); and the xerophilic species *A. amstelodami* have been shown to produce multiple toxins, including Patulin, OTA, and Sterigmatocystin. In addition, other species identified here can produce mycotoxins during favorable conditions. For example, *R. oryzae* is a potential producer of Cyclopiazonic acid and Kojic acid, and *Cladosporium cladosporioides* has been reported to produce aflatoxins, Deoxynivalenol, Fumonisins, OTA, T-2 toxin, and Zearalenone.

As observed in Figure 1, ACP treatments affected the fungal community structure in sundried tomatoes, since fewer fungi were isolated, and their diversity was reduced with prolonged treatment of ACP. In fact, although significant differences were not found in colony counts after 5 min of ACP exposure with respect to the control samples, from eight species present in control samples, only three (*A. niger*, *A. tubingensis*, and *R. oryzae*) were detected in the treated samples with similar incidence. Similar behavior was observed even after 10 min of treatment, even if *A. niger* was isolated at a major frequency. After 20 min of treatment, only *A. niger* and *A. tubingensis* were isolated and in similar proportions. An interesting result was observed at 30 min of exposition. In fact, only *A. niger* colonies were isolated after ACP exposure. As stated in Section 2, it is important to underline that the greater length of the treatment implies a longer exposure of cells to $O_3$. Thus, our results point out the efficacy of the treatment to inactivate spores of species that are frequently reported as a contaminant and able to produce important mycotoxins. Trompeter et al. [49] investigated the sporicidal activity of ACP using dielectric barrier discharges (DBDs) on *A. niger* spores, and reported that among different gases or mixtures of gases used (Ar, synthetic dry and moistened air, a mixture of $O_2$ and $O_3$ and $N_2$, and $N_2$ with 1% added $H_2$), the argon discharge was proven to be the most efficient for spores' inactivation. Recently, the efficacy of the high-voltage atmospheric cold plasma (HVACP) technology to reduce *A. flavus* spores was reported. In fact, about 50% spore inactivation was reached after 1 min of treatment [50].

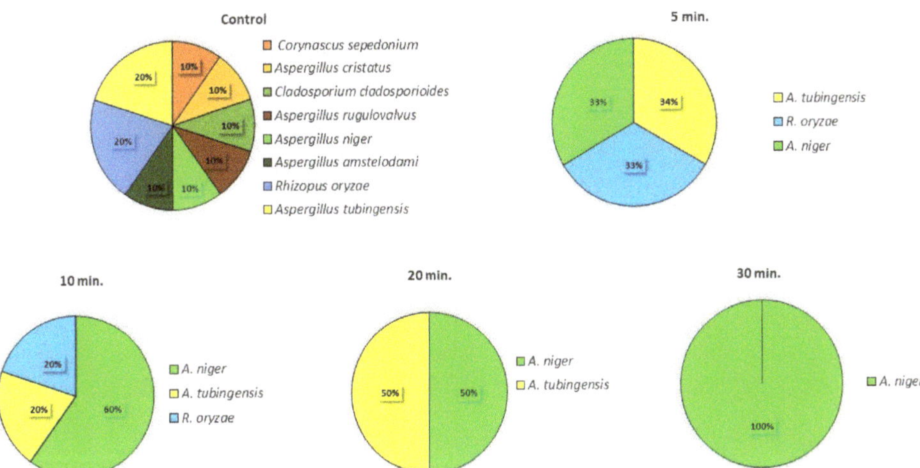


**Figure 1.** Incidence of the fungal species on sundried tomatoes before and after the ACP treatments.

In order to further demonstrate the effect of ACP on spore inactivation, spores of *A. niger* (the most resistant species) and *A. rugulovalvus* (most sensitive species) were subjected to the ACP treatment for 30 min. The spores were individually inoculated in vitro, and the germination percentage and vitality were measured 18 h after treatment. The results, presented in Figure 2a, show that the percentage of non-germinated spores was greater in *A. rugulovalvus* than in *A. niger*, confirming the higher resistance of *A. niger* spores to the stress induced by ACP. It is well known that multiple factors are able to inhibit fungal spores' germination; in the present study, the disruption of the membrane permeability of the fungal spores was evaluated using confocal laser scanning microscopy coupled with CFDA and PI dyes. This methodology was already reported to be suitable for fungal spores analysis [51]. Our results showed that the inhibition of *A. rugulovalvus* spore germination was mainly due to the rupture of the spore permeability barrier after treatment. In fact, the use of nucleic acid stain labels such as propidium iodide (Figure 2b,c), which is unable to enter the nucleus in the intact spore, revealed a decrease in the amount of metabolically active spores after ACP. On the contrary, in treated spores of *A. niger* (Figure 2d), a small percentage was stained red, suggesting that this type of spore possess particular characteristics that allow their resistance against the oxidative stress generated by ozone and other ROS generated by ACP. In this context, some authors reported that the spore pigment melanin increases the firmness of spore cell walls, thus protecting the cell from stressors such as temperature and UV-irradiation, and increases the resistance to ROS, which are also produced in ACP [52]. Similarly, [53] suggested that melanization is a protection mechanism, due to the optical and antioxidant properties of melanin, which allow for limiting and repairing photodamage, due to electromagnetic energy or radiation that could remove electrons from water and other molecules (DNA and proteins) in the fungal cells. Other plausible hypotheses regarding the resistance of spores to ACP have been suggested by Montie et al. [54] who attributed it to their extremely thick polysaccharide cell walls that could represent an adaptation to environmental stresses [55]. On the other hand, as a result of the action of charged particles in ACP, the surface of the cell wall could be subjected to erosion that might cause damage to the cell wall structure and thickness [56].

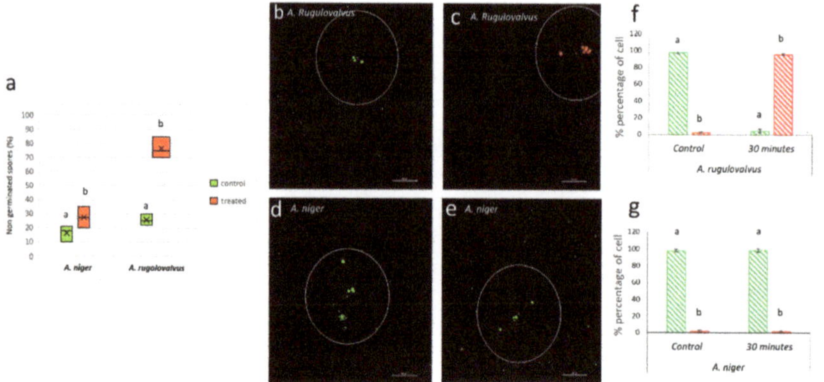

**Figure 2.** Effect of ACP on spore germination (**a**) of *Aspergillus niger* and *A. rugulovalvus* in malt extract medium after 30 min of ACP treatment. Confocal laser scanning microscopy analysis of cell viability in *A. rugulovalvus* (**b**,**c**) *A. niger* (**d**,**e**) after treatment and stained with green fluorescence CFDA (carboxyfuorescein diacetate) and red propidium iodide (PI) dyes. Bars indicate percentage of cell live (green) and death (red) spores in *A. rugulovalvus* (**f**) and *A. niger* (**g**), respectively. Different letters represent significant differences between the sample ($p < 0.05$; Tukey HSD post-hoc test).

### 3.4. Effects on Physicochemical Characteristics of Sundried Tomatoes

The physico-chemical characteristics of sundried tomatoes are presented in Table 3. As can be observed, after the ACP treatments, sundried tomatoes showed pH and $a_w$ values similar ($p \geq 0.05$) to the control sample, while slight differences were observed in the moisture content, which decreased (loss < 6%) after 10 and 20 min of ACP treatment. This moisture loss can be attributed to the evaporation of superficial water due to convective flows triggered by the difference of temperature between the plasma source, which is about 30–35 °C, and the treatment chamber at room temperature (26 °C), and possibly promoted by cellular damages and surface modifications caused by plasma. In particular, ACP treatments were found to decrease the water contact angle of tomato peels and to accelerate the drying of tomatoes. Plasma-reactive species, in fact, can degrade the cuticle making its surface more hydrophilic, therefore increasing the water permeability of the plant surfaces and thus water diffusion from the plant interior [57]. Concerning the slight moisture increase highlighted after 30 min of treatment, it could be dependent on the accidental rehydration of the sample after ACP treatment rather than due to a higher initial moisture content of the sundried tomatoes sampled for this specific treatment.

**Table 3.** Effect of ACP treatment on the physicochemical and antioxidant characteristics of sundried tomatoes. Treatments were carried out at ambient temperatures of 26 ± 1 °C.

| Time of Treatment (min.) | pH | Moisture Content % | $a_w$ | Lycopene (g/g dw) | FC (mg GAE/g dw) | TEAC (μmol TE/g dw) |
|---|---|---|---|---|---|---|
| 0 | 4.36 ± 0.059 $^a$ | 34.25 ± 1.57 $^a$ | 0.608 ± 0.006 $^a$ | 3.22 ± 0.05 $^d$ | 3.35 ± 0.16 $^a$ | 8.35 ± 0.55 $^{ab}$ |
| 5 | 4.34 ± 0.053 $^a$ | 33.29 ± 1.37 $^a$ | 0.610 ± 0.005 $^a$ | 3.58 ± 0.21 $^c$ | 3.58 ± 0.24 $^a$ | 7.58 ± 0.42 $^b$ |
| 10 | 4.30 ± 0.075 $^a$ | 29.00 ± 0.28 $^b$ | 0.610 ± 0.004 $^a$ | 4.13 ± 0.26 $^b$ | 3.57 ± 0.24 $^a$ | 9.28 ± 0.66 $^a$ |
| 20 | 4.25 ± 0.079 $^a$ | 29.86 ± 0.29 $^b$ | 0.611 ± 0.004 $^a$ | 4.57 ± 0.03 $^a$ | 3.35 ± 0.05 $^a$ | 7.51 ± 0.91 $^b$ |
| 30 | 4.29 ± 0.076 $^a$ | 32.43 ± 0.36 $^a$ | 0.613 ± 0.007 $^a$ | 4.14 ± 0.11 $^b$ | 3.48 ± 0.13 $^a$ | 7.89 ± 0.71 $^{ab}$ |

Mean and standard deviation of three repetitions in two different experiments. Different letters in the same line mean significant differences between the treatments ($p < 0.05$; Tukey HSD post-hoc test).

As lycopene is one of the predominant carotenoids found in tomatoes [58], the effect of ACP on this compound was investigated. As shown in Table 3, ACP treatment positively affected the lycopene content leading to an increase of up to 42% after 20 min, while the extension of the treatment to 30 min resulted in a slight reduction. The increase in lycopene content upon ACP treatments was also observed by [58] and could be related to the enhancement of its extractability from the tomato skins as an effect of the cell damage induced by plasma-reactive species. On the other hand, the increase in the concentration of long-lifetime oxidizing species such as ozone in the treatment chamber could have been responsible for the lycopene reduction highlighted after 30 min of treatment. Overall, these results show that, after the previous optimization of the process parameters, cold plasma may be potentially used to enhance the extraction of lycopene from plant tissues.

As regards the antioxidant properties, no significant ($p > 0.05$) variation of the reducing power and of the radical scavenging activity was highlighted after the ACP treatments. Similar results were also reported by [59] on kiwifruit treated by cold plasma at 15 kV for 20 min; conversely, other studies showed a positive influence of ACP on the antioxidant properties of plant products, ascribing this effect to the formation of phenolic compounds with higher antioxidant capacity [60], to the increase in the extractability of polyphenols and other antioxidants such as carotenoids and vitamin C [61], and to the accumulation of secondary metabolites that include phenolic compounds as defense responses to UV irradiation [62,63]. It is important to consider that the instrument used in our study did not produce UV radiation. As highlighted by [64], different effects observed on the antioxidant properties of plant foods treated by cold plasma depend on the food matrix and on the plasma type, applied voltage, working gas, treatment time, and relative humidity. These factors, in fact, affect the types of reactive species generated, UV radiation, energetic ions, and charged particles that may induce physicochemical reactions with treated materials [65].

As color is probably the first quality factor judged by consumers on tomato products, the effect of ACP treatments on this attribute was also evaluated. As shown in Table 4, ACP did not affect the a* and b* values of tomato peels as well as the h°, C*, and a*/b* values, while we determined, irrespectively of the treatment time, a slight reduction (about 11%) of lightness (L*). This darkening could be due to possible melting of cutaneous wax on the peel surface [66].

**Table 4.** Instrumental color in the CIE L*a*b* space.

| Time of Treatment (min) | L* | a* | b* | C* | h° |
|---|---|---|---|---|---|
| 0 | 29.20± 3.64 $^{a}$ | 19.74 ± 3.75 $^{a}$ | 15.20 ± 3.29 $^{a}$ | 24.94 ± 4.81 $^{a}$ | 37.48 $^{a}$ ± 3.14 $^{a}$ |
| 5 | 26.03 ± 5.79 $^{b}$ | 16.98 ± 4.76 $^{a}$ | 14.49 ± 5.25 $^{a}$ | 22.37 ± 6.92 $^{a}$ | 39.54 $^{a}$ ±4.47 $^{a}$ |
| 10 | 25.10 ± 3.27 $^{b}$ | 18.23 ± 4.68 $^{a}$ | 14.38 ± 3.94 $^{a}$ | 23.27 ± 5.92 $^{a}$ | 38.16 $^{a}$ ± 3.79 $^{a}$ |
| 20 | 24.07 ± 4.52 $^{b}$ | 17.45 ± 2.94 $^{a}$ | 12.91 ± 2.69 $^{a}$ | 21.71 $^{a}$ ± 3.93 $^{a}$ | 36.35 $^{a}$ ± 1.83 $^{a}$ |
| 30 | 25.92 ± 4.76 $^{b}$ | 18.63 ± 2.46 $^{a}$ | 14.72 ± 3.27 $^{a}$ | 23.78 $^{a}$ ± 3.81 $^{a}$ | 38.01 $^{a}$ ± 3.42 $^{a}$ |

L*: Luminosity index; a*: Green–red color coordinate; b*: Blue–yellow color coordinate; C*: Croma; h°: Hue. Data on the same column marked with different letters are significantly different at a $p < 0.05$ level.

## 4. Conclusions

The results obtained in the present study highlighted that the application of ACP has potential for the fungal spore decontamination of sundried tomatoes. As demonstrated, it exhibited several levels of the sporicidal effect depending on the types of spores and fungal species. In addition, this study demonstrated that the inactivation processes depend on the exposure time, which is related to the increase in the ozone in the ACP chamber, which reached about 900 ppm at the end of the treatment (30 min). In addition, fluorescence probes revealed that ACP affects spore viability considerably in *A. rugulovalvus*, which was associated with strong damage of the wall and cellular membrane of the spores. For the first time, the sporicidal effect of ACP against *A. rugulovalvus*, a species that produces the antifungal lipopeptide caspofungin B, has been reported. Concerning the physicochemical properties, ACP slightly affected the water content and the color of sundried tomatoes, but did not influence the antioxidant properties, while it increased the extractability of lycopene. In conclusion, the obtained results showed good potentiality of this technology for dry product decontamination and highlighted the importance of process optimization considering the effect on quality and nutritional properties.

**Supplementary Materials:** The following supporting information can be downloaded at: https://www.mdpi.com/article/10.3390/foods11020210/s1, Table S1. Identification of the fungi isolates from dry tomatoes, determined by sequences of the ITS, β-tubulin, and Calmodulin genes.

**Author Contributions:** Conceptualization, C.C.-L. and L.N.; methodology, A.R., L.N. and F.C.; software, J.B.M.-H., J.L. and Y.P.-R.; formal analysis: J.B.M.-H., J.L., Y.P.-R. and V.K.P.; investigation, J.B.M.-H., J.L. and Y.P.-R.; resources, C.C.-L.; data curation, J.B.M.-H., J.L. and L.N.; writing—original draft preparation, C.C.-L. and L.N.; writing—review and editing, C.C.-L., L.N., J.B.M.-H. and S.T.; supervision, C.C.-L.; project administration, C.C.-L.; funding acquisition, C.C.-L. All authors have read and agreed to the published version of the manuscript.

**Funding:** The present work is part of the research activities developed within the project "PLASMAFOOD—Study and optimization of cold atmospheric plasma treatment for food safety and quality improvement" founded by MIUR—Ministero dell'Istruzione dell'Università e della Ricerca—PRIN: Progetti di Ricerca di Rilevante Interesse Nazionale, Bando 2017.

**Institutional Review Board Statement:** Not applicable.

**Informed Consent Statement:** Not applicable.

**Data Availability Statement:** Not applicable.

**Acknowledgments:** The authors thank Luca Valbonetti for their precious technical support in confocal analysis.

**Conflicts of Interest:** Filippo Cappelli is an employee in Company Alma Plasma s.r.l., Bologna, Italy, and contribute to the measurements of the ozone during the CP treatments. The remaining authors have no conflicts of interest to declare.

## References

1. FAOSTAT. News Archive 2011. Available online: https://www.fao.org/news/archive/news-by-date/2011/en/ (accessed on 1 October 2021).
2. Sharma, R.R.; Singh, D.; Singh, R. Biological control of postharvest diseases of fruits and vegetables by microbial antagonists: A review. *Biol. Control* **2009**, *50*, 205–221. [CrossRef]
3. Hegazy, E.M. Mycotoxin and fungal contamination of fresh and dried tomato. *Annu. Res. Rev. Biol.* **2017**, *17*, 1–9. [CrossRef]
4. FAOSTAT. Food and Agriculture Statistics. Available online: https://www.fao.org/food-agriculture-statistics/en/ (accessed on 15 October 2021).
5. FAOSTAT. The State of Food and Agriculture. Available online: https://www.fao.org/3/ca6030en/ca6030en.pdf (accessed on 15 October 2021).
6. Sanzani, S.M.; Gallone, T.; Garganese, F.; Caruso, A.G.; Amenduni, M.; Ippolito, A. Contamination of fresh and dried tomato by Alternaria toxins in southern Italy. *Food Addit. Contam. Part A* **2019**, *36*, 789–799. [CrossRef] [PubMed]
7. Doymaz, I. Air-drying characteristics of tomatoes. *J. Food Eng.* **2007**, *78*, 1291–1297. [CrossRef]
8. Kakde, U.B.; Kakde, H.U. Incidence of post-harvest disease and airborne fungal spores in a vegetable market. *Acta Bot. Croat.* **2012**, *71*, 147–157. [CrossRef]
9. Sephton-Clark, P.C.S.; Muñoz, J.F.; Ballou, E.R.; Cuomo, C.A.; Voelz, K. Pathways of Pathogenicity: Transcriptional Stages of Germination in the Fatal Fungal Pathogen *Rhizopus delemar*. *mSphere* **2018**, *3*, e00403-18. [CrossRef] [PubMed]
10. Dwivedy, A.K.; Prakash, B.; Chanotiya, C.S.; Bisht, D.; Dubey, N.K. Chemically characterized *Mentha cardiaca* L. essential oil as plant based preservative in view of efficacy against biodeteriorating fungi of dry fruits, aflatoxin secretion, lipid peroxidation and safety profile assessment. *Food Chem. Toxicol.* **2017**, *106*, 175–184. [CrossRef]
11. Ndagijimana, M.; Chaves-López, C.; Corsetti, A.; Tofalo, R.; Sergi, M.; Paparella, A.; Guerzoni, M.E.; Suzzi, G. Growth and metabolites production by *Penicillium brevicompactum* in yoghurt. *Int. J. Food Microbiol.* **2008**, *127*, 276–283. [CrossRef]
12. Cappato, L.P.; Ferreira, M.V.S.; Guimaraes, J.T.; Portela, J.B.; Costa, A.L.R.; Freitas, M.Q.; Cunha, R.L.; Oliveira, C.A.F.; Mercali, G.D.; Marzack, L.D.F.; et al. Ohmic heating in dairy processing: Relevant aspects for safety and quality. *Trends Food Sci. Technol.* **2017**, *62*, 104–112. [CrossRef]
13. Amaral, G.V.; Silva, E.K.; Cavalcanti, R.N.; Cappato, L.P.; Guimaraes, J.T.; Alvarenga, V.O.; Esmerino, E.A.; Portela, J.B.; Sant' Ana, A.S.; Freitas, M.Q.; et al. Dairy processing using supercritical carbon dioxide technology: Theoretical fundamentals, quality and safety aspects. *Trends Food Sci. Technol.* **2017**, *64*, 94–101. [CrossRef]
14. Alvarez, I.; Niemira, B.A.; Fan, X.; Sommers, C.H. Inactivation of *Salmonella serovars* in liquid whole egg by heat following irradiation treatments. *J. Food Prot.* **2006**, *69*, 2066–2074. [CrossRef] [PubMed]

15. Olanya, O.M.; Niemira, B.A.; Phillips, J.G. Effects of gamma irradiation on the survival of *Pseudomonas fluorescens* inoculated on romaine lettuce and baby spinach. *LWT* **2015**, *62*, 55–61. [CrossRef]
16. Berrios-Rodriguez, A.; Olanya, O.M.; Annous, B.A.; Cassidy, J.M.; Orellana, L.; Niemira, B.A. Survival of *Salmonella Typhimurium* on soybean sprouts following treatments with gaseous chlorine dioxide and biocontrol Pseudomonas bacteria. *Food Sci. Biotechnol.* **2017**, *26*, 513–520. [CrossRef]
17. Misra, N.N.; Pankaj, S.K.; Segat, A.; Ishikawa, K. Cold plasma interactions with enzymes in foods and model systems. *Trends Food Sci. Technol.* **2016**, *55*, 39–47. [CrossRef]
18. Misra, N.N.; Jo, C. Applications of cold plasma technology for microbiological safety in meat industry. *Trends Food Sci. Technol.* **2017**, *64*, 74–86. [CrossRef]
19. Pankaj, S.K.; Keener, K.M. Cold plasma: Background, applications and current trends. *Curr. Opin. Food Sci.* **2017**, *16*, 49–52. [CrossRef]
20. Munitz, M.S.; Garrido, C.E.; Gonzalez, H.H.L.; Resnik, S.L.; Salas, P.M.; Montti, M.I.T. Mycoflora and Potential Mycotoxin Production of Freshly Harvested Blueberry in Concordia, Entre Ríos Province, Argentina. *Int. J. Fruit Sci.* **2013**, *13*, 312–325. [CrossRef]
21. Delgado-Ospina, J.; Molina-Hernandez, J.B.; Maggio, F.; Fernández-Daza, F.; Sciarra, P.; Paparella, A.; Chaves-López, C. Advances in understanding the enzymatic potential and production of ochratoxin A of moulds isolated from cocoa fermented beans. *Food Microbiol.* 2022, *in press*.
22. Glass, N.L.; Donaldson, G.C. Development of primer sets designed for use with the PCR to amplify conserved genes from filamentous ascomycetes. *Appl. Environ. Microbiol.* **1995**, *61*, 1323–1330. [CrossRef]
23. Makhlouf, J.; Carvajal-Campos, A.; Querin, A.; Tadrist, S.; Puel, O.; Lorber, S.; Oswald, I.P.; Hamze, M.; Bailly, J.D.; Bailly, S. Morphologic, molecular and metabolic characterization of *Aspergillus* section *Flavi* in spices marketed in Lebanon. *Sci. Rep.* **2019**, *9*, 5263. [CrossRef]
24. Ali, A.; Muhammad, M.T.M.; Sijam, K.; Siddiqui, Y. Potential of chitosan coating in delaying the postharvest anthracnose (*Colletotrichum gloeosporioides* Penz.) of Eksotika II papaya. *Int. J. Food Sci. Technol.* **2010**, *45*, 2134–2140. [CrossRef]
25. Molina-Hernandez, J.B.; Aceto, A.; Bucciarelli, T.; Paludi, D.; Valbonetti, L.; Zilli, K.; Scotti, L.; Chaves-López, C. The membrane depolarization and increase intracellular calcium level produced by silver nanoclusters are responsible for bacterial death. *Sci. Rep.* **2021**, *11*, 21557. [CrossRef]
26. Horwitz, W. *Official Methods of Analysis International, Agricultural Chemicals, Contaminants, Drugs*, 15th ed.; AOAC 925.10; AOAC: Rockville, MD, USA, 1990.
27. Salder, G.; Davis, J.; Dezman, D. Rapid Extraction of Lycopene and β-Carotene from Reconstituted Tomato Paste and Pink Grapefruit Homogenates. *J. Food Sci.* **1990**, *55*, 1460–1461. [CrossRef]
28. Souza, A.L.R.; Hidalgo-Chávez, D.W.; Pontes, S.M.; Gomes, F.S.; Cabral, L.M.C.; Tonon, R.V. Microencapsulation by spray drying of a lycopene-rich tomato concentrate: Characterization and stability. *LWT—Food Sci. Technol.* **2018**, *91*, 286–292. [CrossRef]
29. Singleton, V.L.; Rossi, J.A.J. Colorimetry to total phenolics with phosphomolybdic acid reagents. *Am. J. Enol. Vinic.* **1965**, *16*, 144–158.
30. Re, R.; Pellegrini, N.; Proteggente, A.; Pannala, A.; Min Yang, A.; Rice-Evans, C. Antioxidant activity applying an improved abts radical cation decolorization assay. *Free Radic. Biol. Med.* **1999**, *26*, 1231–1237. [CrossRef]
31. Shimizu, T.; Sakiyama, Y.; Graves, D.B.; Zimmermann, J.L.; Morfill, G.E. The dynamics of ozone generation and mode transition in air surface micro-discharge plasma at atmospheric pressure. *New J. Phys.* **2012**, *14*, 103028. [CrossRef]
32. Pavlovich, M.J.; Clarck, D.S.; Graves, D.B. Quantification of air plasma chemistry for surface disinfection. *Plasma Sources Sci. Technol.* **2014**, *23*, 065036. [CrossRef]
33. Eliasson, B.; Hirth, M.; Kogelschatz1, U. Ozone synthesis from oxygen in dielectric barrier discharges. *J. Phys. D Appl. Phys.* **1987**, *20*, 1421. [CrossRef]
34. Kogelschatz, U.; Eliasson, B.; Hirth, M. Ozone generation from oxygen and air: Discharge physics and reactions mechanisms. *Ozone Sci. Eng.* **1987**, *10*, 367–368. [CrossRef]
35. Finn, S.; Condell, O.; McClure, P.; Amézquita, A.; Fanning, S. Mechanisms of survival, responses, and sources of salmonella in low-moisture environments. *Front. Microbiol.* **2013**, *4*, 331. [CrossRef] [PubMed]
36. Puligundla, P.; Mok, C. Inactivation of spores by nonthermal plasmas. *World J. Microbiol. Biotechnol.* **2018**, *34*, 143. [CrossRef]
37. Dobrynin, D.; Fridman, G.; Mukhin, Y.V.; Wynosky-Dolfi, M.A.; Rieger, J.; Rest, R.F.; Gutsol, A.F.; Fridman, A. Cold plasma inactivation of *Bacillus cereus* and *Bacillus anthracis* (anthrax) spores. *IEEE Trans. Plasma Sci.* **2010**, *38*, 1878–1884. [CrossRef]
38. Hertwig, C.; Reineke, K.; Rauh, C.; Schlüter, O. Factors involved in *Bacillus* spore's resistance to cold atmospheric pressure plasma. *Innov. Food Sci. Emerg. Technol.* **2017**, *43*, 173–181. [CrossRef]
39. Young, S.B.; Setlow, P. Mechanisms of *Bacillus subtilis* spore resistance to and killing by aqueous ozone. *J. Appl. Microbiol.* **2004**, *96*, 1133–1142. [CrossRef]
40. Patil, S.; Moiseev, T.; Misra, N.N.; Cullen, P.J.; Mosnier, J.P.; Keener, K.M.; Bourke, P. Influence of high voltage atmospheric cold plasma process parameters and role of relative humidity on inactivation of *Bacillus atrophaeus* spores inside a sealed package. *J. Hosp. Infect.* **2014**, *88*, 162–169. [CrossRef] [PubMed]
41. Gavahian, M.; Peng, H.J.; Chu, Y.H. Efficacy of cold plasma in producing Salmonella-free duck eggs: Effects on physical characteristics, lipid oxidation, and fatty acid profile. *J. Food Sci. Technol.* **2019**, *56*, 5271–5281. [CrossRef] [PubMed]

42. Lacombe, A.; Niemira, B.A.; Gurtler, J.B.; Fan, X.; Sites, J.; Boyd, G.; Chen, H. Atmospheric cold plasma inactivation of aerobic microorganisms on blueberries and effects on quality attributes. *Food Microbiol.* **2015**, *46*, 479–484. [CrossRef] [PubMed]
43. Ulbin-Figlewicz, N.; Jarmoluk, A. Effect of low-pressure plasma treatment on the color and oxidative stability of raw pork during refrigerated storage. *Food Sci. Technol. Int.* **2016**, *22*, 313–324. [CrossRef]
44. Ziuzina, D.; Patil, S.; Cullen, P.J.; Keener, K.M.; Bourke, P. Atmospheric cold plasma inactivation of *Escherichia coli, Salmonella enterica* serovar Typhimurium and *Listeria monocytogenes* inoculated on fresh produce. *Food Microbiol.* **2014**, *42*, 109–116. [CrossRef] [PubMed]
45. Butscher, D.; Zimmermann, D.; Schuppler, M.; von Rohr, P.R. Plasma inactivation of bacterial endospores on wheat grains and polymeric model substrates in a dielectric barrier discharge. *Food Control* **2016**, *60*, 636–645. [CrossRef]
46. Kim, J.E.; Lee, D.U.; Min, S.C. Microbial decontamination of red pepper powder by cold plasma. *Food Microbiol.* **2014**, *38*, 128–136. [CrossRef]
47. Iseki, S.; Ohta, T.; Aomatsu, A.; Ito, M.; Kano, H.; Higashijima, Y.; Hori, M. Rapid inactivation of *Penicillium digitatum* spores using high-density nonequilibrium atmospheric pressure plasma. *Appl. Phys. Lett.* **2010**, *96*, 153704. [CrossRef]
48. Yaguchi, T.; Horie, Y.; Tanaka, R.; Matsuzawa, T.; Ito, J.; Nishimura, K. Molecular phylogenetics of multiple genes on *Aspergillus* section *Fumigati* isolated from clinical specimens in Japan. *Jpn. J. Med. Mycol.* **2007**, *48*, 37–46. [CrossRef]
49. Trompeter, F.J.; Neff, W.J.; Franken, O.; Heise, M.; Neiger, M.; Liu, S.; Pietsch, G.J.; Saveljew, A.B. Reduction of *Bacillus Subtilis* and *Aspergillus Niger* spores using nonthermal atmospheric gas discharges. *IEEE Trans. Plasma Sci.* **2002**, *30*, 1416–1423. [CrossRef]
50. Ott, L.C.; Appleton, H.J.; Shi, H.; Keener, K.; Mellata, M. High voltage atmospheric cold plasma treatment inactivates *Aspergillus flavus* spores and deoxynivalenol toxin. *Food Microbiol.* **2021**, *95*, 103669. [CrossRef]
51. Oliveira, B.R.; Marques, A.P.; Ressurreição, M.; Moreira, C.J.S.; Pereira, C.S.; Crespo, M.T.B.; Pereira, V.J. Inactivation of *Aspergillus* species in real water matrices using medium pressure mercury lamps. *J. Photochem. Photobiol. B Biol.* **2021**, *221*, 112242. [CrossRef] [PubMed]
52. Misra, N.N.; Yadav, B.; Roopesh, M.S.; Jo, C. Cold Plasma for Effective Fungal and Mycotoxin Control in Foods: Mechanisms, Inactivation Effects, and Applications. *Compr. Rev. Food Sci. Food Saf.* **2019**, *18*, 106–120. [CrossRef] [PubMed]
53. Cordero, R.J.B.; Casadevall, A. Functions of fungal melanin beyond virulence. *Fungal Biol. Rev.* **2017**, *31*, 99–112. [CrossRef] [PubMed]
54. Montie, T.C.; Kelly-Wintenberg, K.; Roth, J.R. An overview of research using the one atmosphere uniform glow discharge plasma (OAUGDP) for sterilization of surfaces and materials. *IEEE Trans. Plasma Sci.* **2000**, *28*, 41–50. [CrossRef]
55. Hopke, A.; Brown, A.J.P.; Hall, R.A.; Wheeler, R.T. Dynamic Fungal Cell Wall Architecture in Stress Adaptation and Immune Evasion. *Trends Microbiol.* **2018**, *26*, 284–295. [CrossRef]
56. Ambrico, P.F.; Šimek, M.; Rotolo, C.; Morano, M.; Minafra, A.; Ambrico, M.; Pollastro, S.; Gerin, D.; Faretra, F.; De Miccolis Angelini, R.M. Surface Dielectric Barrier Discharge plasma: A suitable measure against fungal plant pathogens. *Sci. Rep.* **2020**, *10*, 3673. [CrossRef]
57. Bao, Y.; Reddivari, L.; Huang, J.Y. Development of cold plasma pretreatment for improving phenolics extractability from tomato pomace. *Innov. Food Sci. Emerg. Technol.* **2020**, *65*, 102445. [CrossRef]
58. Starek, A.; Pawłat, J.; Chudzik, B.; Kwiatkowski, M.; Terebun, P.; Sagan, A.; Andrejko, D. Evaluation of selected microbial and physicochemical parameters of fresh tomato juice after cold atmospheric pressure plasma treatment during refrigerated storage. *Sci. Rep.* **2019**, *9*, 8407. [CrossRef]
59. Ramazzina, I.; Berardinelli, A.; Rizzi, F.; Tappi, S.; Ragni, L.; Sacchetti, G.; Rocculi, P. Effect of cold plasma treatment on physico-chemical parameters and antioxidant activity of minimally processed kiwifruit. *Postharvest Biol. Technol.* **2015**, *107*, 55–65. [CrossRef]
60. Tappi, S.; Ramazzina, I.; Rizzi, F.; Sacchetti, G.; Ragni, L.; Rocculi, P. Effect of plasma exposure time on the polyphenolic profile and antioxidant activity of fresh-cut apples. *Appl. Sci.* **2018**, *8*, 1939. [CrossRef]
61. Dasan, B.G.; Boyaci, I.H. Effect of Cold Atmospheric Plasma on Inactivation of *Escherichia coli* and Physicochemical Properties of Apple, Orange, Tomato Juices, and Sour Cherry Nectar. *Food Bioprocess Technol.* **2018**, *11*, 334–343. [CrossRef]
62. Li, X.; Li, M.; Ji, N.; Jin, P.; Zhang, J.; Zheng, Y.; Zhang, X.; Li, F. Cold plasma treatment induces phenolic accumulation and enhances antioxidant activity in fresh-cut pitaya (*Hylocereus undatus*) fruit. *LWT* **2019**, *115*, 108447. [CrossRef]
63. Alothman, M.; Bhat, R.; Karim, A.A. UV radiation-induced changes of antioxidant capacity of fresh-cut tropical fruits. *Innov. Food Sci. Emerg. Technol.* **2009**, *10*, 512–516. [CrossRef]
64. Ozen, E.; Singh, R.K. Atmospheric cold plasma treatment of fruit juices: A review. *Trends Food Sci. Technol.* **2020**, *103*, 144–151. [CrossRef]
65. Hoffmann, A.M.; Noga, G.; Hunsche, M. High blue light improves acclimation and photosynthetic recovery of pepper plants exposed to UV stress. *Environ. Exp. Bot.* **2015**, *109*, 254–263. [CrossRef]
66. Sarangapani, C.; O'Toole, G.; Cullen, P.J.; Bourke, P. Atmospheric cold plasma dissipation efficiency of agrochemicals on blueberries. *Innov. Food Sci. Emerg. Technol.* **2017**, *44*, 235–241. [CrossRef]

*Article*

# The Antibacterial Efficacy and Mechanism of Plasma-Activated Water Against *Salmonella* Enteritidis (ATCC 13076) on Shell Eggs

**Chia-Min Lin [1], Chun-Ping Hsiao [2], Hong-Siou Lin [1], Jian Sin Liou [2], Chang-Wei Hsieh [3], Jong-Shinn Wu [2,*,†] and Chih-Yao Hou [1,*,†]**

1 Department of Seafood Science, National Kaohsiung University of Science and Technology, Kaohsiung 811, Taiwan; cmlin@nkust.edu.tw (C.-M.L.); darksevens728@gmail.com (H.-S.L.)
2 Department of Mechanical Engineering, National Chiao Tung University, Hsinchu 300, Taiwan; pandarugby@nctu.edu.tw (C.-P.H.); rinknew@gmail.com (J.S.L.)
3 Department of Food Science and Biotechnology, National Chung Hsing University, Taichung 402, Taiwan; welson@nchu.edu.tw
* Correspondence: chongsin@faculty.nctu.edu.tw (J.-S.W.); chihyaohou@gmail.com (C.-Y.H.); Tel.: +886-985-300-345 (C.-Y.H)
† Both authors contributed equally to this work.

Received: 2 September 2020; Accepted: 16 October 2020; Published: 19 October 2020

**Abstract:** Eggs are one of the most commonly consumed food items. Currently, chlorine washing is the most common method used to sanitize shell eggs. However, chlorine could react with organic matters to form a potential carcinogen, trihalomethanes, which can have a negative impact on human health. Plasma-activated water (PAW) has been demonstrated to inactivate microorganisms effectively without compromising the sensory qualities of shell eggs. For this study, various amounts (250, 500, 750, or 1000 mL) of PAW were generated by using one or two plasma jet(s) at 60 watts for 20 min with an air flow rate at 6 or 10 standard liters per minute (slm). After being inoculated with 7.0 log CFU *Salmonella* Enteritidis, one shell egg was placed into PAW for 30, 60, or 90 s with 1 or 2 acting plasma jet(s). When 2 plasma jets were used in a large amount of water (1000 mL), populations of S. Enteritidis were reduced from 7.92 log CFU/egg to 2.84 CFU/egg after 60 s of treatment. In addition, concentrations of ozone, hydrogen peroxide, nitrate, and nitrite in the PAW were correlated with the levels of antibacterial efficacy. The highest concentrations of ozone (1.22 ppm) and nitrate (55.5 ppm) were obtained with a larger water amount and lower air flow rate. High oxidation reduction potential (ORP) and low pH values were obtained with longer activation time, more plasma jet, and a lower air flow rate. Electron paramagnetic resonance (EPR) analyses demonstrated that reactive oxygen species (ROS) were generated in the PAW. The observation under the scanning electron microscope (SEM) revealed that bacterial cells were swollen, or even erupted after treatment with PAW. These results indicate that the bacterial cells lost control of cell permeability after the PAW treatment. This study shows that PAW is effective against S. Enteritidis on shell eggs in a large amount of water. Ozone, nitrate, and ROS could be the main causes for the inactivation of bacterial cells.

**Keywords:** plasma-activated water (PAW); *Salmonella*; egg

## 1. Introduction

Eggs are a popular food choice in daily life and an excellent source of nutrients [1,2]. The major pathogenic microorganism associated with eggs is *Salmonella* spp. Salmonellosis associated with eggs is frequently reported in many countries [3]. In Taiwan, 77% of confirmed *Salmonella* cases were associated with eggs (TCDC, 2011). Among more than 2000 serovars, *S. enterica* serovar Enteritidis (S. Enteritidis) is the most frequently occurring serovar in eggs and egg products [4–6]. Thus, inactivating *Salmonella* on egg surfaces is a necessary process. Currently, the most common practice used to disinfect and clean the surface of shell eggs is washing with 100–200 ppm chlorinated water, alkaline detergent (such as quaternary ammonium salts), hot water, or steaming at 5–10 °C above the egg's surface temperature. Chlorinated water is the most commonly used method, because of its relatively low cost and high efficacy. However, chlorine is not stable in high temperatures, and easily reacts with organic substances to form a potential carcinogen, trihalomethane [7]. In addition, modern consumers perceive chemical food treatments negatively. Thus, a novel technique to disinfect and clean shell eggs without chemical residue is strongly needed.

Non-thermal plasma is a novel physical method to disinfect microorganisms on foods without compromising sensory characteristics. Plasma ionizes the gas to generate charged particles, electric fields, ultraviolet (UV) photons, reactive nitrogen-oxygen (RNS), and reactive oxygen species (ROS), which are able to inactivate microorganisms [8]. Applying the plasma jet inside water, the charged particles, ROS, and RNS generated from gas ionization describe those which were previously easily dissolved in water. In addition, plasma also ionized the water to produce charged particles and ROS. These substances and the reaction products of these substances were effective antimicrobial agents. The substances are relatively non-stable and degrade rapidly, thus leaving no chemical residue [9,10]. This type of water is called plasma-activated water (PAW), which has been reported to inactivate foodborne pathogens on food-contact surfaces and food items without a negative impact to the environment and human health [8]. Several examples of using PAW to inactivate microorganisms on food have been reported, which include mushroom [11], strawberry [12], bean sprout [13], Korean rice cake [14], egg [15], and cabbage [16].

In past studies, several substances generated by the reactions of RNS and ROS in PAW were suggested to be the key agents in inactivating bacteria [17]. These agents included hydrogen peroxide ($H_2O_2$), singlet oxygen (O), nitrite ($NO_2^-$), and nitrate ($NO_3^-$). However, these potential antibacterial agents were not measured in the previous examples of using PAW on food items. Additionally, the amount of PAW used was relatively small (100–200 mL). Increasing the amount of PAW is a critical step towards industrial applications. Therefore, the objectives of this research were to evaluate the efficacy of larger amounts of PAW on shell eggs and investigate the potential antibacterial mechanism of PAW. To achieve these goals, different quantities of plasma generators and air flow rates were used in various amounts of water to obtain the optimal condition to inactivate *S.* Enteritidis on eggs. The concentrations of hydrogen peroxide ($H_2O_2$), ozone ($O_3$), nitrite ($NO_2^-$), and nitrate ($NO_3^-$) in PAW was determined. The spectrum of plasma and the existence of singlet oxygen (O) were measured. Lastly, damage of the PAW-treated bacterial cells was observed under a scanning electron microscope.

## 2. Material and Methods

### 2.1. Preparation of Bacterial Suspension

The most commonly occurring pathogen associated with eggs, *Salmonella enterica* subsp. *enterica* serovar Enteritidis (ATCC 13076, Bioresource Collection and Research Centre, Taiwan), was used. This bacterium was maintained on tryptic soy agar (TSA) and confirmed by the selective media, xylose lysine deoxycholate agar (XLD). Fresh working culture was prepared by inoculating in tryptic soy broth (TSB), then incubated at 37 °C for 18–20 h consecutively, twice. Bacterial population density was maintained at a 1.0–1.5 optical density at 600 nm ($OD_{600}$), which was approximately 9 log CFU/mL. The media used in this study were all purchased from Difco Laboratories (Detroit, MI, USA).

### 2.2. Preparation of Plasma-Activated Water (PAW)

The device of the non-thermal atmospheric plasma generator was assembled by the Aerothermal and Plasma Physics Lab. (APPL), Department of Mechanical Engineering, National Chiao Tung University (Hsingchu, Taiwan). The major components of the system included a high-voltage power supply, an air pump, and an atmospheric pressure plasma jet (APPJ) (patent, US10,121,638B1) generator.

During the study, single or duplicate system(s) were used to evaluate the effect of plasma jet quantities. This system was specifically designed to operate an electrode beneath the water surface to generate PAW. Atmospheric air was pumped into the electrode by the air pump, which also controlled the air flow rate. As per the previous study [15], reverse osmotic (RO) water was used as the water source. The voltage, frequency, pre-activation time, and power were set at 3.0 kV, 16 kHz, 20 min, and 60 Watts, respectively. The duplicate system used in this study is shown in Figure 1.

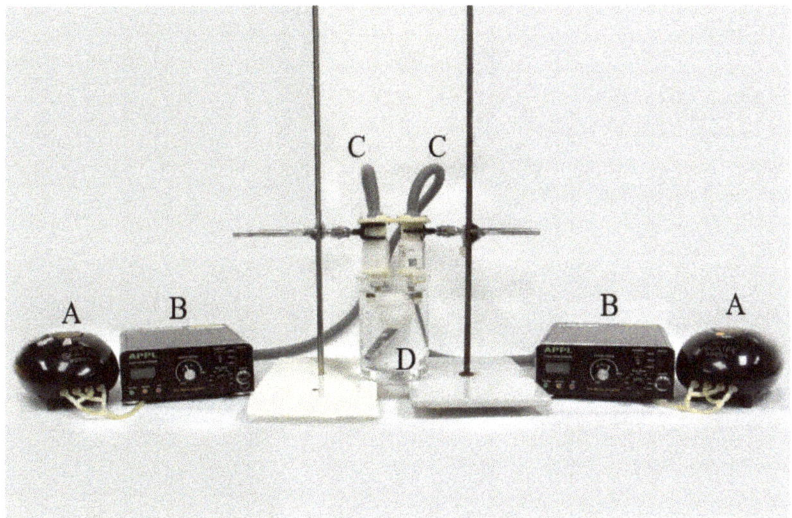


**Figure 1.** The devices of the air plasma-activated water system, including (**A**) air pumps, (**B**) high-voltage power supply, (**C**) plasma jet generators, and (**D**) the beaker containing one egg.

### 2.3. *Inactivation of Salmonella* spp. *on Eggs by PAW*

Eggs were obtained from a commercial egg farm with its own brand name. All testing eggs were unwashed and laid less than 2 days before testing. Only the eggs without surface debris, such as feces or other litter, were used. The chosen eggs were thoroughly washed by sterilized water, wiped by sterile paper, and then air-dried in a laminar hood. The washed egg was placed into a sampling bag with 100 mL phosphate buffer saline (PBS, pH 7.2). After being completely hand-rubbed for 2 min, decimally serial dilution and spreading on plate count agar (PCA) were used to check the presence of bacteria. Absence of *Salmonella* was confirmed based on the official method of the US Food and Drug Administration [18], in which the washed egg was placed into a sampling bag containing 225 mL lactose broth, then incubated at 37°C for 24 h. Tetrathionate (TT) broth and Rappaport-Vassiliadis (RV) broth were used as selective enrichment broth following the incubation of lactose broth. Selective agar media were xylose lysine deoxycholate (XLD), Hektoen enteric (HE), and bismuth sulfite (BS), which were used after the incubation of the selective enrichment broth. Before testing, S. Enteritidis (SE) culture was centrifuged at 5000× *g* for 5 min at 4 °C and resuspended in new TSB to achieve a population at 2–3 × $10^9$ CFU/mL. One hundred µL bacterial suspension was placed onto the surface of each egg in 33–35 drops, then air dried for 20–30 min. The inoculated eggs of the treatment group were transferred into a beaker containing pre-activated PAW. Different combinations of the operating parameters, such as air flow rate, water amount, treatment time, and plasma generator quantity, were tested to obtain the optimal conditions (Table 1). The treated egg was transferred into a sterile sampling bag with 100 mL of PBS, then gently rubbed by hand for 2 min. After being decimally serial-diluted, 0.1 mL of the PBS was spread onto PCA. The plates were incubated at 37 °C for 18–24 h, and triplicate plates were used for each dilution. Two to three colonies on a PCA plate which were randomly re-streaked onto XLD agar to confirm the recovered bacteria were *Salmonella*. Inoculated eggs washed with the same amount of sterile RO water with the same time were used as the water washing control. Inoculated eggs without washing were used as the negative control [15].

**Table 1.** The operating parameters of plasma-activated water (PAW) to inactivate *S.* Enteritidis on shell eggs.

| Parameters | Variance | Unit |
|---|---|---|
| Air flow rate | 6, 10 | standard liter per min (slm) |
| Water amount | 250, 500, 750, 1000 | mL |
| Treatment time | 30, 60, 90 | sec |
| Plasma jet number | 1, 2 | |

### *2.4. Measurement of pH, Oxidative-Reductive Potential (ORP), Hydrogen Peroxide ($H_2O_2$), Ozone ($O_3$), Nitrite ($NO^{2-}$), and Nitrate ($NO^{3-}$) of PAW*

The values of pH, oxidative-reductive potential (ORP), conductivity, hydrogen peroxide ($H_2O_2$), ozone ($O_3$), nitrite ($NO^{2-}$), and nitrate ($NO^{3-}$) were determined to investigate the physio-chemical characteristics of PAW.

#### 2.4.1. Measurement of pH and Oxidative-Reductive Potential (ORP)

The values of pH and oxidative-reductive potential (ORP) were measured by a pH/ORP meter (PC-200, Cole-Parmer, Vernon Hills, IL, USA) connected with a pH or ORP probe (serial 100 probe, Cole-Parmer).

#### 2.4.2. Measurements of Hydrogen Peroxide ($H_2O_2$)

The concentration of $H_2O_2$ at each treatment time, 30, 60, and 90 s was measured according to the method of the Department of Health and Welfare, Taipei City, Taiwan (2013). Five mL of PAW was collected immediately after treatment and mixed with 0.1% *o*-phenylenediamine (OPD) in citric acid buffer (pH 5.0). After reacting for 10 min, 1 mL 10 N sulfuric acid was added, and absorbance at 490 nm was recorded. The $H_2O_2$ concentrations were determined by comparing with the standard curve ranging from 0 to 6.0 mg/L. The absorbance at 490 nm was adjusted by using sterile RO water as the testing sample.

#### 2.4.3. Measurement of Ozone ($O_3$)

Ozone concentrations at each treatment time, 30, 60, and 90 s, were determined by Ozone AccuVac® Ampuls (0–1.5 mg/L) in a handheld colorimeter (DR900, HACH, Loveland, CO, USA). Following manufactural instruction, 40 mL PAW was sampled immediately after treatment and the ampule containing the reagents was placed into the tilted PAW sample. After the ampule was filled with PAW, it was sealed by a rubber stopper and shaken until the reagents totally dissolved. After reacting for 1 min, the concentrations were measured in the colorimeter. Sterile RO water was used as the blank.

#### 2.4.4. Measurement of Nitrite ($NO^{2-}$), and Nitrate ($NO^{3-}$)

The concentrations of nitrate and nitrite at each treatment time, 30, 60, and 90 s were measured by Nitrite Reagent Powder Pillows (2–250 mg/L $NO^{2-}$, and 0.3–30 mg/L $NO^{3-}$) in a handheld colorimeter (DR900, HACH, Loveland, CO, USA). As per the manufactural instructions, 10 mL of PAW was sampled and mixed with the powder pillows. After reacting for 5 min, the concentrations of $NO^{2-}$ and $NO^{3-}$ were determined in the colorimeter. Sterile RO water was used as the blank.

### 2.5. Measurement of the Optical Characteristics of Plasma by Optical Emission Spectroscopy (OES)

The optical characteristics of the plasma jet in the water was measured by an emission spectroscope (SP-2500, Acton Research, Acton, MA, USA). A plasma jet was placed beneath the water surface in a quartz tube. Plasma was discharged beneath the water. A fiber optics cable connected with the spectroscope was used to measure the optical signals with wavelengths between 200 to 850 nm.

### 2.6. Detection of Reactive Oxygen Species (ROS) by Electron Paramagnetic Resonance (EPR)

The generation of ROS in the PAW was determined by the reaction of superoxide, as well as superoxide-related formations of hydroxyl radicals within the spin trap of DMPO (5,5-dimethylpyrroline-N-oxide). Magnetic fields strength (G) of the sin trap was detected by an electron paramagnetic resonance (EPR) spectrometer (EMXplus-10/12/P/L system, Bruker Inc., Billerica, MA, USA). PAW was generated on site, and 1 L RO water was activated for 20 min using two plasma jets with an air flow rate at 10 slm. After 20 min activation, 1 mL PAW was collected at 30, 60, and 90 s with the plasma jets were still acting. After mixing with 20 µL of 1 M DMPO, the samples were immediately measured by EPR. RO water added with the same amount of DMPO was used as the control [19].

### 2.7. Observation Under Scanning Electron Microscope (SEM)

The inoculated eggs were treated with PAW for 60 or 90 s in 1 L PAW, as described previously, and unwashed inoculated eggs were used as the control. After treatment, eggshells were collected, and the inner membrane was removed. The shells were immersed in a phosphate buffer containing 2.5% glutaraldehyde at 4 °C for 2 h, then further washed three times with phosphate buffer and deionized water. The washed shells were soaked in gradually increasing concentrations of ethanol solutions (50%, 70%, 80%, 90%, and 95%). The dehydrated eggshell was further treated at −20 °C for 2 h, then −80 °C for 12 h. Finally, the shell samples were freeze-dried for 4 h, then stored in a desiccator before SEM observation. The samples were observed under a Quanta 200 Environmental Scanning Electron Microscope (Thermo Fisher Scientific, Waltham, MA, USA).

### 2.8. Statistical Analyses

Each treatment had triplicate samples, and all experiments were conducted at least twice. Data were analyzed by using one-way ANOVA and Duncan's test was used for post hoc analysis. The significant differences between treatments were set at $p < 0.05$. All statistical analyses were conducted by using an IBM SPSS program (Version 22.0, St. Armonk, NY, USA).

## 3. Results

### 3.1. The Effects of Plasma Jet Quantity, Treatment Time, and Air Flow Rates

While one plasma jet was used, bactericidal activities increased with longer treatment time ($p < 0.05$) and reduced significantly when large water amounts were tested ($p < 0.05$). When compared with egg samples washed by sterile water under the same treatment time, 2.82–3.07 and 1.92–2.89 log CFU/egg reductions were obtained from the PAW at 250 and 500 mL, respectively. However, only 0.53–0.92 log and 0–0.08 CFU/egg reductions were obtained at 750 and 1000 mL (Table 2). Bactericidal activities increased greatly when two plasma jets were used. More than five log reductions were obtained when eggs were treated for 90 s in 750 and 1000 mL PAW (Table 2). Longer treatment time still resulted in greater reductions ($p < 0.05$). However, no significant difference was observed between different water volumes at the same treatment time ($p > 0.05$).

**Table 2.** The populations of *Salmonella* Enteritidis (log CFU/egg) on eggs treated by the plasma-activated water (PAW) generated with one or two plasma jets at 6 slm air flow with different treatment times and volumes.

| Treatments/Volume | 250 mL | 500 mL | 750 mL | 1000 mL |
|---|---|---|---|---|
| Unwashed | 7.22 ± 1.04 $^{aA}$ | 7.37 ± 0.60 $^{aA}$ | 8.06 ± 0.06 $^{aA}$ | 7.92 ± 0.25 $^{aA}$ |
| Sterile water-30s | 6.85 ± 0.23 $^{bA}$ | 6.69 ± 0.42 $^{bA}$ | 7.12 ± 0.21 $^{bA}$ | 7.06 ± 0.01 $^{bA}$ |
| Sterile water-60s | 6.66 ± 0.06 $^{bA}$ | 6.60 ± 0.06 $^{bA}$ | 6.63 ± 0.16 $^{bA}$ | 6.54 ± 0.34 $^{bcA}$ |
| Sterile water-90s | 6.24 ± 0.27 $^{bA}$ | 6.49 ± 0.27 $^{bA}$ | 6.43 ± 0.27 $^{bA}$ | 6.36 ± 0.16 $^{bcA}$ |
| One plasma jet | | | | |
| PAW-30s | 3.78 ± 0.64 $^{cdA}$ | 4.77 ± 0.92 $^{cB}$ | 6.49 ± 0.13 $^{bC}$ | 6.98 ± 0.12 $^{bcC}$ |
| PAW-60s | 4.27 ± 0.63 $^{cA}$ | 4.26 ± 0.67 $^{cdA}$ | 6.10 ± 0.11 $^{bcB}$ | 6.59 ± 0.27 $^{bcB}$ |
| PAW-90s | 3.42 ± 0.37 $^{cdA}$ | 3.60 ± 0.65 $^{dA}$ | 5.51 ± 0.27 $^{cB}$ | 6.28 ± 0.19 $^{cC}$ |
| Two plasma jet | | | | |
| PAW-30s | 4.55 ± 0.60 $^{cA}$ | 4.33 ± 0.15 $^{cdA}$ | 4.36 ± 0.41 $^{dA}$ | 4.67 ± 0.32 $^{dA}$ |
| PAW-60s | 3.48 ± 0.66 $^{cdA}$ | 3.87 ± 0.39 $^{dA}$ | 3.73 ± 0.48 $^{dfA}$ | 2.84 ± 1.34 $^{eA}$ |
| PAW-90s | 3.16 ± 0.10 $^{dA}$ | 3.38 ± 0.32 $^{dA}$ | 2.69 ± 0.37 $^{fA}$ | 2.78 ± 0.56 $^{eA}$ |

Data are presented as average ± STD. Average in the same column with different lower-case letters are significantly different ($p < 0.05$). Average in the same row with different upper-case letters are significantly different ($p < 0.05$).

Since the PAW generated with two plasma jets was more effective and a large water volume was the objective, the effects of different air flow rates were only tested in the two plasma jet systems at the 1000 mL water amount. Significantly lower reduction ($p < 0.05$) was obtained from the samples using 10 slm rather than 6 slm air flow rate for all three treatment time-periods. When treatment time was extended from 60 s to 90 s, no significant decrease ($p < 0.05$) in bacterial populations were obtained at both air flow rates (Table 3).

**Table 3.** The populations of *Salmonella* Enteritidis (log CFU/egg) on eggs treated in 1000 mL plasma-activated water (PAW) generated with two jets at 6 slm or 10 slm.

| Treatments/Flow Rate | 6 slm | 10 slm |
|---|---|---|
| Unwashed | 7.24 ± 0.23 $^{Aa}$ | 7.71 ± 0.11 $^{Aa}$ |
| Sterile water-30s | 7.06 ± 0.01 $^{Aa}$ | 6.69 ± 0.17 $^{Ab}$ |
| Sterile water-60s | 6.54 ± 0.34 $^{Aa}$ | 6.46 ± 0.06 $^{Ab}$ |
| Sterile water-90s | 6.23 ± 0.16 $^{Aa}$ | 6.34 ± 0.13 $^{Ab}$ |
| PAW-30s | 4.67 ± 0.32 $^{Ab}$ | 5.20 ± 0.16 $^{Bc}$ |
| PAW-60s | 2.84 ± 1.34 $^{Ac}$ | 4.53 ± 0.17 $^{Bd}$ |
| PAW-90s | 2.78 ± 0.56 $^{Ac}$ | 3.83 ± 0.07 $^{Bd}$ |

Data are presented as average ± STD. Average in the same column with different lower-case letters are significantly different ($p < 0.05$). Average in the same row with different upper-case letters are significantly different ($p < 0.05$).

### 3.2. The pH and Oxidative-Reductive Potential (ORP) Values of PAW

Within the PAW generated by one plasma jet, the values of pH decrease and ORP increase were reduced significantly ($p < 0.05$) when a larger water volume was used. However, when adding one more plasma jet, the pH and ORP values of the PAW were not significantly different ($p > 0.05$) between various water amounts (Table 4). However, when the PAW was generated with two plasma jets, significantly lower ORP values were obtained with the high air flow rate (10 slm) (Table 5).

**Table 4.** The pH and oxidative-reduction potential (ORP) values of the plasma-activated water (PAW) generated with one plasma jet and two plasma jets at various treatment times and water volumes.

| Treatments/Volume | | 250 mL | 500 mL | 750 mL | 1000 mL |
|---|---|---|---|---|---|
| RO water | pH | 6.23 ± 0.20 $^{aA}$ | 6.27 ± 0.28 $^{aA}$ | 6.25 ± 0.21 $^{aA}$ | 6.13 ± 0.13 $^{aA}$ |
| | ORP | 200.9 ± 3.5 $^{dA}$ | 249.4 ± 6.1 $^{dA}$ | 254.6 ± 28.7 $^{dA}$ | 220.4 ± 13.5 $^{dA}$ |
| One plasma jet | | | | | |
| PAW-30s | pH | 3.36 ± 0.07 $^{bA}$ | 3.43 ± 0.17 $^{bA}$ | 3.53 ± 0.15 $^{bA}$ | 3.95 ± 0.11 $^{bB}$ |
| | ORP | 492.5 ± 14.1 $^{cA}$ | 499.8 ± 7.1 $^{cA}$ | 476.2 ± 8.6 $^{cB}$ | 467.8 ± 7.5 $^{cB}$ |
| PAW-60s | pH | 3.29 ± 0.10 $^{bA}$ | 3.46 ± 0.16 $^{bA}$ | 4.01 ± 0.04 $^{bA}$ | 3.87 ± 0.03 $^{bA}$ |
| | ORP | 503.2 ± 18.6 $^{bA}$ | 513.9 ± 11.2 $^{bA}$ | 505.5 ± 7.2 $^{bA}$ | 491.1 ± 11.3 $^{bA}$ |
| PAW-90s | pH | 3.41 ± 0.10 $^{bA}$ | 3.55 ± 0.14 $^{bA}$ | 3.71 ± 0.08 $^{bB}$ | 3.81 ± 0.06 $^{bB}$ |
| | ORP | 531.6 ± 3.4 $^{aA}$ | 536.7 ± 10.3 $^{aA}$ | 496.8 ± 17.3 $^{aB}$ | 509.6 ± 2.2 $^{aB}$ |
| Two plasma jet | | | | | |
| PAW-30s | pH | 3.20 ± 0.08 $^{bA}$ | 3.12 ± 0.02 $^{bA}$ | 3.24 ± 0.08 $^{bA}$ | 3.29 ± 0.02 $^{bA}$ |
| | ORP | 540.5 ± 0.4 $^{aA}$ | 543.20 ± 0.1 $^{aA}$ | 530.8 ± 12.7 $^{aA}$ | 505.95 ± 1.2 $^{aB}$ |
| PAW-60s | pH | 3.10 ± 0.02 $^{bA}$ | 3.17 ± 0.09 $^{bA}$ | 3.16 ± 0.03 $^{bA}$ | 3.20 ± 0.02 $^{bA}$ |
| | ORP | 510.2 ± 8.1 $^{bA}$ | 534.0 ± 1.4 $^{bA}$ | 523.6 ± 7.6 $^{bA}$ | 515.7 ± 2.1 $^{bA}$ |
| PAW-90s | pH | 3.06 ± 0.01 $^{cA}$ | 3.08 ± 0.05 $^{cA}$ | 3.12 ± 0.06 $^{cA}$ | 3.25 ± 0.08 $^{cA}$ |
| | ORP | 529.5 ± 6.2 $^{bA}$ | 533.7 ± 10.7 $^{bA}$ | 520.7 ± 23.3 $^{bA}$ | 527.1 ± 9.7 $^{bA}$ |

Data are presented as average ± STD. Average in the same column for the same testing with different lower-case letters are significantly different ($p < 0.05$). Average in the same row with different upper-case letters are significantly different ($p < 0.05$).

**Table 5.** The values of pH and oxidative-reduction potential (ORP) of 1000 mL PAW generated by two plasma jets at different treatment times and flow rates.

| Treatments/Flow Rate | | 6 slm | 10 slm |
|---|---|---|---|
| RO water | pH | 7.91 ± 0.12 $^{aA}$ | 8.05 ± 0.23 $^{aA}$ |
| | ORP | 239.6 ± 18.10 $^{cA}$ | 210.1 ± 12.5 $^{bA}$ |
| PAW-30s | pH | 3.29 ± 0.02 $^{bA}$ | 3.31 ± 0.10 $^{bA}$ |
| | ORP | 505.95 ± 1.2 $^{aA}$ | 465.2 ± 4.1 $^{aB}$ |
| PAW-60s | pH | 3.20 ± 0.02 $^{bA}$ | 3.40 ± 0.02 $^{bA}$ |
| | ORP | 515.70 ± 2.12 $^{bA}$ | 488.8 ± 7.4 $^{aB}$ |
| PAW-90s | pH | 3.25 ± 0.08 $^{cA}$ | 3.32 ± 0.03 $^{bA}$ |
| | ORP | 527 ± 9.76 $^{bA}$ | 501.2 ± 24.6 $^{aB}$ |

Data are presented as average ± STD. Average in the same column for the same testing with different lower-case letters are significantly different ($p < 0.05$). Average in the same row with different upper-case letters are significantly different ($p < 0.05$).

### *3.3. The Concentrations of Ozone ($O_3$), Hydrogen Peroxide ($H_2O_2$), Nitrate ($NO_3^-$), and Nitrile ($NO^{2-}$) in PAW*

Both the concentrations of $O_3$ and $H_2O_2$ were lower at 10 slm compared to 6 slm, particularly $H_2O_2$, whose values were significantly lower ($p < 0.05$) at 10 slm. When the water amount increased, the variation of $O_3$ concentrations was insignificant ($p > 0.05$). However, $H_2O_2$ concentrations reached the highest at 750 mL, but decreased markedly at 1000 mL (Table 6).

The $NO_2^-$ concentrations were higher than $NO_3^-$ in PAW (Table 7). Both $NO_3^-$ and $NO_2^-$ concentrations increased with longer treatment time. At different air flow rates, $NO_3^-$ concentrations increased at 10 slm, but $NO_2^-$ concentrations decreased at 10 slm. However, $NO_3^-$ concentrations decreased when the water volume increased, but $NO_2^-$ concentrations increased along with the water volume.

**Table 6.** The concentrations of ozone ($O_3$) and hydrogen peroxide ($H_2O_2$) of the PAW generated by two plasma jets at 6 or 10 slm with different treatment times and water volumes.

| slm | Time/Volume | 250 mL | 500 mL | 750 mL | 1000 mL |
|---|---|---|---|---|---|
| $O_3$ | | | | | |
| | 30 s | 1.03 ± 0.05 $^{bA}$ | 0.92 ± 0.03 $^{bA}$ | 0.84 ± 0.02 $^{aA}$ | 0.90 ± 0.06 $^{aA}$ |
| 6 | 60 s | 1.13 ± 0.12 $^{aA}$ | 1.18 ± 0.05 $^{aA}$ | 0.93 ± 0.09 $^{aA}$ | 0.96 ± 0.04 $^{aA}$ |
| | 90 s | 1.12 ± 0.02 $^{aA}$ | 1.22 ± 0.02 $^{aA}$ | 0.96 ± 0.52 $^{aA}$ | 1.16 ± 0.18 $^{aA}$ |
| | 30 s | 0.91 ± 0.04 $^{aA}$ | 0.94 ± 0.01 $^{aA}$ | 0.87 ± 0.01 $^{aA}$ | 0.75 ± 0.06 $^{bA}$ |
| 10 | 60 s | 0.99 ± 0.05 $^{aA}$ | 0.90 ± 0.08 $^{aA}$ | 0.93 ± 0.04 $^{aA}$ | 0.60 ± 0.02 $^{bA}$ |
| | 90 s | 0.97 ± 0.02 $^{aA}$ | 1.02 ± 0.02 $^{aA}$ | 0.91 ± 0.60 $^{aA}$ | 0.71 ± 0.54 $^{bA}$ |
| $H_2O_2$ | | | | | |
| 6 | 30 s | 11.78 ± 0.73 $^{a}$ | 17.84 ± 0.44 $^{a}$ | 21.16 ± 0.70 $^{a}$ | 18.03 ± 0.71 $^{a}$ |
| | 60 s | 14.95 ± 1.02 $^{a}$ | 17.52 ± 0.70 $^{a}$ | 20.83 ± 0.68 $^{a}$ | 18.27 ± 0.75 $^{a}$ |
| | 90 s | 11.42 ± 0.77 $^{a}$ | 16.73 ± 0.25 $^{a}$ | 20.22 ± 0.86 $^{a}$ | 17.49 ± 0.39 $^{a}$ |
| 10 | 30 s | 2.96 ± 0.03 $^{b}$ | 2.91 ± 0.06 $^{b}$ | 2.89 ± 0.11 $^{b}$ | 1.88 ± 0.15 $^{b}$ |
| | 60 s | 2.76 ± 0.01 $^{b}$ | 2.89 ± 0.08 $^{b}$ | 3.06 ± 0.07 $^{b}$ | 1.91 ± 0.49 $^{b}$ |
| | 90 s | 2.72 ± 0.04 $^{b}$ | 2.78 ± 0.07 $^{b}$ | 2.75 ± 0.17 $^{b}$ | 1.65 ± 0.14 $^{b}$ |

Data are presented as average ± STD. Average in the same column for the same testing with different lower-case letters are significantly different ($p < 0.05$). Average in the same row with different upper-case letters are significantly different ($p < 0.05$).

**Table 7.** The concentrations of nitrate ($NO_3^-$) and nitrite ($NO_2^-$) of the PAW generated by two plasma jets at 6 or 10 slm with different treatment times and water volumes.

| slm | Time/Volume | 250 mL | 500 mL | 750 mL | 1000 mL |
|---|---|---|---|---|---|
| $NO_3^-$ | | | | | |
| | 30 s | 5.20 ± 0.60 $^{bA}$ | 4.56 ± 0.20 $^{bB}$ | 4.60 ± 1.25 $^{aB}$ | 5.01 ± 0.17 $^{aAB}$ |
| 6 | 60 s | 4.96 ± 0.68 $^{bA}$ | 4.53 ± 0.37 $^{aA}$ | 4.46 ± 0.21 $^{cA}$ | 4.71 ± 0.11 $^{aA}$ |
| | 90 s | 7.26 ± 0.61 $^{aA}$ | 6.70 ± 0.05 $^{bA}$ | 6.06 ± 0.20 $^{bB}$ | 4.70 ± 0.25 $^{aC}$ |
| | 30 s | 5.46 ± 0.40 $^{bB}$ | 5.70 ± 0.10 $^{aB}$ | 6.53 ± 0.05 $^{bA}$ | 4.70 ± 0.30 $^{bC}$ |
| 10 | 60 s | 5.40 ± 0.36 $^{bA}$ | 5.90 ± 0.34 $^{aA}$ | 5.33 ± 0.15 $^{cA}$ | 5.56 ± 0.15 $^{aA}$ |
| | 90 s | 9.66 ± 0.72 $^{aA}$ | 5.63 ± 0.10 $^{aB}$ | 8.10 ± 0.23 $^{aA}$ | 5.93 ± 0.26 $^{aB}$ |
| $NO_2^-$ | | | | | |
| 6 | 30 s | 40.5 ± 0.70 $^{aB}$ | 43.5 ± 0.90 $^{aB}$ | 49.5 ± 0.40 $^{aA}$ | 55.5 ± 0.37 $^{aA}$ |
| | 60 s | 42.5 ± 0.60 $^{aB}$ | 44.5 ± 0.50 $^{aB}$ | 50.5 ± 1.53 $^{aA}$ | 53.0 ± 0.80 $^{aA}$ |
| | 90 s | 40.0 ± 0.24 $^{aB}$ | 43.5 ± 1.06 $^{aB}$ | 46.5 ± 1.76 $^{aB}$ | 54.5 ± 0.35 $^{aA}$ |
| 10 | 30 s | 34.1 ± 1.41 $^{bB}$ | 31.5 ± 0.32 $^{bB}$ | 44.5 ± 0.40 $^{bA}$ | 47.5 ± 0.30 $^{bA}$ |
| | 60 s | 35.5 ± 0.60 $^{bB}$ | 31.0 ± 1.41 $^{bB}$ | 44.0 ± 1.41 $^{bA}$ | 46.5 ± 0.70 $^{bA}$ |
| | 90 s | 35.4 ± 0.10 $^{bB}$ | 36.0 ± 0.35 $^{bB}$ | 45.0 ± 1.10 $^{bA}$ | 48.0 ± 0.51 $^{bA}$ |

Data are presented as average ± STD. Average in the same column for the same testing with different lower-case letters are significantly different ($p < 0.05$). Average in the same row with different upper-case letters are significantly different ($p < 0.05$).

### 3.4. *The Optical Characteristics of Plasma and the Reactive Oxygen Species (ROS) in PAW*

The spectrum of the plasma jet in the PAW was from 190 to 460 nm. Most of the spectrum was located in the wavelength range of ultraviolet light (UV) (91.25%). Further examination showed UVA (320 nm–400 nm) and UVB (275-320 nm) were the major parts (66.6%). The shorter wavelength UVC (100–275 nm) that possesses higher antibacterial ability was the minor part (25.19%) (Figure 2).

Compared with the water control, several higher absorbance peaks, from between 810 to 1040 G, were observed in the PAW. These results indicate the existence of unpaired electrons in the PAW, which were predominantly generated by the activation of plasma in the water (Figure 3).

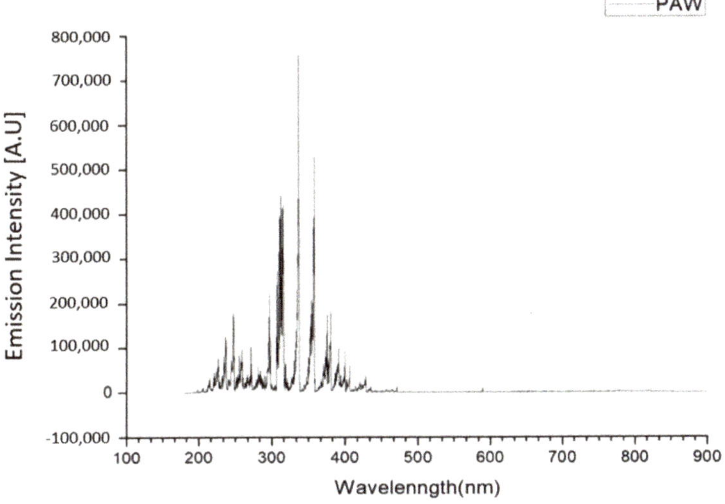


**Figure 2.** Space-integrated optical spectra of the plasma for water activation.

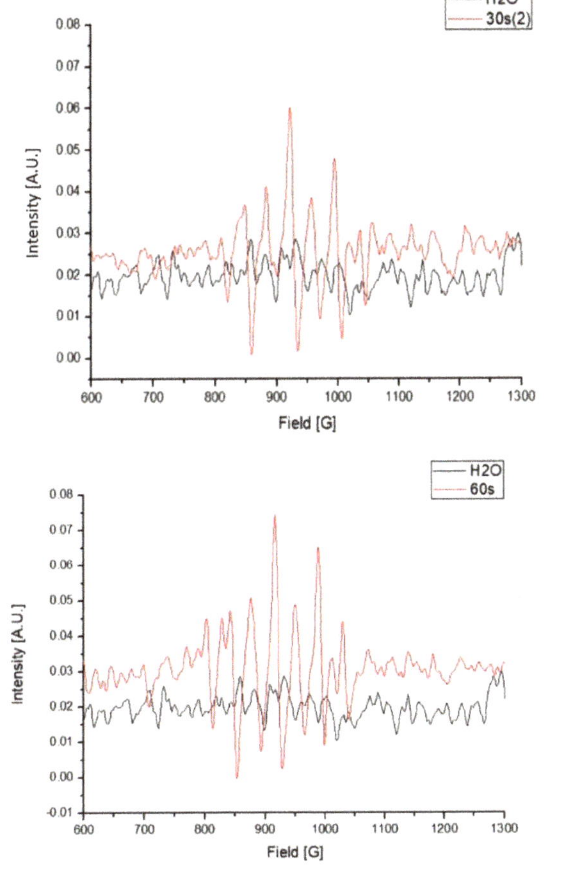


**Figure 3.** *Cont.*

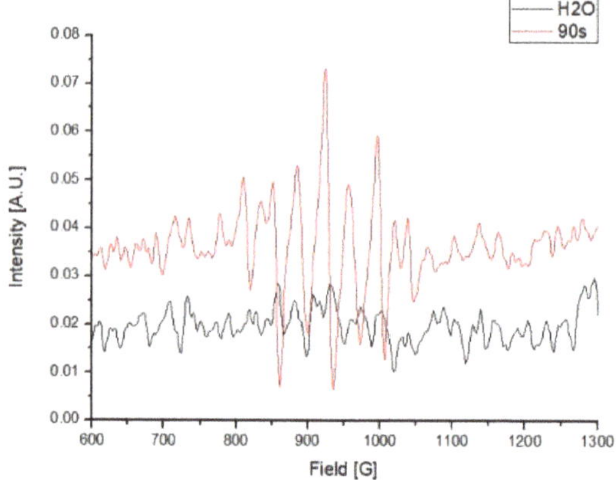


**Figure 3.** Electron paramagnetic resonance (EPR) of the plasma-activated water (PAW) in 1000 mL.

### 3.5. *The Observation of Scanning Electronic Microscope (SEM)*

After PAW treatment, most *Salmonella* cells on the egg surface were much larger than the cells on the untreated eggs. In addition, some bacterial cells were even ruptured (Figure 4).

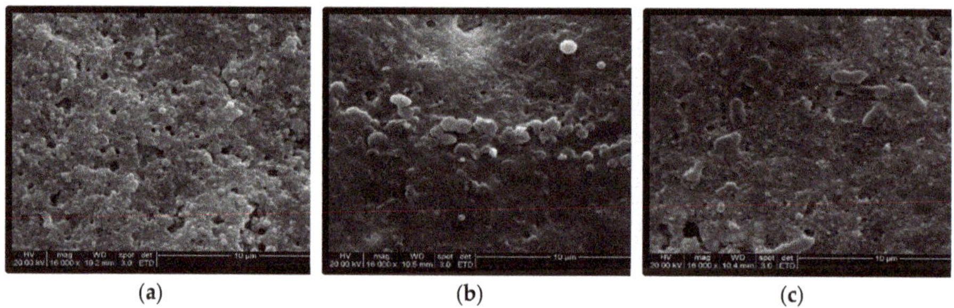

(a) (b) (c)

**Figure 4.** SEM observation of the inoculated eggshells (×16,000 magnification): (**a**) Unwashed, (**b**) PAW 60 s, (**c**) PAW 90 s.

## 4. Discussion

Increasing the water volume is an essential step for practical use of PAW. In this study, the water volume increased to 1 L which was larger than most studies using PAW to inactivate foodborne pathogens on foods [15,20–23]. Among the reports that used 1 L of water, Han et al. (2020) used two plasma jets in 1 L of distilled water for 20 min, then treated Korean rice cakes for 20 min [14]. Choi et al. (2019) used a device containing 12 pins of discharged plasma over the surface of 1 L of distilled water for 30, 60, and 120 min, then treated cabbages for 10 min [16]. Lastly, Liu et al. (2020) submerged 28 acting plasma jets in 1 L of distilled water for 10 min, then treated fresh-cut apples for 5 min [24]. In this study, two plasma jets were used to activate 1 L of PAW that was able to inactivate more than 3 log SE on egg surfaces after 60 s treatment. The number of plasma jets used was less, and the sample treatment time was also shorter than previous studies at the same water volume. Thus, the use of electrical energy in this study was more efficient.

Several PAW studies combined PAW treatment with mild heat (50–60 °C) to enhance antibacterial activity. In the study on cabbages [16], treatment with the PAW activated under 120 min (PAW-120) for 10 min was able to reduce 2.0, 2.2, 1.8, and 0.9 log for mesophilic bacteria, lactic acid bacteria, yeast and mold, and coliform, respectively. Subsequent treatment in distilled water at 60 °C for 5 min resulted in greater pathogen reduction. When the cabbage was inoculated with *L. monocytogenes* and

*Staphylococcus aureus*, 3.4 and 3.7 log reduction was obtained, respectively, after the combined treatment of PAW-120 and 60 °C water. Xiang et al. (2020) applied 30 mL of APW at 50–55 °C for 30 min to treat grapes and obtained more than 5 log reduction of *Saccharomyces cerevisiae*. During this research, the PAW temperature raised naturally to 40–42 °C for all treatments. Thus, the treatment temperatures of this study were lower than the previous studies, in which 50–60 °C [16] and 50–55 °C [21] were used for treating cabbage and grapes, respectively. The PAW temperature was mainly affected by the type of plasma, number of plasma jets, and treatment time. Generally, the temperature was lower when using noble gas, such as argon, instead of regular air [9]. Longer treatment time and a higher number of plasma jets induced higher temperatures [15,17]. In this study, the temperature of smaller-volume PAW (250 mL) was slightly higher than large-volume PAW (1000 mL). However, the temperature difference was not significant, and all were within the range of 40–42 °C. Unlike many other food items, the temperature of the egg washing solution must be higher than the temperature of the egg surface to prevent the infiltration of surface bacteria into the egg [25]. Therefore, the PAW temperature used in this research was in the same temperature range used in commercial egg washing. The RO water used in the controls were also raised to the same temperature as the PAW. The RO control groups obtained a SE reduction of around 1 log when washed at 40–42 °C. In a preliminary study, the same amount of SE reduction was achieved when 25–28 °C RO water was used. Thus, the SE reduction obtained from PAW treatments was caused by the PAW, and not the elevated temperature at 40–42 °C.

The main mechanisms of PAW to inactivate microorganisms were reported to be the creation of reactive oxygen species (ROS), reactive nitrogen species (RNS), and UV light [26]. In this study, the existence of ROS was detected by EPR. The photo spectrum showed that the majority of the light generated by the plasma was within the range of UV light. It has been reported that the UV light generated from the acting plasma under water contributed to the bacterial inactivation [27]. The ROS generated during the PAW activation was also confirmed by previous studies [19]. Moreover, Wu et al. (2017) demonstrated that adding $H_2O_2$ simultaneously enhanced the bactericidal activity and the intensity of the magnetic field, which provided a direct connection between the existence of ROS and the bactericidal ability of PAW [19]. In our study, the EPR results also showed that higher intensity was obtained from 60 and 90 s of treatment time compared to 30 s, which corresponded with the antibacterial results. The majority of ROS are hydrogen peroxide ($H_2O_2$) and oxygen molecules containing unpaired electrons, such as atomic oxygen (O), ozone ($O_3$), superoxide ($O_2^-$), hydroxyl radicals ($OH^\bullet$), and singlet oxygen ($^1O^2$). Nitrate and nitrile were the major products of the reaction between RNS and these active oxygen molecules [28]. The production of these substances result in the increase of ORP and the decrease of pH [28,29]. Thus, ORP and pH values were used in many studies [17,19,28,30] to monitor the antibacterial activity of PAW, including our previous study [15]. In this study, the increase of ORP values and decrease of pH values became less as more water was added in PAW activated by one plasma jet. In contrast, the ORP and pH values were not significantly different between various water volumes in the PAW activated by two plasma jets. These results corresponded with the efficacy of SE reduction. When the PAW was activated with two jets in 1 L of water at 6 or 10 slm, the higher air flow rate not only diminished the ORP values, but also the $O_3$, $H_2O_2$, and $NO_2^-$ concentrations, particularly the $H_2O_2$ concentrations. Our results also showed the reductions of SE populations were less when PAW was activated at 10 slm than 6 slm. Thus, the concentrations of $O_3$, $H_2O_2$, and $NO_2^-$ were related with the bactericidal activity of PAW, which was also shown in previous studies [17,19,31,32]. Though $NO_3^-$ concentrations slightly increased at 10 slm, its concentrations were far less than the $NO_2^-$ concentrations.

SEM observation also revealed that the inoculated SE cells inflated, or even erupted. The possible reason for this could be that the cell membrane lost the ability to maintain osmotic balance, which resulted in the extracellular water infiltrating into the cell continuously. Several reports have described that the impairment of the cell membrane was the major antibacterial mechanism of PAW and resulted in higher permeability, which caused the leakage of intracellular substances [13,33]. However, this study was the first one to demonstrate the enlargement of the treated bacterial cells after PAW treatment.

Since the main components of the cell membrane are phospholipids, the UV light, ROS, and the acidic substances generated by PAW were all able to injure phospholipids and cause the loss of the cell membrane's permeability control [13,33]. Besides the impairment of the cell membrane, damage to bacterial DNA and their metabolism were also reported [9,32,34].

The results of this study demonstrate that a large-scale PAW generation system is achievable, and provide a foundation for future practical use, in which a larger and faster production of PAW is essential. In addition, the antibacterial efficacy of PAW obtained in this study was greater than the study using sodium chlorine and chlorine dioxide [35]. Furthermore, it offered a direct correlation between the PAW products described previously and its antibacterial capacity, which also presented a clearer road map to understanding the antibacterial mechanisms of PAW.

## 5. Conclusions

A patented air plasma jet system was used to generate 1 L of PAW under various conditions to inactivate SE on egg surfaces. The results showed that using two plasma jets at 6 slm was the optimal condition. The values of ORP and pH were related with the antibacterial efficacy. The products generated during PAW activation, such as UV light, unpaired oxygen electrons, $O_3$, $H_2O_2$, nitrile, and nitrate were detected. Furthermore, the relationship between the quantitative amount of these products and the antibacterial efficacy of PAW was demonstrated. The observation under SEM revealed SE cells were inflated, which was mainly caused by the loss permeability of the cytomembrane. The results of this study indicated that air-generated PAW could be a practical option to clean and disinfect shell eggs. More evidence regarding the connection between the PAW products and PAW's antibacterial activity was also provided.

**Author Contributions:** C.-M.L. and C.-Y.H.; Data curation, C.-M.L. and C.-W.H.; Funding acquisition, J.-S.W. and C.-Y.H.; Investigation, C.-P.H., J.-S.W. and C.-Y.H.; Methodology, C.-P.H., H.-S.L., C.-W.H. and C.-Y.H.; Project administration, J.-S.W.; Resources, C.-P.H. and H.-S.L.; Software, H.-S.L. and J.S.L.; Supervision, J.S.L. and C.-W.H.; Validation, C.-P.H., H.-S.L. and J.S.L.; Visualization, H.-S.L. and J.S.L.; Writing—Original draft, C.-M.L., J.-S.W. and C.-Y.H.; Writing—Review & editing, C.-M.L. All authors have read and agreed to the published version of the manuscript.

**Funding:** This research was funded by the Ministry of Science and Technology (MOST) under grant numbers MOST-107-2622-E-009-009-CC1, MOST-108-2622-E-009-003-CC1, MOST-109-2622-E-009-001-CC1. and the APC was funded by MOST-107-2622-E-009-009-CC1, MOST-108-2622-E-009-003-CC1, MOST-109-2622-E-009-001-CC1. The authors would like to thank all of the individuals who volunteered for this study.

**Conflicts of Interest:** The authors declare no conflict of interest.

## References

1. Réhault-Godbert, S. The Golden Egg: Nutritional Value, Bioactivities, and Emerging Benefits for Human Health. *Nutrients* **2019**, *11*, 684. [CrossRef] [PubMed]
2. Sass, C.A.B.; Kuriya, S.; Da Silva, G.V.; Silva, H.L.A.; Da Cruz, A.G.; Esmerino, E.A.; Freitas, M.Q. Completion task to uncover consumer's perception: A case study using distinct types of hen's eggs. *Poult. Sci.* **2018**, *97*, 2591–2599. [CrossRef] [PubMed]
3. Okorie-Kanu, O.J.; Ezenduka, E.V.; Okorie-Kanu, C.O.; Ugwu, L.C.; Nnamani, U.J. Occurrence and antimicrobial resistance of pathogenic *Escherichia coli* and *Salmonella* spp. in retail raw table eggs sold for human consumption in Enugu state, Nigeria. *Vet. World* **2016**, *9*, 1312–1319. [CrossRef]
4. Han, J.; Gokulan, K.; Barnette, D.; Khare, S.; Rooney, A.W.; Deck, J.; Nayak, R.; Stefanova, R.; Hart, M.E.; Foley, S.L. Evaluation of Virulence and Antimicrobial Resistance in Salmonella enterica Serovar Enteritidis Isolates from Humans and Chicken- and Egg-Associated Sources. *Foodborne Pathog. Dis.* **2013**, *10*, 1008–1015. [CrossRef] [PubMed]
5. Hu, L.; Deng, X.; Brown, E.W.; Hammack, T.S.; Ma, L.M.; Zhang, G. Evaluation of Roka Atlas *Salmonella* method for the detection of *Salmonella* in egg products in comparison with culture method, real-time PCR and isothermal amplification assays. *Food Control* **2018**, *94*, 123–131. [CrossRef]

6. Jakočiūnė, D.; Herrero-Fresno, A.; Jelsbak, L.; Olsen, J.E. Highly expressed amino acid biosynthesis genes revealed by global gene expression analysis of *Salmonella enterica* serovar Enteritidis during growth in whole egg are not essential for this growth. *Int. J. Food Microbiol.* **2016**, *224*, 40–46. [CrossRef]
7. Yu-Chung, L.; Hund-Der, Y. Trihalomethane Species Forecast Using Optimization Methods: Genetic Algorithms and Simulated Annealing. *J. Comput. Civ. Eng.* **2005**, *19*, 248–257. [CrossRef]
8. Joshi, S.G.; Cooper, M.; Yost, A.; Paff, M.; Ercan, U.K.; Fridman, G.; Friedman, G.; Fridman, A.; Brooks, A.D. Nonthermal Dielectric-Barrier Discharge Plasma-Induced Inactivation Involves Oxidative DNA Damage and Membrane Lipid Peroxidation in *Escherichia coli*. *Antimicrob. Agents Chemother.* **2011**, *55*, 1053–1062. [CrossRef]
9. Thirumdas, R.; Kothakota, A.; Annapure, U.; Siliveru, K.; Blundell, R.; Gatt, R.; Valdramidis, V.P. Plasma activated water (PAW): Chemistry, physico-chemical properties, applications in food and agriculture. *Trends Food Sci. Technol.* **2018**, *77*, 21–31. [CrossRef]
10. Ma, S.; Kim, K.; Huh, J.; Hong, Y. Characteristics of microdischarge plasma jet in water and its application to water purification by bacterial inactivation. *Sep. Purif. Technol.* **2017**, *188*, 147–154. [CrossRef]
11. Ma, R.; Wang, G.; Tian, Y.; Wang, K.; Zhang, J.; Fang, J. Non-thermal plasma-activated water inactivation of food-borne pathogen on fresh produce. *J. Hazard. Mater.* **2015**, *300*, 643–651. [CrossRef] [PubMed]
12. Qian, J.; Zhuang, H.; Nasiru, M.M.; Muhammad, U.; Zhang, J.; Yan, W. Action of plasma-activated lactic acid on the inactivation of inoculated Salmonella Enteritidis and quality of beef. *Innov. Food Sci. Emerg. Technol.* **2019**, *57*, 102196. [CrossRef]
13. Xiang, Q.; Liu, X.; Liu, S.; Ma, Y.; Xu, C.; Bai, Y. Effect of plasma-activated water on microbial quality and physicochemical characteristics of mung bean sprouts. *Innov. Food Sci. Emerg. Technol.* **2019**, *52*, 49–56. [CrossRef]
14. Han, J.-Y.; Song, W.-J.; Kang, J.H.; Min, S.C.; Eom, S.; Hong, E.J.; Ryu, S.; Kim, S.; Cho, S.; Kang, D.-H. Effect of cold atmospheric pressure plasma-activated water on the microbial safety of Korean rice cake. *LWT* **2020**, *120*, 108918. [CrossRef]
15. Lin, C.-M.; Chu, Y.-C.; Hsiao, C.-P.; Wu, J.-S.; Hsieh, C.-W.; Hou, C.-Y. The Optimization of Plasma-Activated Water Treatments to Inactivate *Salmonella enteritidis* (ATCC 13076) on Shell Eggs. *Foods* **2019**, *8*, 520. [CrossRef] [PubMed]
16. Choi, E.J.; Park, H.W.; Kim, S.B.; Ryu, S.; Lim, J.; Hong, E.J.; Byeon, Y.S.; Chun, H.H. Sequential application of plasma-activated water and mild heating improves microbiological quality of ready-to-use shredded salted kimchi cabbage (*Brassica pekinensis* L.). *Food Control* **2019**, *98*, 501–509. [CrossRef]
17. Royintarat, T.; Seesuriyachan, P.; Boonyawan, D.; Choi, E.H.; Wattanutchariya, W. Mechanism and optimization of non-thermal plasma-activated water for bacterial inactivation by underwater plasma jet and delivery of reactive species underwater by cylindrical DBD plasma. *Curr. Appl. Phys.* **2019**, *19*, 1006–1014. [CrossRef]
18. Andrews, W.; Wang, H.; Jacobson, A.; Ge, B.; Zhang, G.; Hammack, T. BAM: *Salmonella*. In *A Bacteriological Analytical Manual (BAM)*; FDA: Hampton, VI, USA, 2009.
19. Sun, P.; Wu, H.; Bai, N.; Zhou, H.; Wang, R.; Feng, H.; Zhu, W.; Zhang, J.; Fang, J. Inactivation of *Bacillus subtilis* Spores in Water by a Direct-Current, Cold Atmospheric-Pressure Air Plasma Microjet. *Plasma Process. Polym.* **2012**, *9*, 157–164. [CrossRef]
20. Khan, M.S.I.; Kim, Y.-J. Inactivation mechanism of *Salmonella typhimurium* on the surface of lettuce and physicochemical quality assessment of samples treated by micro-plasma discharged water. *Innov. Food Sci. Emerg. Technol.* **2019**, *52*, 17–24. [CrossRef]
21. Xiang, Q.; Kang, C.; Zhao, D.; Niu, L.; Liu, X.; Bai, Y. Influence of organic matters on the inactivation efficacy of plasma-activated water against *E. coli* O157:H7 and *S. aureus*. *Food Control* **2019**, *99*, 28–33. [CrossRef]
22. Xiang, Q.; Zhang, R.; Fan, L.; Ma, Y.; Wu, D.; Li, K.; Bai, Y. Microbial inactivation and quality of grapes treated by plasma-activated water combined with mild heat. *LWT* **2020**, *126*, 109336. [CrossRef]
23. Liao, X.; Xiang, Q.; Cullen, P.J.; Su, Y.; Chen, S.; Ye, X.; Liu, D.; Ding, T. Plasma-activated water (PAW) and slightly acidic electrolyzed water (SAEW) as beef thawing media for enhancing microbiological safety. *LWT* **2020**, *117*, 108649. [CrossRef]
24. Liu, C.; Chen, C.; Jiang, A.; Sun, X.; Guan, Q.; Hu, W. Effects of plasma-activated water on microbial growth and storage quality of fresh-cut apple. *Innov. Food Sci. Emerg. Technol.* **2020**, *59*, 102256. [CrossRef]

25. Caudill, A.B.; Curtis, P.A.; Anderson, K.E.; Kerth, L.K.; Oyarazabal, O.; Jones, D.R.; Musgrove, M.T. The effects of commercial cool water washing of shell eggs on Haugh unit, vitelline membrane strength, aerobic microorganisms, and fungi. *Poult. Sci.* **2010**, *89*, 160–168. [CrossRef] [PubMed]
26. Vatansever, F.; de Melo, W.C.M.A.; Avci, P.; Vecchio, D.; Sadasivam, M.; Gupta, A.; Chandran, R.; Karimi, M.; Parizotto, N.A.; Yin, R.; et al. Antimicrobial strategies centered around reactive oxygen species—bactericidal antibiotics, photodynamic therapy, and beyond. *FEMS Microbiol. Rev.* **2013**, *37*, 955–989. [CrossRef] [PubMed]
27. Spetlikova, E.; Janda, V.; Lukes, P.; Clupek, M. Role of UV Radiation, Solution Conductivity and Pulse Repetition Frequency in the Bactericidal Effects During Pulse Corona Discharges. In Proceedings of the WDS'10 Contributed Papers, Prague, Czech Republic, 1–4 June 2010; pp. 96–100.
28. Chen, T.-P.; Liang, J.; Su, T.-L. Plasma-activated water: Antibacterial activity and artifacts? *Environ. Sci. Pollut. Res.* **2018**, *25*, 26699–26706. [CrossRef] [PubMed]
29. Zhang, Q.; Liang, Y.; Feng, H.; Ma, R.; Tian, Y.; Zhang, J.; Fang, J. A study of oxidative stress induced by non-thermal plasma-activated water for bacterial damage. *Appl. Phys. Lett.* **2013**, *102*, 203701. [CrossRef]
30. Bai, Y.; Idris Muhammad, A.; Hu, Y.; Koseki, S.; Liao, X.; Chen, S.; Ye, X.; Liu, D.; Ding, T. Inactivation kinetics of *Bacillus cereus* spores by plasma activated water (PAW). *Food Res. Int.* **2020**, *131*, 109041. [CrossRef]
31. Oh, J.-S.; Szili, E.J.; Ogawa, K.; Short, R.D.; Ito, M.; Furuta, H.; Hatta, A. UV-vis spectroscopy study of plasma-activated water: Dependence of the chemical composition on plasma exposure time and treatment distance. *JPN J. Appl. Phys.* **2017**, *57*, 0102B9. [CrossRef]
32. Thirumdas, R.; Sarangapani, C.; Annapure, U.S. Cold Plasma: A novel non-thermal technology for food processing. *Food Biophys.* **2015**, *10*, 1–11. [CrossRef]
33. Lukes, P.; Locke, B.R.; Brisset, J.-L. Aqueous-phase chemistry of electrical discharge plasma in water and in gas-liquid environments. *Plasma Chem. Catal. Gases Liq.* **2012**, 243–308.
34. Tian, Y.; Ma, R.; Zhang, Q.; Feng, H.; Liang, Y.; Zhang, J.; Fang, J. Assessment of the Physicochemical Properties and Biological Effects of Water Activated by Non-thermal plasma above and beneath the water surface. *Plasma Process. Polym.* **2015**, *12*, 439–449. [CrossRef]
35. Ni, L.; Cao, W.; Zheng, W.C.; Chen, H.; Li, B.M. Efficacy of slightly acidic electrolyzed water for reduction of foodborne pathogens and natural microflora on shell eggs. *Food Sci. Technol. Res.* **2014**, *20*, 93–100. [CrossRef]


**Publisher's Note:** MDPI stays neutral with regard to jurisdictional claims in published maps and institutional affiliations.

*Article*

# Enhanced Antioxidant Capacity of Puffed Turmeric (*Curcuma longa* L.) by High Hydrostatic Pressure Extraction (HHPE) of Bioactive Compounds

**Yohan Choi [1,†], Wooki Kim [1,†], Joo-Sung Lee [1], So Jung Youn [2], Hyungjae Lee [2,*] and Moo-Yeol Baik [1,*]**

1 Department of Food Science and Biotechnology, Graduate School of Biotechnology, Institute of Life Science and Resources, Kyung Hee University, Yongin 17104, Korea; yo3247@naver.com (Y.C.); kimw@khu.ac.kr (W.K.); daishi8882@naver.com (J.-S.L.)
2 Department of Food Engineering, Dankook University, Cheonan 31116, Korea; govl1990@naver.com
* Correspondence: lee252@dankook.ac.kr (H.L.); mooyeol@khu.ac.kr (M.-Y.B.); Tel.: +82-41-550-3561 (H.L.); +82-31-201-2625 (M.-Y.B.)
† These authors contributed equally to this work.

Received: 12 October 2020; Accepted: 16 November 2020; Published: 18 November 2020

**Abstract:** Turmeric (*Curcuma longa* L.) is known for its health benefits. Several previous studies revealed that curcumin, the main active compound in turmeric, has antioxidant capacity. It has been previously demonstrated that puffing, the physical processing using high heat and pressure, of turmeric increases the antioxidant and anti-inflammatory activities by increasing phenolic compounds in the extract. The current study sought to determine if high hydrostatic pressure extraction (HHPE), a non-thermal extraction at over 100 MPa, aids in the chemical changes and antioxidant functioning of turmeric. 2,2-diphenyl-1-picrylhydrazyl (DPPH), 2,2′-azino-bis (3-ethylbenzothiazoline-6-sulphonic acid) (ABTS), and ferric reducing antioxidant power (FRAP) analyses were conducted and assessed the content of total phenol compounds in the extract. The chemical changes of curcuminoids were also determined by high performance liquid chromatography (HPLC). Among the three variables of ethanol concentration, pressure level, and treatment time, ethanol concentration was the most influential factor for the HHPE of turmeric. HHPE at 400 MPa for 20 min with 70% EtOH was the optimal extraction condition for the highest antioxidant activity. Compositional analysis revealed that 2-methoxy-4-vinylphenol was produced by puffing. Vanillic acid and ferulic acid content increased with increasing HHPE time. Synergistic effect was not observed on antioxidant activity when the turmeric was sequentially processed using puffing and HHPE.

**Keywords:** turmeric; curcumin; puffing; high hydrostatic pressure; antioxidant

---

## 1. Introduction

Turmeric (*Curcuma longa* L.) has been used widely as a spice, but its applications are limited due to its pungent flavor [? ]. Many traditional medicines have utilized turmeric as a beneficial ingredient, and recent studies identified the major functional components, which are curcuminoids, essential oils, fixed oils, and various volatile oils including turmerone, alantone, and zingiberone [? ]. Among these bioactive components, curcuminoids and their derivatives have been thoroughly investigated and reported to have anticancer, antibacterial, antioxidant, and anti-inflammatory properties [? ]. Curcumin (1,7-bis-(4-hydroxy-3-methoxyphenyl-1,6-hepatadiene-3,5-dione), the major curcuminoid component, is a yellowish polyphenolic compound and has been used in many drugs and foods [? ]. Curcuminoid degradation products, i.e., ferulic acid, vanillic acid, and vanillin, are also present in turmeric and have demonstrated antioxidant activity [? ? ].

Various extraction methods, such as high hydrostatic pressure extraction (HHPE), microwave extraction (ME), supercritical fluid extraction (SFE), ultrasound, pulsed electric field (PEF), moderate electric field (MEF), etc., have been developed to enhance the yield or efficacy of bioactive components [? ]. The application of supercritical carbon dioxide extraction on turmeric was optimized and turmerone was determined to be the major component [? ]. In the case of supercritical carbon dioxide extraction of turmeric, high pressure (20–40 MPa) and low temperature (313–333 K) were the best extraction conditions for turmerone. Dandekar and Gaikar also reported a novel hydrotropy-based extraction method for selective extraction of curcuminoids from turmeric [? ]. It was demonstrated that application of pressurized liquids at 10 MPa is an economic method for augmented extraction of curcuminoids [? ]. Puffing, a simple processing method, was recently applied to turmeric resulting in increased antioxidant capacity in vitro [? ]. Puffing uses relatively high heat and pressure and reportedly allows for increased antioxidant activity of natural resources, such as doraji, cacao beans, coffee beans, turmeric, and ginseng [? ? ? ? ? ? ? ? ? ]. In addition, the Maillard reaction during the puffing process imparts a unique flavor by producing volatile substances, such as formic acid, acetaldehyde, formaldehyde, and glyoxal acid [? ].

Among recent advances in food processing methods, high hydrostatic pressure (HHP) treatment is a non-thermal pasteurization method that uses pressure over 100 MPa [? ? ? ? ]. HHPE is an effective method for extraction of heat sensitive materials. HHPE increases the yields and mass transfer rates of herbal products by cell wall breakdown as compared to the conventional extraction methods such as soxhlet, heat reflux, ultrasonic, microwave, and supercritical $CO_2$ extractions [? ? ? ? ? ].

While a significant number of papers has indicated the beneficial roles of turmeric in health as well as its active components of curcuminoids, its applications in processed foods are limited due to its insolubility in water resulting in low bioavailability and bioaccessibility [? ? ? ? ? ]. In this regard, many approaches, including nano-particlization, encapsulation, emulsion formation, and bioconversion, to enhance the biofunctions of turmeric have been reported [? ? ? ? ]. Since there is no report on the HHPE of turmeric, this study investigated the best HHPE conditions for extraction of turmeric to maximize the antioxidant activity. In addition, the combinational effect of puffing and HHPE on antioxidant activity and bioactive compounds of turmeric was also investigated.

## 2. Materials and Methods

### 2.1. Materials and Chemicals

Dried turmeric (moisture content 15.7 ± 0.1%, crude fat 6.4 ± 0.1, crude protein 7.7 ± 0.1, ash 6.5 ± 0.4) was kindly donated from Bibong Herb (Yangju-si, Korea) and kept in a deep freezer before use. Fermented ethanol, an extraction solvent, was provided by ethanol Supplies World Co. (Jeonju-si, Korea). Methyl alcohol, sodium carbonate, sodium hydroxide, and hydrochloric acid were purchased from Daejung Chemicals & Metals Co. (Siheung-si, Korea). Folin-Ciocalteu's phenol reagent, 2,2′-azino-bis (3-ethylbenzthiazoline-6-sulfonic acid) diammonium salt (ABTS), 2,2′-azobis (2-amidino-propane) dihydrochloride (AAPH), 1,1-diphenyl-2-picrylhydrazyl (DPPH), gallic acid, catechin, ascorbic acid, HPLC analytical standard curcumin, demethoxycurcumin, bis-demethoxycurcumin, ferulic acid, and 2-Methoxy-4-vinylphenol were purchased from Sigma-Aldrich Co. (St. Louis, MO, USA). Aluminum chloride and sodium nitrite were purchased from Junsei Chemical Co., LTD (Tokyo, Japan). HPLC grade water, acetonitrile, and methanol were purchased from Honeywell Burdick & Jackson Inc. (Charlotte, NC, USA).

### 2.2. HHPE

HHPE was carried out using a high hydrostatic pressure unit (Suflux, Ilshin Autoclave Co., Daejeon, Korea). Extraction pressure, treatment time, and ethanol concentration were considered as the variable factors in this experiment according to Lee et al. [? ]. Ground turmeric (5 g) and solvent (100 mL) were placed into a plastic pouch and hermetically heat-sealed. Following an application of high hydrostatic pressure at a specific condition, the solutions were vacuum filtered by using Whatman no. 2 filter paper (Cytiva, Marlborough, MA, USA) on a funnel and Kimble-filtering flask. The filtrate was subsequently stored at −20 °C for further experiments. It has been reported that pressure level and treatment time did not have a big impact on the HHPE of puffed ginseng [? ]. Therefore, we wanted to confirm whether pressure level and treatment time had an impact on the HHPE of turmeric. In order to determine the best HHPE conditions for the highest antioxidant activity of turmeric, pressure, treatment time, and ethanol concentration were varied. First, pressure was varied (0.1 (atmospheric), 100, 250, 400, 550 MPa) with fixed treatment time (15 min) and ethanol concentration (70% ethanol) [? ? ]. Second, treatment time was varied (5, 10, 15, 20, 30 min) with the best fixed pressure as previously determined (400 MPa) and ethanol concentration (70% ethanol). Last, ethanol concentration was varied (0, 20, 40, 70, 95% *v/v*) with the previously determined best pressure (400 MPa) and treatment time (20 min). Consequently, the best HHPE conditions were applied for the combinational effect of puffing and HHPE.

### 2.3. Combination of HHPE and Puffing

Puffing of turmeric was carried out at 980 kPa as previously reported [? ]. Briefly, 150 g of sliced and dried turmeric was mixed with 600 g of rice (1:4 *w/w*), which served as an excessive carbonization preventer at a high temperature and a heat-transfer medium [? ? ? ]. The mixture was placed in a rotary gun puffing machine chamber and subjected to an incremental increase of internal pressure to 980 kPa by heating with a gas burner. Subsequently, puffing of the mixture was induced with a sudden pressure release by opening of the chamber door. The puffed turmeric was ground and HHPE was applied at 400 MPa for 20 min using a 70% ethanol concentration. The non-puffed with no HHPE sample was considered as the control sample. For extraction of non-puffed (control) and puffed turmeric, ground samples (5 g) in a 70% ethanol solvent (200 mL) were agitated by a magnet stirrer for 30 min at room temperature.

### 2.4. Extraction Yield

The extract was dried at 105 °C using a drying oven (HB-502M, HanBeak Scientific Co., Bucheon, Korea), and the extraction yield was determined by the following Equation (1) [? ]:

$$\text{Extraction yield } (\%) = \frac{(w_2 - w_1)}{A} \times \frac{E}{E'} \times 100 \quad (1)$$

where,

$A$ = Weight of used sample (g)
$E$ = Total volume of extract (mL)
$E'$ = Used volume of extract to be dried (mL)
$w_1$ = Initial weight of aluminum dish (g)
$w_2$ = Weight of aluminum dish and dried sample (g)

### 2.5. Radical-Scavenging and Ferric-Reducing Activities

Antioxidant activity of all turmeric extracts (ground turmeric, HHPE turmeric, puffed turmeric, puffed and HHPE turmeric) was determined using three different methods. In order to assess the radical scavenging activity, DPPH and ABTS radical scavenging activity were carried out. A FRAP assay was performed to determine the ferric-reducing activity. Ascorbic acid served as a standard, and the antioxidant activity of the extracts were determined in the unit of mg vitamin C equivalent (VCE)/g dried turmeric [? ]. The radical scavenging and ferric-reducing activities of each extract were measured in a 96-well plate using a microplate reader (Bio-Rad, Hercules, CA, USA).

Specifically, the DPPH radical scavenging activity of the extracts was determined using the method of Brand-Williams, Cuvelier and Berset [? ]. Briefly, 0.1 mM DPPH solution was prepared using DPPH and 80% methanol followed by a normalization at an absorbance of 0.650 ± 0.020 at 517 nm. Extract (0.05 mL) was added to 2.95 mL of the DPPH solution and reacted at room temperature for 30 min. The absorbance at 517 nm subtracted from a blank was determined.

The ABTS radical scavenging activity of the extracts was determined by the method of Van den Berg et al. [? ]. A phosphate buffer saline solution of AAPH (1.0 mM) was reacted with ABTS (2.5 mM) for 30 min at 70 °C. Subsequently, the reaction solution was filtrated through a 0.45 μm syringe filter followed by an adjustment of an absorbance to 0.650 ± 0.020 at 517 nm. ABTS reagent (980 μL) was applied to 20 μL of extract and reacted at 37 °C for 10 min. The absorbance at 517 nm subtracted from a blank was determined.

A FRAP assay was performed according to the method of Benzie and Strain [? ]. Briefly, 300 mM acetate buffer (pH 3.6), made with 3.1 g $C_2H_3NaO_2{\cdot}3H_2O$ with 16 mL $C_2H_4O_2$ and 10 mM TPTZ (2,4,6- tripyridyl-s-triazine) solution in 40 mM HCl, and 20 mM $FeCl_3{\cdot}6H_2O$ solution was prepared and stocked. A fresh working solution of acetate buffer (25 mL), TPTZ solution (2.5 mL), and $FeCl_3{\cdot}6H_2O$ solution (2.5 mL) was stored at 37 °C until application of 2850 mL FRAP solution to 150 mL of extract. The mixture was reacted for 30 min in a dark condition, and the colorimetric changes by the ferrous tripyridyltriazine complex was determined at 593 nm.

### 2.6. *Total Polyphenol Content (TPC)*

Total polyphenol content (TPC) of all turmeric extracts (ground turmeric, HHPE turmeric, puffed turmeric, puffed and HHPE turmeric) was determined using Folin and Ciocalteu's assay [? ]. The extract (200 μL), distilled water (2.6 mL), and Folin-Ciocalteu solution (200 μL) were left to react for 6 min. After 6 min, 2 mL of $Na_2CO_3$ was added and absorbance was measured at 750 nm after another 90 min. Gallic acid was employed as the standard and results were expressed as mg gallic acid equivalents (GAE)/g dried turmeric.

### 2.7. *Quantitative Analysis of Bioactive Compounds by HPLC*

Bioactive compounds in the extracts, i.e., curcumin (CUR), demethoxycurcumin (DMC), bis-demethoxycurcumin (BDMD), ferulic acid (FA), 4-vinyl guaiacol (4VG), vanillin (VN), and vanillic aicd (VA), were quantitatively determined by using an HPLC method [? ]. In brief, 10 μL of extract, filtered through a 0.45 μm Millipore filter, was analyzed using the Agilent 1260 Infinity II HPLC system (Agilent Technologies, Santa Clara, CA, USA) equipped with a Zorbax SB-C18 column (4.6 × 250 mm, 5 μm, Agilent Technologies, Santa Clara, CA, USA), monitored at 260 nm. The mobile phase was run at 1 mL/min with the following differential gradient of 0.4% acetic acid in deionized water (A) and acetonitrile (B): 0–30 min, 92–9% A; 30–39 min, 9–0% A; 39–44 min, 0% A; 44–47 min, 0–92% A; 47–50 min, 92% A. The standard molecules of the curcuminoids, as well as their degradation products, were obtained from Sigma-Aldrich Co. (St. Louis, MO, USA), whose calibration curves served for the identification and quantification of the bioactive molecules in the extracts.

### 2.8. *Statistical Analysis*

Data are representative of three repeated experiments with three replicates. All experimental data were analyzed using one-way analysis of variance (ANOVA) and are expressed as the mean ± standard deviation (SD). Duncan's multiple range test was conducted to assess significant differences among experimental mean values for extraction yield, antioxidant capacity, and total phenolic content using SAS software (version 8.2, SAS Institute, Inc., Cary, NC, USA). Tukey's post-hoc multiple comparisons test was applied to the HPLC analyses using GraphPad Prism software (version 5, La Jolla, CA, USA). For all statistical analyses, ($p < 0.05$) was considered statistically significant.

## 3. Results

### 3.1. High Hydrostatic Pressure Extraction (HHPE) Yield

The effects of various HHPE parameters, i.e., pressure, time, and ethanol concentration, on extraction yield of ground turmeric are shown in (Table ??). Pressure did not have a positive impact on extraction yield in this case. Moreover, a slight decrease of extraction yield was observed with increasing pressure level. Extraction time also exhibited only a small impact on extraction yield. A slight increase in extraction yield was observed with increasing extraction time. On the other hand, ethanol concentration greatly influenced the extraction yield of the HHPE of turmeric. The highest extraction yield was observed with 0% ethanol in the solution, while a 95% ethanol solution exhibited about an 87% reduced extraction yield. It has been reported that extraction yield was greatly influenced by ethanol concentration in puffed ginseng [? ]. Those authors reported that extraction yield decreased with increasing ethanol concentration because water soluble polymers are efficiently extracted in polar solvent conditions.

**Table 1.** High hydrostatic pressure extraction (HHPE) yield of ground turmeric.

| Pressure (MPa) | Extraction Yield (%) | Time (min) | Extraction Yield (%) | Ethanol Concentration (%) | Extraction Yield (%) |
|---|---|---|---|---|---|
| 0.1 | 10.77 ± 0.24 [a,*] | 5 | 8.46 ± 0.38 [b] | 0 | 20.04 ± 0.82 [a] |
| 100 | 10.60 ± 0.30 [a] | 10 | 9.99 ± 0.54 [a] | 20 | 17.24 ± 0.37 [b] |
| 250 | 10.55 ± 0.27 [a] | 15 | 10.23 ± 0.24 [a] | 40 | 16.85 ± 0.40 [b] |
| 400 | 10.23 ± 0.24 [b] | 20 | 10.60 ± 0.71 [a] | 70 | 10.60 ± 0.71 [d] |
| 550 | 10.12 ± 0.22 [b] | 30 | 10.77 ± 0.75 [a] | 95 | 2.52 ± 0.01 [e] |

* Values with the same letters in the same column are not significantly different ($p < 0.05$).

### 3.2. Factors Affecting Antioxidant Activity and Total Phenolic Content of HHPE of Ground Turmeric

#### 3.2.1. Effect of Pressure

The effects of HHPE variables, i.e., pressure, extraction time, and ethanol concentration, on antioxidant activity of the extracts were investigated (Figure ??). Pressure demonstrated very minor effects (Figure ??A): Only a slight increment of antioxidant activities of turmeric was observed in extracts at 400 MPa (DPPH, ABTS, and FRAP values at 7.35 ± 0.30, 13.82 ± 0.82, and 7.61 ± 0.21 mg VCE/g dried turmeric, respectively) as compared to 0.1 MPa atmospheric extraction (6.61 ± 0.26, 11.76 ± 1.01, and 7.17 ± 0.23 mg VCE/g dried turmeric, respectively) (Figure ??A). Total phenolic content (TPC) was not affected by the pressure level.

#### 3.2.2. Effect of Extraction Time

Antioxidant activities and TPC were analyzed in the HHPE at various extraction times (5–30 min) at 400 MPa using a 70% ethanol solvent. Following a 15 min extraction, DPPH, ABTS, and FRAP significantly increased to 7.35 ± 0.30, 13.82 ± 0.82, and 7.61 ± 0.21 mg VCE/g dried turmeric, respectively, as compared to a 5 min extraction (6.97 ± 0.52, 12.80 ± 0.73, and 7.27 ± 0.37 mg VCE/g dried turmeric, respectively). Further extension of extraction time reached the maximal increment of all antioxidant activities in the HHPE turmeric at 20 min ($p < 0.05$) (Figure ??B). TPC also significantly increased to its maximum in a 20 min as compared to a 5 min extraction (6.12 ± 0.10 vs. 5.62 ± 0.49 mg gallic acid equivalents (GAE)/g dried turmeric, respectively) ($p < 0.05$).

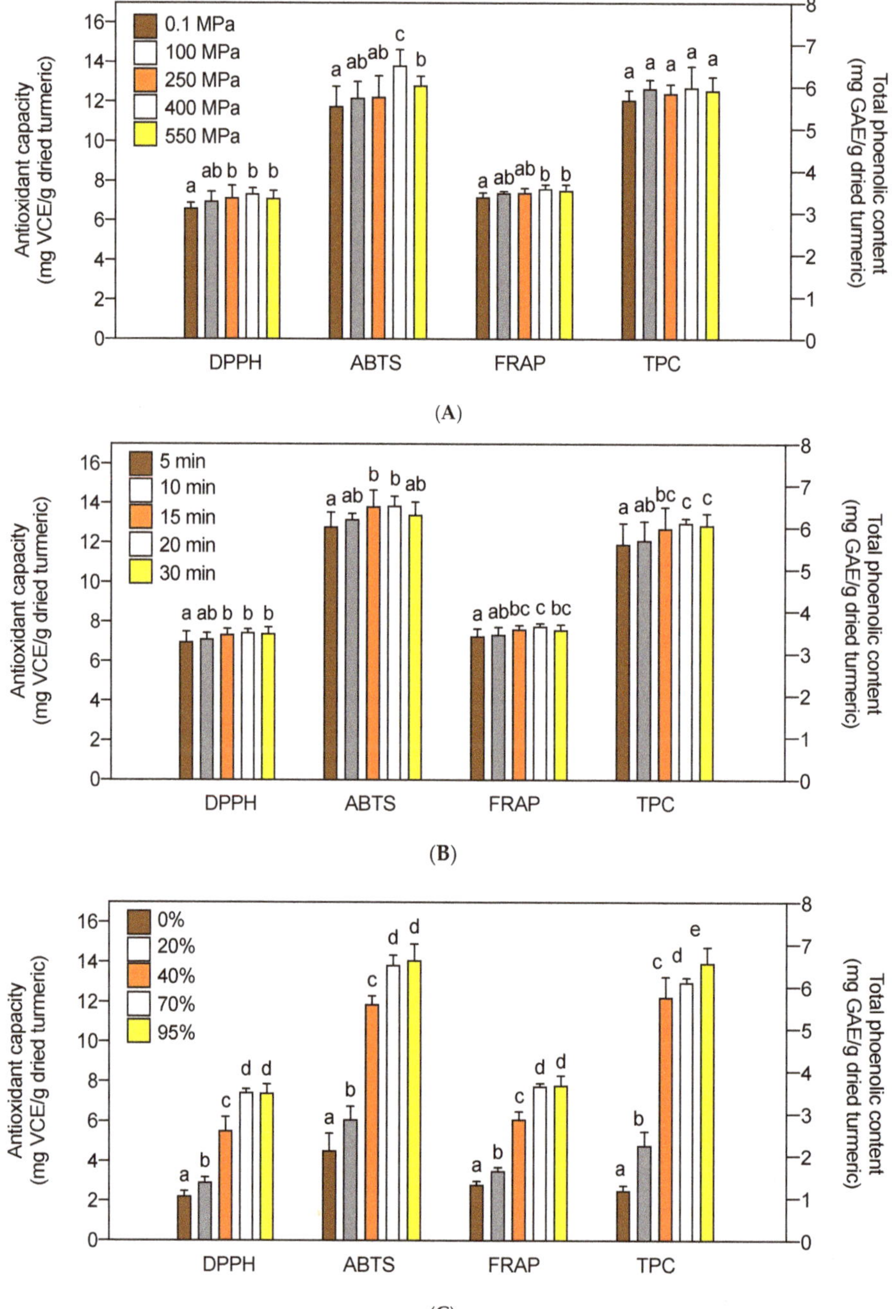


**Figure 1.** Effects of pressure (**A**), extraction time (**B**), and ethanol concentration (**C**) on antioxidant activities and total phenolic contents (TPC) of the HHPE turmeric. Same letters in each group indicates that they are not statistically different at $p < 0.05$.

#### 3.2.3. Effect of Ethanol Concentration

Ethanol concentration greatly influenced the antioxidant activity of HHPE turmeric (Figure ??C). All antioxidant activities increased with increasing ethanol concentration up to 70% ($p < 0.05$) at which the maximal effects were observed (DPPH 7.45 ± 0.17, ABTS 13.85 ± 0.49, and FRAP 7.77 ± 0.13 mg VCE/g dried turmeric, respectively). TPC also significantly increased with increasing ethanol concentration giving 6.57 ± 0.37 mg GAE/g dried turmeric in 95% ethanol up from 1.19 ± 0.11 mg GAE/g dried turmeric in 0% ethanol water.

### 3.3. *HPLC Analysis of HHPE of Ground Turmeric*

The bioactive compounds in the HHPE turmeric were analyzed using HPLC (Table **??**). Curcumin (CUR), demethoxycurcumin (DMC), and bis-demethoxycurcumin (BDMC) were determined to be the major curcuminoids in turmeric extract, and their degradation products, ferulic acid (FA), vanillin (VN), and vanillic acid (VA) [**?** ], were also found in the HHPE turmeric. Pressure level did not greatly influence the amount of the bioactive compounds of HHPE turmeric even though a slight but significant decrement of FA and VN was observed (Table **??**). In contrast, extraction time significantly influenced the composition of bioactive compounds in the HHPE turmeric. The amount of curcuminoids decreased and the amount of the minor bioactive compounds (VA and FA) increased with increasing extraction time. Ethanol concentration also greatly influenced the composition of bioactive compounds in the HHPE turmeric. CUR content dramatically increased with increasing ethanol concentration from 5.31 (0% ethanol) to 407.65 µg/ g dried turmeric (70% ethanol). Additionally, DMC and BDMC were not detected in up to 20% and 40% ethanol concentrations, respectively.

**Table 2.** Effects of pressure, extraction time, and ethanol concentration on bioactive compounds (µg/g dried turmeric) in the HHPE turmeric analyzed by HPLC.

| Pressure | VA ** | VN | FA | 4VG | BDMC | DMC | CUR |
|---|---|---|---|---|---|---|---|
| 0.1 MPa | 3.46 ± 0.32 $^{a,*}$ | 32.12 ± 2.85 $^{a}$ | 20.46 ± 0.44 $^{a}$ | N/D *** | 23.29 ± 0.12 $^{b,c}$ | 100.91 ± 1.72 $^{a}$ | 399.96 ± 2.49 $^{a}$ |
| 100 MPa | 2.67 ± 0.40 $^{b,c}$ | 28.96 ± 0.89 $^{b}$ | 19.05 ± 0.21 $^{a,b}$ | N/D | 21.87 ± 0.22 $^{c}$ | 95.44 ± 3.19 $^{a}$ | 395.65 ± 5.07 $^{a}$ |
| 250 MPa | 2.40 ± 0.28 $^{b,c}$ | 28.16 ± 0.46 $^{b}$ | 18.50 ± 0.18 $^{b}$ | N/D | 20.85 ± 0.58 $^{d}$ | 98.24 ± 1.56 $^{a}$ | 396.81 ± 5.39 $^{a}$ |
| 400 MPa | 2.23 ± 0.18 $^{c}$ | 19.87 ± 1.04 $^{c}$ | 16.26 ± 0.52 $^{c}$ | N/D | 24.70 ± 0.25 $^{a,b}$ | 95.65 ± 2.64 $^{a}$ | 390.72 ± 1.97 $^{a}$ |
| 550 MPa | 3.04 ± 0.45 $^{a,b}$ | 17.79 ± 0.30 $^{c}$ | 14.52 ± 0.38 $^{d}$ | N/D | 25.89 ± 0.28 $^{a}$ | 95.97 ± 0.75 $^{a}$ | 399.57 ± 3.79 $^{a}$ |
| **Extraction Time** | | | | | | | |
| 5 min | 1.86 ± 0.00 $^{d}$ | 13.71 ± 0.11 $^{d}$ | 21.82 ± 1.05 $^{a}$ | N/D | 29.37 ± 0.04 $^{a}$ | 126.6 ± 0.21 $^{a}$ | 438.83 ± 1.14 $^{a}$ |
| 10 min | 2.94 ± 0.21 $^{b,c}$ | 13.64 ± 0.77 $^{d}$ | 20.27 ± 2.55 $^{b}$ | N/D | 25.73 ± 0.75 $^{a}$ | 115.7 ± 2.60 $^{b}$ | 400.32 ± 8.23 $^{b,c}$ |
| 15 min | 2.23 ± 0.18 $^{c,d}$ | 19.87 ± 1.04 $^{c}$ | 16.26 ± 0.52 $^{c}$ | N/D | 24.70 ± 0.25 $^{a}$ | 95.65 ± 2.64 $^{c}$ | 390.72 ± 1.97 $^{b,c}$ |
| 20 min | 3.51 ± 0.65 $^{a,b}$ | 28.02 ± 2.36 $^{b}$ | 20.36 ± 1.04 $^{b}$ | N/D | 24.40 ± 2.35 $^{a}$ | 100.1 ± 2.03 $^{c}$ | 407.65 ± 2.63 $^{b}$ |
| 30 min | 4.20 ± 0.93 $^{a}$ | 30.92 ± 1.06 $^{a}$ | 19.47 ± 1.01 $^{b}$ | N/D | 23.66 ± 2.49 $^{a}$ | 100.9 ± 1.21 $^{c}$ | 408.51 ± 3.26 $^{b}$ |
| **EtOH Concentration** | | | | | | | |
| 0% | 7.23 ± 0.43 $^{a}$ | 19.49 ± 2.69 $^{b}$ | N/D | N/D | N/D | N/D | 5.31 ± 0.06 $^{d}$ |
| 20% | 4.31 ± 0.27 $^{b}$ | 20.23 ± 1.43 $^{b}$ | N/D | N/D | N/D | N/D | 9.65 ± 0.05 $^{d}$ |
| 40% | 4.37 ± 1.40 $^{b}$ | 20.96 ± 0.84 $^{b}$ | 1.25 ± 0.14 $^{c}$ | N/D | N/D | 4.12 ± 0.11 $^{b}$ | 110.74 ± 3.98 $^{c}$ |
| 70% | 3.51 ± 0.65 $^{c}$ | 28.02 ± 2.36 $^{a}$ | 20.36 ± 1.04 $^{a}$ | N/D | 24.40 ± 2.35 $^{a}$ | 100.12 ± 2.03 $^{a}$ | 407.65 ± 2.63 $^{a}$ |
| 95% | N/D | 20.64 ± 0.10 $^{b}$ | 26.96 ± 0.68 $^{a}$ | N/D | 22.29 ± 0.38 $^{a}$ | 94.50 ± 1.44 $^{a}$ | 389.67 ± 5.57 $^{a}$ |

* Values designated by different letters (a–d) are statistically different at $p < 0.05$ within the column. ** VA, vanillic acid; VN, vanillin; FA, ferulic acid; 4VG, 4–vinyl guaiacol; BDMC, bisdemethoxycurcumin; DMC, demethoxycurcumin; CUR, curcumin. *** N/D, not detected.

### 3.4. *Combinatory Effect of Puffing and HHPE on Turmeric*

Following the investigation on the optimal conditions for HHPE turmeric, which were determined to be at 400 MPa for 20 min using 70% ethanol, the combinatory effect of puffing at 980 kPa and HHPE on turmeric was determined by assessing extraction yield, antioxidant activity, TPC, and the bioactive compounds profile. Neither puffing nor HHPE alone showed any effect in the extraction yield of puffed and HHPE turmeric (Table **??**). Puffing of turmeric followed by HHPE, however, demonstrated significantly increased extraction yield to 11.53% as compared to non-puffed turmeric extracted at atmospheric pressure (0.1 MPa) ($p < 0.05$).

**Table 3.** Effect of the combination of puffing and HHPE on extraction yield of turmeric.

| Treatment | Extraction Yield (%) |
|---|---|
| Control | 10.77 ± 0.24 $^{b,*}$ |
| HHPE | 10.60 ± 0.71 $^{b}$ |
| Puffing | 10.84 ± 0.94 $^{b}$ |
| Puffing + HHPE | 11.53 ± 0.73 $^{a}$ |

* Values designated by different letters (a,b) are statistically different at $p < 0.05$ within the column.

On the other hand, puffing and HHPE did not show any synergistic effect on antioxidant activity and TPC (Figure ??). Puffing showed the highest antioxidant activity (11.89, 21.24, 12.98 mg VCE/g dried turmeric in DPPH, ABTS, and FRAP, respectively) and TPC at 13.49 mg GAE/g dried turmeric, and puffing followed by HHPE showed slightly lower antioxidant activity (11.58, 21.04, 12.75 mg VCE/g dried turmeric in DPPH, ABTS, FRAP, respectively) and TPC at 12.58 mg GAE/g dried turmeric. Consequently, the combination of puffing and HHPE is not beneficial to antioxidant activity and total phenolic content compared to the single treatment of puffing. Overall, puffing is the best way to increase the antioxidant activity and TPC of turmeric.

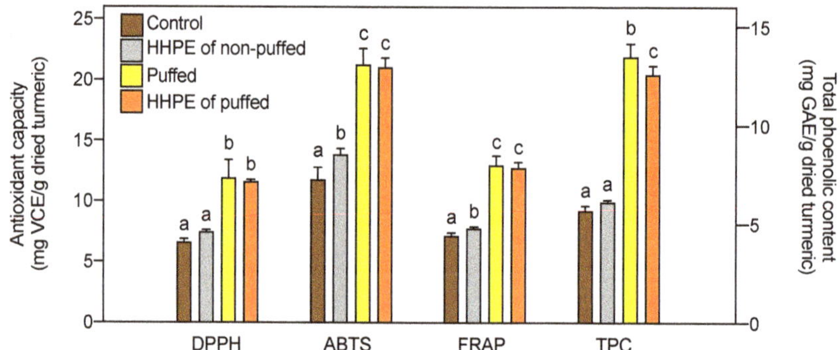


**Figure 2.** Combination effect of puffing and HHPE on antioxidant activities and total phenolic contents (TPC) of turmeric. Same letters in each group indicates that they are not statistically different at $p < 0.05$.

The combinatory effect of puffing and HHPE on bioactive compounds of turmeric was analyzed using HPLC (Figure ??) and quantitatively analyzed (Table ??). Puffing showed an increase in minor bioactive compounds and decrease in curcuminoids. 4VG was not observed in non-puffed turmeric, but detected in all puffed turmeric and increased with HHPE.

**Table 4.** Effect of the combination of puffing and HHPE on bioactive compounds (g/g dried turmeric) of turmeric.

| Treatment | VA ** | VN | FA | 4VG | BDMC | DMC | CUR |
|---|---|---|---|---|---|---|---|
| Control | 3.46 ± 0.32 $^{c,*}$ | 32.12 ± 2.85 $^{c}$ | 20.46 ± 0.44 $^{a}$ | N/D *** | 23.29 ± 0.12 $^{a}$ | 100.91 ± 1.72 $^{a}$ | 399.96 ± 2.49 $^{b}$ |
| HHPE | 3.51 ± 0.65 $^{c}$ | 28.02 ± 2.36 $^{d}$ | 20.36 ± 1.04 $^{a}$ | N/D | 24.40 ± 2.35 $^{a}$ | 100.12 ± 2.03 $^{a}$ | 407.65 ± 2.63 $^{a}$ |
| Puffing | 13.66 ± 0.83 $^{b}$ | 79.22 ± 6.00 $^{a}$ | 6.07 ± 1.29 $^{b}$ | 77.61 ± 5.30 $^{b}$ | 10.46 ± 1.32 $^{c}$ | 63.58 ± 3.80 $^{c}$ | 306.53 ± 18.96 $^{d}$ |
| Puffing + HHPE | 16.25 ± 1.16 $^{a}$ | 72.55 ± 1.99 $^{b}$ | 5.78 ± 0.42 $^{b}$ | 140.00 ± 3.14 $^{a}$ | 16.54 ± 1.10 $^{b}$ | 70.04 ± 1.29 $^{b}$ | 352.41 ± 3.47 $^{c}$ |

* Values designated by different letters (a–d) are statistically different at $p < 0.05$ within the col umn. ** VA, vanillic acid; VN, vanillin; FA, ferulic acid; 4VG, 4–vinyl guaiacol; BDMC, bisdemethoxycurcumin; DMC, demethoxycurcumin; CUR, curcumin. *** N/D, not detected.

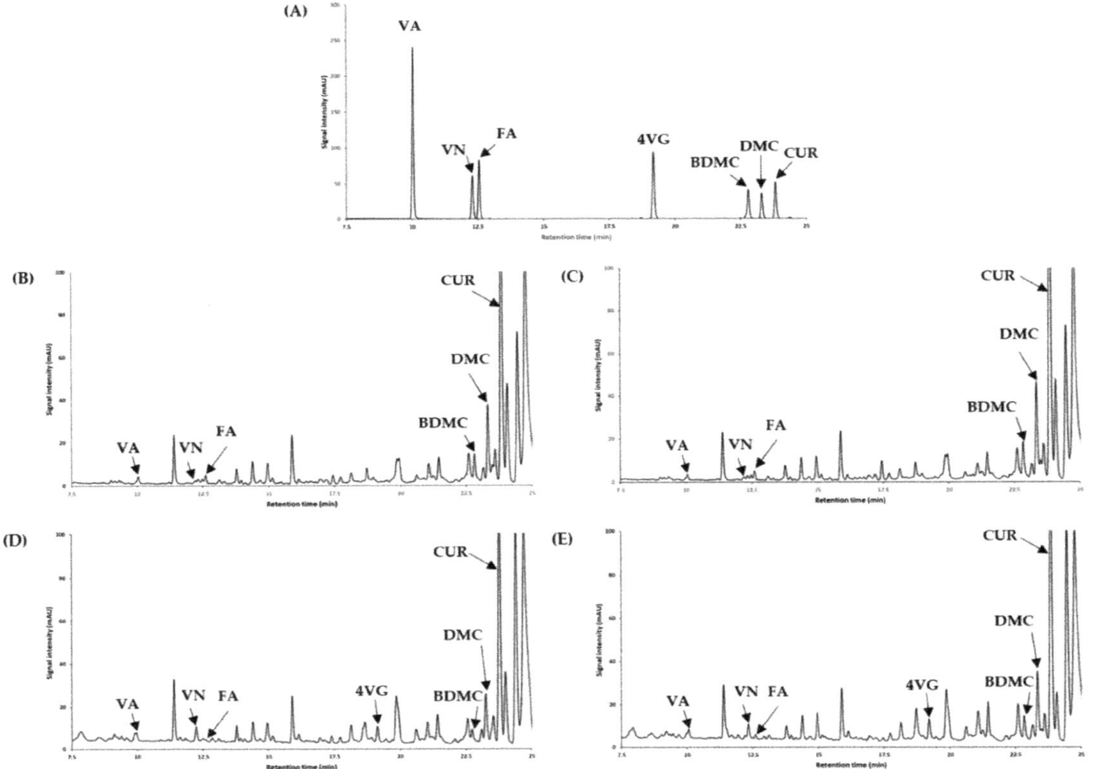


**Figure 3.** HPLC chromatograms of seven standard materials including curcuminoids and their degradation products (**A**), turmeric extract (**B**), HHPE turmeric (**C**), puffed turmeric (**D**), and puffed and HHPE turmeric (**E**) for quantification. VA, vanillic acid; VN, vanillin; FA, ferulic acid; 4VG, 4-vinyl guaiacol; BDMC, bisdemethoxycurcumin; DMC, demethoxycurcumin; CUR, curcumin.

## 4. Discussion

The current study sought to find the best condition of HHPE for the augmentation of turmeric's antioxidant properties. In addition, the combination of the puffing process with HHPE, for which the enhanced antioxidant capacity was clearly demonstrated [? ], was investigated for clarification of any synergistic and/or additive effects.

The main parameters of turmeric HHPE, i.e., pressure, time, and ethanol concentration, were investigated. Pressure and extraction time in HHPE have a very marginal influence on the extraction yield of turmeric. In contrast, HHPE of ginseng powder exhibited improved extraction yield in a pressure and time dependent manner [? ]. This discrepancy may come from the difference in soluble components between turmeric vs. ginseng powder. Of interest, ethanol concentration in HHPE greatly affected the extraction yield and the amount of major bioactive compounds (CUR, DMC, and BDMC) of turmeric. The hydrophobicity of the major curcuminoids (CUR, DMC, and BDMC) made them not well solubilized in water. Consequently, BDMC revealed the highest hydrophobicity followed by DMC and CUR. At 0% ethanol concentration the highest yield was observed, but only a very small amount of the major curcuminoids was detected, suggesting that most of the extracts from turmeric are water soluble. Most of the water-soluble components in turmeric, mainly dietary fibers, are not extracted at higher ethanol concentrations, resulting in a drastic decrease in extraction yield at a 95% ethanol concentration. Similarly, it has been reported that increased ethanol concentration negatively affected the extraction yield of red-ginseng [? ] and that the high HHPE yield of ginseng was shown in a lower ethanol concentration, suggesting that water soluble molecules including starches were the main components in the extraction process [? ? ].

Extraction pressure and time were also found to have very marginal to no impacts on antioxidant activity and TPC. This apparent discrepancy may demonstrate the hydrophobicity of the antioxidant components in turmeric. A significant change in bioactive compounds of turmeric in HHPE was not observed in this study. High pressure level (550 MPa) and long treatment time (30 min) had no impact on the HHPE of bioactive compounds. Consequently, the optimal conditions based on antioxidant activity for the HHPE of non-puffed turmeric were determined to be 400 MPa for 20 min using 70% ethanol.

HHPE increased the curcuminoids and minor bioactive compounds except FA in puffed turmeric. This suggests that curcuminoids are degraded to minor bioactive compounds by puffing. At first, curcuminoids decompose to FA and then the FA degrades further to produce 4VG. Due to the differences in degradation rates of curcuminoids-to-FA and FA-to-4VG, FA may be continuously accumulated despite its decomposition to 4VG [? ? ]. It has been reported that curcumin is degraded and transformed to volatile phenolic compounds such as vanillin, guaiacol, and isoeugenol [? ]. Curcumin may be further degraded into ferulic acid, 4-vynilguaiacol, vanillic acid, and vanillin in thermal treatment. The degradation rate of curcumin to ferulic acid may be faster than the degradation rate of ferulic acid to 4-vynilguaiacol. Finally, degradation to volatile compounds, such as vanillic acid or vanillin, occurred slowly [? ]. The HPLC analysis in the current study suggests that curcuminoids are highly insoluble in water, and BDMC has the highest insolubility to water followed by DMC and CUR. In contrast to the trend seen in curcuminoids, VA content decreased with increasing ethanol concentration, indicating that VA was more water soluble than ethanol. While previous studies showed an increased amount of bioactive compounds using HHPE for anthocyanin in grapes, carotenoids in carrot puree, and lycopene in tomato puree [? ? ], no significant increase in bioactive compounds in HHPE of ground turmeric was observed in this study. Moreover, intense conditions such as high pressure level (550 MPa) and long treatment time (30 min) were not necessary for better extraction of bioactive compounds in HHPE of ground turmeric. This is in line with the results of antioxidant activities of HHPE in this study and the results of Casquete et al. [? ]. Overall, other bioactive compounds except VA had the highest amounts in the HHPE using 70% ethanol. Similarly, 60% ethanol extraction for phenolic compounds in flaxseed, and 70% ethanol extraction for high antioxidant activity in ginseng have been previously recommended [? ? ]. On the other hand, combination of puffing and HHPE showed an increase in extraction yield and bioactive compounds. Assessment of antioxidant capacity further revealed that puffing is an effective method to increase antioxidant capacity, but HHPE after puffing did not show any synergistic nor additive impacts in antioxidant capacity. Consequently, HHPE was found to be a more effective method than conventional extraction in the case of the porous and weak-structured materials created by puffing.

## 5. Conclusions

The best HHPE conditions for extraction of turmeric to maximize the antioxidant activity were investigated. The synergistic effect of puffing and HHPE on antioxidant activity and bioactive compounds of turmeric was also investigated. Ethanol concentration was the most effective variable for HHPE of ground turmeric among the three studied variables of ethanol concentration, pressure level, and treatment time. HHPE of ground turmeric at 400 MPa for 20 min with 70% EtOH was the best extraction condition for the highest antioxidant activity. Extraction of bioactive compounds in the HHPE of ground turmeric was also greatly influenced by ethanol concentration, possibly due to their hydrophilic or hydrophobic characteristics. Although puffing and HHPE showed an increase in extraction yield and bioactive compounds, a synergistic effect was not observed on antioxidant activity.

**Author Contributions:** Conceptualization, W.K., H.L., and M.-Y.B.; methodology, Y.C., J.-S.L., W.K., S.J.Y., H.L., and M.-Y.B.; HPLC method development and analysis, S.J.Y. and H.L; investigation, Y.C., W.K., H.L., and M.-Y.B.; data curation, Y.C., J.-S.L., W.K., S.J.Y., H.L., and M.-Y.B.; writing—original draft preparation, Y.C., W.K., H.L., and M.-Y.B.; writing—review and editing, W.K., H.L., and M.-Y.B.; project administration, W.K., H.L., and M.B.; funding acquisition, W.K., H.L., and M.-Y.B. All authors have read and agreed to the published version of the manuscript.

**Funding:** This work was supported by the Rural Development Administration, Republic of Korea (PJ01314402).

**Conflicts of Interest:** The authors declare no conflict of interest.

## References

1. Jung, S.-H.; Chang, K.-S.; Ko, K.-H. Physiological effects of curcumin extracted by supercritical fluid from turmeric (*Curcuma longa* L.). *Korean J. Food Sci. Technol.* **2004**, *36*, 317–320.
2. Palombo, E.A. Traditional medicinal plant extracts and natural products with activity against oral bacteria: Potential application in the prevention and treatment of oral diseases. *Evid.-Based Complement. Altern. Med.* **2011**, *2011*, 680354. [CrossRef]
3. Choi, Y.; Ban, I.; Lee, H.; Baik, M.-Y.; Kim, W.-K. Puffing as a novel process to enhance the antioxidant and anti-inflammatory properties of *Curcuma longa* L. (Turmeric). *Antioxidants* **2019**, *8*, 506. [CrossRef] [PubMed]
4. Huang, M. Effect of dietary curcumin and dibenzoylmethane on formation of 7,12-dimethylbenz[a]anthracene-induced mammary tumors and lymphomas/leukemias in Sencar mice. *Carcinogenesis* **1998**, *19*, 1697–1700. [CrossRef] [PubMed]
5. Fiddler, W.; Parker, W.E.; Wasserman, A.E.; Doerr, R.C. Thermal decomposition of ferulic acid. *J. Agric. Food Chem.* **1967**, *15*, 757–761. [CrossRef]
6. Kikuzaki, H.; Hisamoto, M.; Hirose, K.; Akiyama, K.; Taniguchi, H. Antioxidant properties of ferulic acid and its related compounds. *J. Agric. Food Chem.* **2002**, *50*, 2161–2168. [CrossRef]
7. Gavahian, M.; Chu, Y.-H.; Sastry, S.K. Extraction from food and natural products by moderate electric field: Mechanisms, benefits, and potential industrial applications. *Compr. Rev. Food Sci. Food Saf.* **2018**, *17*, 1040–1052. [CrossRef]
8. Gopalan, B.; Goto, M.; Kodama, A.; Hirose, T. Supercritical carbon dioxide extraction of turmeric (*Curcuma longa*). *J. Agric. Food Chem.* **2000**, *48*, 2189–2192. [CrossRef]
9. Dandekar, D.V.; Gaikar, V.G. Hydrotropic extraction of curcuminoids from turmeric. *Sep. Sci. Technol.* **2003**, *38*, 1185–1215. [CrossRef]
10. Osorio-Tobón, J.F.; Carvalho, P.I.; Rostagno, M.A.; Petenate, A.J.; Meireles, M.A.A. Extraction of curcuminoids from deflavored turmeric (*Curcuma longa* L.) using pressurized liquids: Process integration and economic evaluation. *J. Supercrit. Fluids* **2014**, *95*, 167–174. [CrossRef]

11. Kim, J.-H.; Ahn, S.-C.; Choi, S.-W.; Hur, N.-Y.; Kim, B.-Y.; Baik, M.-Y. Changes in effective components of ginseng by puffing. *Appl. Biol. Chem.* **2008**, *51*, 188–193.
12. Kim, S.-T.; Jang, J.-H.; Kwon, J.-H.; Moon, K.-D. Changes in the chemical components of red and white ginseng after puffing. *Korean J. Food Preserv.* **2009**, *16*, 355–361.
13. An, Y.-E.; Ahn, S.-C.; Yang, D.-C.; Park, S.-J.; Kim, B.-Y.; Baik, M.-Y. Chemical conversion of ginsenosides in puffed red ginseng. *LWT* **2011**, *44*, 370–374. [CrossRef]
14. Park, K.-N.; Park, L.-Y.; Kim, D.-G.; Park, G.-S.; Lee, S.-H. Effect of turmeric (Curcuma aromatica Salab.) on shelf life of tofu. *Korean J. Food Preserv.* **2007**, *14*, 136–141.
15. Kim, A.-Y.; Kim, W.-K.; Kim, D.-O.; Kim, H.Y.; Kim, B.-Y.; Baik, M.-Y.; Lee, H. Effects of moisture content and puffing pressure on extraction yield and antioxidant activity of puffed 21-year-old *Platycodon grandiflorum* roots. *Food Sci. Biotechnol.* **2015**, *24*, 1293–1299. [CrossRef]
16. Hu, S.; Kim, B.-Y.; Baik, M.-Y. Physicochemical properties and antioxidant capacity of raw, roasted and puffed cacao beans. *Food Chem.* **2016**, *194*, 1089–1094. [CrossRef]
17. Kim, W.-K.; Kim, S.-Y.; Kim, D.-O.; Kim, B.-Y.; Baik, M.-Y. Puffing, a novel coffee bean processing technique for the enhancement of extract yield and antioxidant capacity. *Food Chem.* **2018**, *240*, 594–600. [CrossRef]
18. Shin, J.-H.; Park, Y.J.; Kim, W.; Kim, D.-O.; Kim, B.-Y.; Lee, H.; Baik, M.-Y. Change of ginsenoside profiles in processed ginseng by drying, steaming, and puffing. *J. Microbiol. Biotechnol.* **2019**, *29*, 222–229. [CrossRef]
19. Kim, S.-B.; Do, J.-R.; Lee, Y.-W.; Gu, Y.-S.; Kim, C.-N.; Park, Y.-H. Nitrite-scavenging effects of roasted-barley extracts according to processing conditions. *Korean. J. Food Sci. Technol.* **1990**, *22*, 748–752.
20. Cao, X.; Zhang, Y.; Zhang, F.; Wang, Y.; Yi, J.; Liao, X. Effects of high hydrostatic pressure on enzymes, phenolic compounds, anthocyanins, polymeric color and color of strawberry pulps. *J. Sci. Food Agric.* **2011**, *91*, 877–885. [CrossRef]
21. Chawla, R.; Patil, G.R.; Singh, A.K. High hydrostatic pressure technology in dairy processing: A review. *J. Food Sci. Technol.* **2011**, *48*, 260–268. [CrossRef] [PubMed]
22. Rendueles, E.; Omer, M.K.; Alvseike, O.; Alonsocalleja, C.; Capita, R.; Prieto, M.A.R. Microbiological food safety assessment of high hydrostatic pressure processing: A review. *LWT* **2011**, *44*, 1251–1260. [CrossRef]
23. Rivalain, N.; Roquain, J.; Boiron, J.-M.; Maurel, J.-P.; Largeteau, A.; Ivanovic, Z.; Demazeau, G. High hydrostatic pressure treatment for the inactivation of *Staphylococcus aureus* in human blood plasma. *New Biotechnol.* **2012**, *29*, 409–414. [CrossRef] [PubMed]
24. Ruizhan, C.; Meng, F.; Zhang, S.; Liu, Z. Effects of ultrahigh pressure extraction conditions on yields and antioxidant activity of ginsenoside from ginseng. *Sep. Purif. Technol.* **2009**, *66*, 340–346. [CrossRef]
25. Riahi, E.; Ramaswamy, H.S. High pressure inactivation kinetics of amylase in apple juice. *J. Food Eng.* **2004**, *64*, 151–160. [CrossRef]
26. Shouqin, Z.; Junjie, Z.; Changzhen, W. Novel high pressure extraction technology. *Int. J. Pharm.* **2004**, *278*, 471–474. [CrossRef]
27. Shin, J.-S.; Ahn, S.-C.; Choi, S.-W.; Lee, D.-U.; Kim, B.-Y.; Baik, M.-Y. Ultra high pressure extraction (UHPE) of ginsenosides from Korean *Panax ginseng* powder. *Food Sci. Biotechnol.* **2010**, *19*, 743–748. [CrossRef]
28. Jun, X.; Deji, S.; Ye, L.; Rui, Z. Micromechanism of ultrahigh pressure extraction of active ingredients from green tea leaves. *Food Control.* **2011**, *22*, 1473–1476. [CrossRef]
29. Lee, A.-R.; Choi, S.-H.; Choi, H.-W.; Ko, J.-H.; Kim, W.; Kim, D.-O.; Kim, B.-Y.; Baik, M.-Y. Optimization of ultra high pressure extraction (UHPE) condition for puffed ginseng using response surface methodology. *Food Sci. Biotechnol.* **2014**, *23*, 1151–1157. [CrossRef]
30. Prasad, K.N.; Yang, B.; Zhao, M.; Wei, X.; Jiang, Y.M.; Chen, F. High pressure extraction of corilagin from longan (*Dimocarpus longan* Lour.) fruit pericarp. *Sep. Purif. Technol.* **2009**, *70*, 41–45. [CrossRef]
31. Gil Muñoz, M.I.; Tomás-Barberán, F.A.; Hess-Pierce, B.; Kader, A.A. Antioxidant capacities, phenolic compounds, carotenoids, and vitamin C contents of nectarine, peach, and plum cultivars from California. *J. Agric. Food Chem.* **2002**, *50*, 4976–4982. [CrossRef]
32. Brand-Williams, W.; Cuvelier, M.; Berset, C. Use of a free radical method to evaluate antioxidant activity. *LWT* **1995**, *28*, 25–30. [CrossRef]
33. Berg, R.V.D.; Haenen, G.R.; Berg, H.V.D.; Bast, A. Applicability of an improved Trolox equivalent antioxidant capacity (TEAC) assay for evaluation of antioxidant capacity measurements of mixtures. *Food Chem.* **1999**, *66*, 511–517. [CrossRef]

34. Benzie, I.F.F.; Strain, J.J. The ferric reducing ability of plasma (FRAP) as a measure of "antioxidant power": The FRAP assay. *Anal. Biochem.* **1996**, *239*, 70–76. [CrossRef]
35. Singleton, V.L.; Rossi, J.A. Colorimetry of total phenolics with phosphomolybdic-phosphotungstic acid reagents. *Am. J. Enol. Vitic.* **1965**, *16*, 144–158.
36. Kim, H.; Ban, I.; Choi, Y.; Yu, S.; Youn, S.J.; Baik, M.-Y.; Lee, H.; Kim, W.-K. Puffing of turmeric (*Curcuma longa* L.) enhances its anti-inflammatory effects by upregulating macrophage oxidative phosphorylation. *Antioxidants* **2020**, *9*, 931. [CrossRef]
37. Esatbeyoglu, T.; Ulbrich, K.; Rehberg, C.; Rohn, S.; Rimbach, G. Thermal stability, antioxidant, and anti-inflammatory activity of curcumin and its degradation product 4-vinyl guaiacol. *Food Funct.* **2015**, *6*, 887–893. [CrossRef]
38. Sung, H.; Yoon, S.; Kim, W.; Yang, C. Relationship between chemical components and their yields of red ginseng extract extracted by various extracting conditions. *Korean J. Ginseng Sci.* **1985**, *9*, 170–178.
39. Corrales, M.; Toepfl, S.; Butz, P.; Knorr, D.; Tauscher, B. Extraction of anthocyanins from grape by-products assisted by ultrasonics, high hydrostatic pressure or pulsed electric fields: A comparison. *Innov. Food Sci. Emerg. Technol.* **2008**, *9*, 85–91. [CrossRef]
40. Patras, A.; Brunton, N.P.; Da Pieve, S.; Butler, F. Impact of high pressure processing on total antioxidant activity, phenolic, ascorbic acid, anthocyanin content and colour of strawberry and blackberry purées. *Innov. Food Sci. Emerg. Technol.* **2009**, *10*, 308–313. [CrossRef]
41. Casquete, R.; Castro, S.; Villalobos, M.; Serradilla, M.; Queirós, R.; Saraiva, J.; Córdoba, M.; Teixeira, P. High pressure extraction of phenolic compounds from citrus peels. *High Press. Res.* **2014**, *34*, 447–451. [CrossRef]
42. Waszkowiak, K.; Gliszczyńska-Świgło, A. Binary ethanol–water solvents affect phenolic profile and antioxidant capacity of flaxseed extracts. *Eur. Food Res. Technol.* **2015**, *242*, 777–786. [CrossRef]

 *foods*

*Article*

# Impact of Ultra-High Pressure Homogenization on the Structural Properties of Egg Yolk Granule

**Romuald Gaillard [1], Alice Marciniak [1], Guillaume Brisson [1], Véronique Perreault [1], James D. House [2], Yves Pouliot [1] and Alain Doyen [1,*]**

[1] Department of Food Science, Institute of Nutrition and Functional Foods (INAF), Université Laval, Quebec, QC G1V 0A6, Canada; romuald.gaillard.1@ulaval.ca (R.G.); alice.marciniak.1@ulaval.ca (A.M.); guillaume.brisson@fsaa.ulaval.ca (G.B.); veronique.perreault.5@ulaval.ca (V.P.); yves.pouliot@fsaa.ulaval.ca (Y.P.)

[2] Department of Food and Human Nutritional Sciences, University of Manitoba, Winnipeg, MB R3T 2N2, Canada; james.house@umanitoba.ca

* Correspondence: alain.doyen@fsaa.ulaval.ca; Tel.: +1-418-656-2131 (ext. 405454)

**Abstract:** Ultra-high pressure homogenization (UHPH) is a promising method for destabilizing and potentially improving the techno-functionality of the egg yolk granule. This study's objectives were to determine the impact of pressure level (50, 175 and 300 MPa) and number of passes (1 and 4) on the physico-chemical and structural properties of egg yolk granule and its subsequent fractions. UHPH induced restructuration of the granule through the formation of a large protein network, without impacting the proximate composition and protein profile in a single pass of up to 300 MPa. In addition, UHPH reduced the particle size distribution up to 175 MPa, to eventually form larger particles through enhanced protein–protein interactions at 300 MPa. Phosvitin, apovitellenin and apolipoprotein-B were specifically involved in these interactions. Overall, egg yolk granule remains highly stable during UHPH treatment. However, more investigations are needed to characterize the resulting protein network and to evaluate the techno-functional properties of UHPH-treated granule.

**Keywords:** egg yolk granule; ultra-high pressure homogenization; microstructure; protein aggregation; proteins

**Citation:** Gaillard, R.; Marciniak, A.; Brisson, G.; Perreault, V.; House, J.D.; Pouliot, Y.; Doyen, A. Impact of Ultra-High Pressure Homogenization on the Structural Properties of Egg Yolk Granule. *Foods* **2022**, *11*, 512. https://doi.org/10.3390/foods11040512

Academic Editors: Asgar Farahnaky, Mahsa Majzoobi and Mohsen Gavahian

Received: 16 December 2021
Accepted: 9 February 2022
Published: 10 February 2022



## 1. Introduction

Hen egg yolk is used in a wide variety of food products as an emulsifier and gelling agent due to its nutritional and techno-functional properties. It is particularly interesting because it contains proteins of high biological value, as well as other nutrients (phospholipids, vitamins, minerals, essential fatty acids) [1]. The protein fraction of egg yolk is composed of 68% low-density lipoproteins (LDLs), 16% high-density lipoproteins (HDLs), 10% globular proteins (livetins), 4% phosphoprotein (phosvitin), and 2% other minor proteins [2]. After centrifugation, egg yolk can be fractionated into two fractions called plasma (supernatant) and granule (pellet). The plasma is mainly composed of LDLs (85%) and livetin (15%), whereas the granule consists of HDLs (70%), phosvitin (16%) and LDLs (12%). The granule has interesting nutritional properties related to its low cholesterol (LDLs) content [3–5] and high 5-methyl-tetrahydro-folate (5-MTHF) concentration, compared to the whole egg yolk [6]. Ultimately, destabilizing the egg yolk granule could improve the accessibility of its nutritional and bioactive compounds and eventually improve the digestibility and functionality of the product [7]. Structurally, the granule consists of a circular structure formed by non-soluble complexes involving HDL and phosvitin stabilized by phosphocalcic bridges [8]. The large number of phosphocalcic bridges gives the granule a very compact and poorly hydrated structure, leading to its high resistance to thermal treatment, ultrasound and enzymatic hydrolysis [9,10].

Different strategies have been applied to destabilize the egg yolk granule. Usually, this involves increasing the ionic strength, which induces the disruption of phosphocalcic bridges and improves the granule solubility [7,8]. This high solubility improves the emulsion stability of the granule compared to the egg yolk and plasma [7]. More recently, high hydrostatic pressure (HHP) (400 and 600 MPa, for 5 min) applied to the granule was sufficient to destabilize its structure, probably by disrupting the phosphocalcic bridges between phosvitin and HDLs [11,12]. After centrifugation of the isostatically pressured granule, the recovered insoluble fraction was enriched in folic acid and phosvitin. Consequently, the use of high pressure shows great potential to generate new ingredients with interesting nutritional and functional properties for many industrial applications such as use as a low-cholesterol emulsifying agent in mayonnaise [11–15].

The ultra-high pressure homogenization (UHPH) system, with higher pressure levels of up to 400 MPa, is an emerging technology used to modify the structure of food matrices. Contrary to the isostatic HHP process, UHPH is a dynamic pressure process inducing turbulence, high shear, cavitation, and temperature to help reduce particle size distributions, allowing emulsion stability to be enhanced [16,17]. UHPH also induced modification of protein structures through unfolding of the quaternary and tertiary structure, and generation of inter- and intra-molecular interactions [16,18–20]. However, these modifications depend on the matrices and pressure parameters applied, such as pressure level, process temperature and the number of passes. These particular modifications of protein structure and interactions induced by UHPH were studied to improve the techno-functional properties of proteins [21,22], such as emulsifying [23–26] and foaming [27–32].

To the best of our knowledge, limited data are available regarding the impact of UHPH on granule destabilization. Only Sirvente et al. [10] demonstrated that conventional high pressure homogenization (0.3 to 20 MPa) had little effect on granule destabilization and its techno-functional properties. Thus, the use of UHPH for the destabilization of the compact granule structure represents a promising technology for generating a high value-added egg fraction with interesting functional and nutritional properties for the food industry [28,31]. Consequently, this project aims to study the impact of the pressure level (50, 175 and 300 MPa) and number of passes (1 and 4) on the structural properties of egg yolk granule and its subsequent fractions (plasma and granule from pressure-treated granule).

## 2. Materials and Methods

### 2.1. Preparation of Egg Yolk Granules

Fresh hen eggs were purchased from a local supermarket and stored at 4 °C until preparation according to the protocol described by Naderi et al. [6]. Briefly, fresh hen eggs were manually broken, and the albumen was discarded. Albumen residues were eliminated from yolk by absorption on filter paper (Whatman, MA, USA). The vitelline membrane of yolk was removed with tweezers. Egg yolks were diluted in distilled water (1:1 *w/w*) and centrifuged at 10,000× *g* for 45 min at 4 °C for the recovery of plasma (supernatant) and granule (pellet) fractions. The granule fraction was prefrozen at −30 °C and freeze-dried (pressure of 27 Pa and plate temperature of 20 °C) for 48 h. The freeze-dried granule was kept frozen under vacuum until analysis and treatment.

### 2.2. Ultra-High Pressure Homogenization

The freeze-dried granule was solubilized in distilled water (1% $w/v$) overnight at 4 °C under constant agitation to generate the initial granule solution (G1c). Immediately after solubilization, the G1c was pressure-treated by single or multiple-pass (1 to 4) ultra-high pressure homogenization (Nano DeBEE laboratory homogenizer, Bee International, South Easton, MA, USA) at 50, 175 and 300 MPa at a flow rate of 4 L/h. While it is estimated that the temperature of the sample increased by around 16 to 25 °C per 100 Mpa [31,33,34], the system was equipped with a cooling system allowing a fast temperature decrease after UHPH treatment. The resulting emulsions obtained after pressurization of G1c were labeled G1p. After UHPH treatments, 1 mL of sodium azide solution (0.02% $w/v$) was added to 500 mL of sample for preservation. Next, pressure-treated granule samples (G1p) were centrifuged at 10,000× *g* for 45 min at 4 °C to generate a second granule (G2p) and plasma (P2p). Control granule (G2c) and plasma (P2c) were generated following the same experimental design without pressurization treatment (Figure 1). Secondary granules (G2p and G2c) were resuspended in 10 mL of distilled water for native PAGE analysis. Finally, initial granules (G1p and G1c), secondary granules (G2p and G2c) and plasma (P2p and P2c) were prefrozen at −30 °C and freeze-dried (pressure of 20 Pa and plate temperature of 20 °C) for 48 h before subsequent analysis.

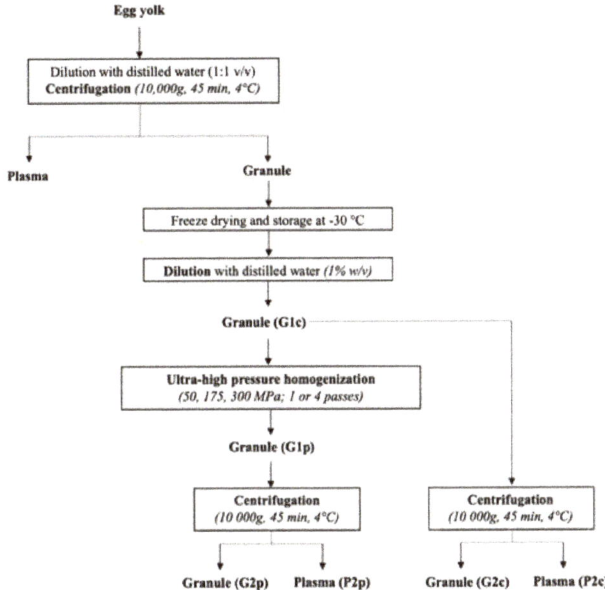


**Figure 1.** Experimental design of the production of the different fractions from egg yolk granule after ultra-high-pressure homogenization treatments. G1c: initial control granule, G1p: initial pressure-treated granule, G2p: granule from initial pressure-treated granule, P2p: plasma from initial pressure-treated granule, G2c: control granule from initial control granule, P2c: control plasma from initial control granule.

### 2.3. Proximate Composition

Dry matter and ash content of the initial control granule (G1c), pressure-treated granule (G1p), granule from pressure-treated granule (G2p) and control granule (G2c) fractions were determined according to the Association of Official Agricultural Chemists (AOAC) 923.03 (dry matter) and 925.09 (ash) methods. The crude protein content of G1c, G2c, G1p, G2p, control plasma (P2c) and plasma originating from pressure-treated granule (P2p) was obtained by the Dumas method (LECO FP-528, Model 601-500, LECO, St. Joseph, MI, USA) using a protein-to-nitrogen conversion factor of 6.25 [6]. The lipid content of G1c, G1p, G2p and G2c fractions was determined by the Mojonnier method (AOAC International 925.32). Phosphorus and iron contents of G1c, G1p, G2p and G2c fractions were obtained by inductively coupled plasma optical emission spectroscopy (ICP-OES) (Agilent 5110 ICP-OES, Agilent Technologies, Inc., Santa Clara, CA, USA) according to the AOAC 923.03 and 925.09 methods. Dry matter, lipid, ash and mineral composition were not determined for P2c and P2p due to limited quantities of those fractions.

### 2.4. Particle Size Distribution

The distribution of particle sizes of G1c and G1p was determined on liquid samples using a laser diffraction system (Malvern Mastersizer 3000, Malvern Instrument Ltd.,

Worcestershire, UK), and the data were analyzed with the software Mastersizer V.3.72. Samples were analyzed without dilution. The sample dispersion particle type was set to non-spherical particle mode and the refractive index was set to 1.45, whereas the dispersant phase's (deionized water) refractive index was set to 1.33. The sample was added to the cell up to an obscuration rate between 8–13%. The mean diameter expressed as volume mean (De Brouckere mean diameter) was expressed as $D_{[4,3]}$. All conditions were analyzed in duplicate with 3 measures for each duplicate.

### 2.5. Microstructure of Egg Yolk Granules and Fractions

The microstructure of control and pressure-treated granule and plasma samples was visualized by transmission electron microscopy (TEM) (JEOL JEM-1230 TEM, Tokyo, Japan) operating at 80 kV and at a magnification factor of 3K. Granule fractions (G1c, G1p, G2c and G2p) were diluted to 1:200 for obtaining good quality images, whereas plasma fractions (P2c and P2p) were used undiluted. A droplet of each sample was mixed with 3% (*w*/*v*) uranyl acetate and placed onto a copper grid/carbon film for 1 min. The excess was removed, and the grid was air-dried. A Gatan camera ultrascan US1000SP1 (Gatan, Inc., Pleasanton, CA, USA) was used for image capture.

### 2.6. Protein Profiles

Protein profiles of control and pressure-treated samples were obtained by native and sodium dodecyl sulphate (SDS)-polyacrylamide gel electrophoresis (PAGE) under reducing conditions according to the protocol of Naderi et al. [6]. First, native PAGE was carried out with fresh liquid samples. For control and pressure-treated granule fractions, 25 μL of diluted sample (1:4 *v*/*v*) was mixed with 25 μL of native sample buffer (62.5 mM Tris-HCl, 40% glycerol, 0.01% bromophenol blue—Bio-Rad, Mississauga, ON, Canada), and 10 μL of each sample was loaded into the well. For control and pressure-treated plasma fractions, 50 μL of sample and 20 μL of native sample buffer were mixed and a final volume of 20 μL was loaded into each well. For SDS PAGE under reducing conditions, a 1% (*w*/*v*) protein solution was prepared for each freeze-dried granule and plasma fraction, and 200 μL of that solution was diluted into 800 μL of deionized water. Samples G2c and G2p were sonicated for 30 min to improve solubility. A volume of 25 μL of sample buffer (5% 2-mercaptoethanol, 95% Laemmli buffer—Bio-Rad, Mississauga, ON, Canada) was added to 25 μL of each diluted sample. Samples were boiled for 10 min and cooled on ice before loading 10 μL into the gel wells. Electrophoresis was performed using 4–20% TGX Stain-Free polyacrylamide gel (Bio-Rad, Mississauga, ON, Canada) and run at 30 mA per gel for a total of 35 min at room temperature. Gels were stained with 0.1 M aluminum nitrate and destained with a solution of 40% *v*/*v* methanol, 10% *v*/*v* acetic acid and 50% *v*/*v* distilled water. The molecular weights of migrated proteins were estimated using molecular weight markers (Precision Plus Protein™ 161-0373 All Blue Prestained Protein Standards, Bio-Rad, Mississauga, ON, Canada). Images of the gels were captured using the ChemiDoc™ MP Imaging System (Bio-Rad Laboratories Ltd., Mississauga, ON, Canada).

### 2.7. Proteomics Analysis

Protein digestion and mass spectrometry analyses were performed by the Proteomics Platform of the Research Center of the Centre Hospitalier Universitaire (CHU) of Quebec (Quebec City, QC, Canada). Specific bands corresponding to high molecular weight aggregates from native gels were excised for G1c and G1p at 300 MPa with 1 and 4 passes and washed with MilliQ water. Tryptic digestion products for each sample were separated by online reversed-phase (RP) nanoscale capillary liquid chromatography (nanoLC) and analyzed by electrospray mass spectrometry (ES MS/MS). Data obtained by mass spectrometry were analyzed using Mascot (Matrix Science, London, UK; version 2.5.1) as described by Duffuler et al. [13]. The Uniprot Gallus 2020 database and Scaffold software (version Scaffold_4.11.1, Proteome Software Inc., Portland, OR) were used for peptide and protein identification. The false discovery rates (FDR) and the minimum number of peptides occurrence were set to <1.0% and 2, respectively.

### 2.8. Statistical Analysis

Data analyses were carried out using Statistical Analysis System (SAS), SAS® Studio software (Copyright © 2020 SAS Institute Inc., Cary, NC, USA.) and three replicates were performed for each experiment. The proximate compositions of the fractions were analyzed by one-way analysis of variance (ANOVA) and the Tukey test ($\alpha = 0.05$) for multiple comparisons. The particle size distribution and the mean diameter expressed as volume mean, $D_{[4,3]}$, were analyzed by multifactorial ANOVA. A 95% confidence interval ($p < 0.05$) was used for all tests. Data were expressed as mean ± standard deviation (SD).

## 3. Results

### 3.1. Proximate Composition of Control and Pressure-Treated Egg Yolk Fractions

Table 1 shows the proximate composition of control and pressure-treated fractions for each pressurization level (50, 175 and 300 MPa) and number of passes (1 and 4). No differences ($p > 0.05$) were observed regarding the content of dry matter and all components (protein, lipid, ash, phosphorus and iron) between control and pressure-treated fractions (granule or plasma), regardless of the pressure applied (50, 175 and 300 MPa) and the number of passes (1 and 4).

**Table 1.** Composition of control and pressure-treated fractions of egg yolk.

| Fraction | Dry Matter | Protein | Lipid (% *w*/*w*, Dry Basis) | Ash | Phosphorus (P) | Iron (Fe) ($\times 10^{-2}$) |
|---|---|---|---|---|---|---|
| G1c (control) | 96.1 ± 1.0 $^{a*}$ | 64.6 ± 0.7 $^{a}$ | 39.7 ± 11.2 $^{a}$ | 6.5 ± 0.2 $^{a}$ | 1.1 ± 0.5 $^{a}$ | 2.4 ± 1.2 $^{a}$ |
| G1p (50, 175 and 300 MPa for 1 and 4 passes) | 96.5 ± 2.2 $^{a}$ | 62.7 ± 0.8 $^{a}$ | 29.7 ± 5.0 $^{a}$ | 6.4 ± 0.2 $^{a}$ | 1.0 ± 0.4 $^{a}$ | 1.5 ± 0.8 $^{a}$ |
| G2c (control) | 99.0 ± 0.6 $^{a}$ | 63.6 ± 1.2 $^{a}$ | 37.3 ± 6.4 $^{a}$ | 5.4 ± 0.6 $^{b}$ | 0.5 ± 0.2 $^{a}$ | 1.9 ± 0.7 $^{a}$ |
| G2p (50, 175 and 300 MPa for 1 and 4 passes) | 99.1 ± 0.4 $^{a}$ | 62.8 ± 2.1 $^{a}$ | 42.0 ± 8.6 $^{a}$ | 5.5 ± 0.2 $^{b}$ | 0.5 ± 0.2 $^{a}$ | 1.5 ± 0.4 $^{a}$ |
| P2c (control) | N.D. | 50.0 ± 10.7 $^{b}$ | N.D. | N.D. | N.D. | N.D. |
| P2p (50, 175 and 300 MPa for 1 and 4 passes) | N.D. | 50.1 ± 3.6 $^{b}$ | N.D. | N.D. | N.D. | N.D. |

* Mean value ± standard deviation. Data with different letters (a–b) within each column are significantly different at $p < 0.05$ (Tukey test, $\alpha = 0.05$, $n = 3$). G1c: initial granule, G1p: pressure-treated granule, G2c: control granule from initial granule, G2p: granule from pressure-treated granule, P2c: control plasma from initial granule, P2p: plasma from pressure-treated granule. N.D.: not determined.

### 3.2. Effect of UHPH Treatments on Particle Size Distribution

Figure 2 shows the particle size distribution of initial control (G1c) and pressure-treated granule (G1p) at 50, 175 and 300 MPa with 1 and 4 passes (Figure 2A), and the factorial effect of pressure level and number of passes (1 and 4) on the two main populations (Figure 2B). The two main heterogenous populations were distinguishable in the non-treated fraction (G1c), with a particle size between 0.33 and 1.36 μm for population 1 and between 1.76 and 625 μm for population 2 (Figure 2A). The application of UHPH treatments decreased the particle size of the initial control granule (G1c) drastically. Particles above about 30 μm that were initially present were reduced to two main populations: population 1 ranging from 0.25 to 1.55 μm and population 2 ranging from 1.76 to 15.4 μm.

To understand the impact of pressure level (50, 175 and 300 MPa) and number of passes (1 and 4) on the size of each population, a multiple factorial analysis was performed for all G1p fractions (Figure 2B). The figure represents the size corresponding to the highest intensity for each population (1—hatched and 2—filled) as a function of pressure (50, 175 and 300 MPa) and passes (1—blue and 4—green). Population 1 had strong interactions between both pressure and passes ($p < 0.0001$). While no effect was observed for the number of passes (1 or 4) at 50 and 175 MPa, at 300 MPa, population 1's particle sizes for 1 (0.63 μm) and 4 passes (0.56 μm) were lower than those obtained at 50 and 175 MPa (mean of 0.72 μm). In contrast, for population 2, the impact of UHPH was solely due to the pressure level ($p < 0.0001$). Indeed, the particle size decreased from 4.04 to 3.66 μm for 50 and 175 MPa, and finally increased from 3.66 to 4.96 μm for 175 and 300 MPa.

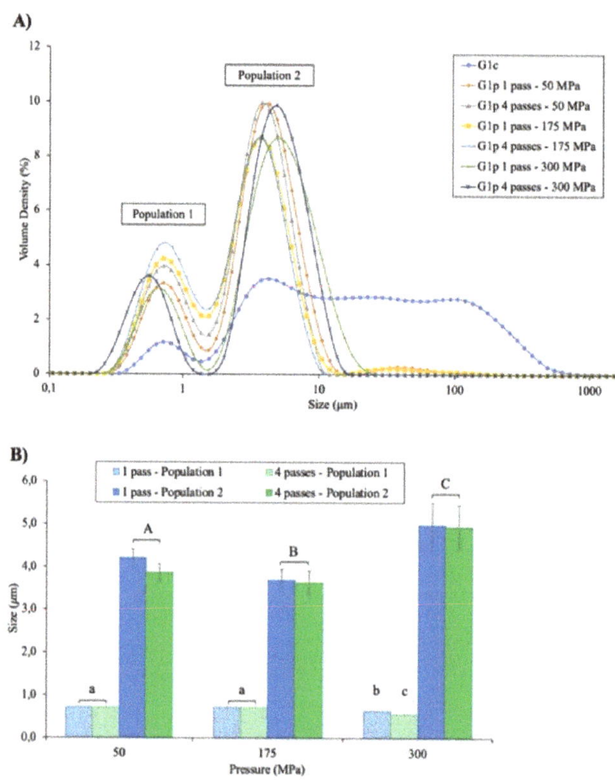


**Figure 2.** Particle size distribution of initial (G1c) and pressure-treated granule (G1p) at 50, 175 and 300 MPa with 1 and 4 passes (**A**), and factorial effect of pressure level (50, 175 and 300 MPa) and number of passes (1—blue and 4—green) on the two populations (1—hatched and 2—filled) (**B**). Different letters, a–c for population 1 and A–C for population 2, indicate significant differences ($p < 0.05$) due to interaction between pressure and passes and solely due to the pressure level (Tukey test, $\alpha = 0.05$, $n = 3$).

Figure 3 represents the mean diameter expressed as volume ($D_{[4,3]}$), which reflects the sizes of particles that constituted the majority of the sample volume of pressure-treated granule (G1p) as a function of pressure (50, 175 and 300 MPa) and number of passes (1 and 4). The $D_{[4,3]}$ value of the initial granule (G1c—data not shown) was higher ($p < 0.0001$) than that of pressure-treated granules, with average values of 58.5 ± 43.2 μm for G1c and 4.2 ± 0.8 μm for all G1p, which corresponded to a decrease of approximately 93%. Within UHPH treatments, the multiple factorial analyses indicated a simple effect of pressure ($p < 0.0001$) and passes ($p < 0.0001$) without interaction between these two parameters (Figure 3). Globally, regardless of the pressure level (50–300 MPa), UHPH at 1 pass increased the $D_{[4,3]}$ value compared to that in samples pressurized with 4 passes ($p < 0.0001$), with average values of 4.8 ± 1.2 μm for G1p at 1 pass and 3.5 ± 0.4 μm for G1p at 4 passes. In addition, regardless of the number of passes, pressure treatment at 300 MPa induced higher $D_{[4,3]}$ compared to pressures of 50 and 175 MPa, with values of 4.8 ± 0.6 μm (average of G1p at 300 MPa at 1 and 4 passes) and 3.9 ± 0.9 μm (average of G1p at 50 and 175 MPa at 1 and 4 passes).

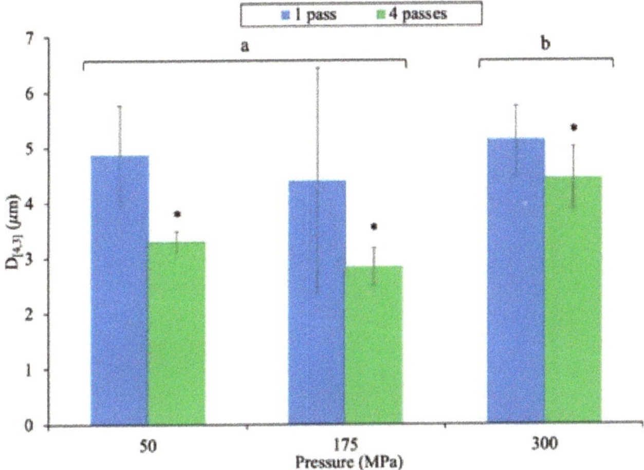


**Figure 3.** Impact of pressure level (50, 175 and 300 MPa) and number of passes (1—blue and 4—green) on the mean diameter expressed as volume ($D_{[4,3]}$) of pressure-treated granule (G1p). Data with different letters (a-b) and * are significantly different ($p < 0.05$), due to a simple effect of pressure and passes (Tukey test, $\alpha = 0.05$, $n = 3$).

### 3.3. Microstructural Modifications Following UHPH Treatments

Figure 4 shows the changes in microstructure associated with the initial granule (G1c) before and after UHPH treatment (G1p—Figure 4A), and their respective secondary granules (G2c and G2p—Figure 4B) and plasmas (P2c and P2p—Figure 4C) using TEM. First, the initial granule (G1c) appeared as large and compact (dark) aggregates with diameters close to 4 μm, whereas a change in the morphology of the particles was observed after UHPH treatment, correlating with the increase in pressure and number of passes (Figure 4A). Specifically, the particles observed for the pressure-treated granule (G1p) at 300 MPa with 4 passes formed a diffuse and thin network compared to the particle shape in the initial control granule (G1c). Second, the control granule from initial granule (G2c) and granule from pressure-treated granule (G2p) (Figure 4B) showed microstructure and effects of pressure and number of passes similar to those observed for G1c and G1p. Last, the microstructure of control plasma from initial granule (P2c) and plasma from pressure-treated granule (P2p) (Figure 4C) was drastically different from that of granules (G1 and G2) (Figure 4A,B). Indeed, TEM images show very particular leaf-like structures but none of the network previously observed in G1p and G2p. It should be noted that no major differences, regardless of pressurization level (50–300) and number of passes (1–4), were observed between plasma samples. Compared with the controls (G1c and G2c), UHPH treatments caused major changes in the microstructure of the particles in the granule fractions that correlated with the increases in pressure (50–300 MPa) and number of passes (1–4). UHPH treatments reduced the particle size and changed the microstructure of the complex, thus forming a thin and highly diffuse network when subject to the most severe parameters (300 MPa, 4 passes).

### 3.4. Effect of UHPH Treatments on Protein Profiles

Figure 5 shows the native (Figure 5A,C,E) and reduced (Figure 5B,D,F) SDS PAGE of G1c and G1p (Figure 5A,B), G2c and G2p (Figure 5C,D) and P2c and P2p (Figure 5E,F) fractions. Protein profiles from native PAGE of the initial (G1c) and pressure-treated granule (G1p) (Figure 5A) were similar except that the intensity of the band in the loading well (X1) decreased as a function of the increase in pressure level, mainly at 300 MPa and passes. Two other bands (X2) with similar intensities were detected in native PAGE gels through all pressure levels and numbers of passes. Native PAGE of the granule obtained after centrifugation of initial granule (G2c) and pressure-treated granule (G2p) (Figure 5C) showed that the intensity in the loading well (X3) decreased as a function of UHPH treatments, mainly for 300 MPa with 4 passes, as observed for G1c and G1p (Figure 5A). Similarly to G1c and G1p, no differences were observed in the protein profile (X4) for all G2c and G2p fractions (Figure 5B). Native PAGE of plasma (P2c and P2p) fractions (Figure 5E) showed no remarkable modifications in the protein profiles between P2c and P2p at 50 and 175 MPa, regardless of the number of passes. However, at 300 MPa, differences in the protein profiles (X5) were observed after 1 and 4 passes.

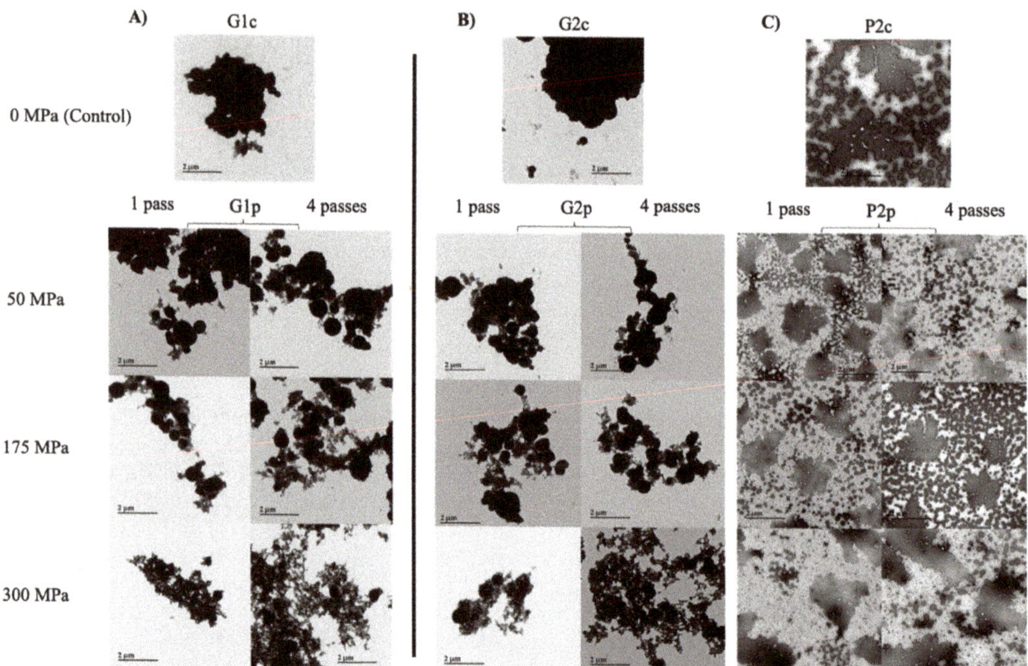


**Figure 4.** Microstructures of (**A**) initial control granule (G1c) and pressure-treated granule (G1p), (**B**) control granule from initial control granule (G2c) and granule from pressure-treated granule (G2p), and (**C**) control plasma from initial control granule (P2c) and plasma from pressure-treated granule (P2p) observed by transmission electron microscopy with a magnification of 3K.

The protein profiles of control and pressure-treated granule fractions (Figure 5B,D), obtained from SDS PAGE under reducing conditions showed 6 major protein bands from 31 kDa to 110 kDa that were also identified by Guilmineau et al. [35], Naderi et al. [5] and [12]. Their intensity and distribution remained similar, regardless of the treatment and granule sample. The protein profiles of plasma from pressure-treated granule (P2p) (Figure 5F) obtained from SDS PAGE under reducing conditions were similar to those of the control for samples at 50 and 175 MPa with 1 or 4 passes. However, and as observed in native PAGE (Figure 5E), at 300 MPa, the intensity of the protein bands at 68 and 78 kDa (X6) decreased as a function of the number of passes. In addition, for P2p at 300 MPa and 4 passes, a band with a molecular weight of 45 kDa appeared that was not present for other treatments and control (P2c).

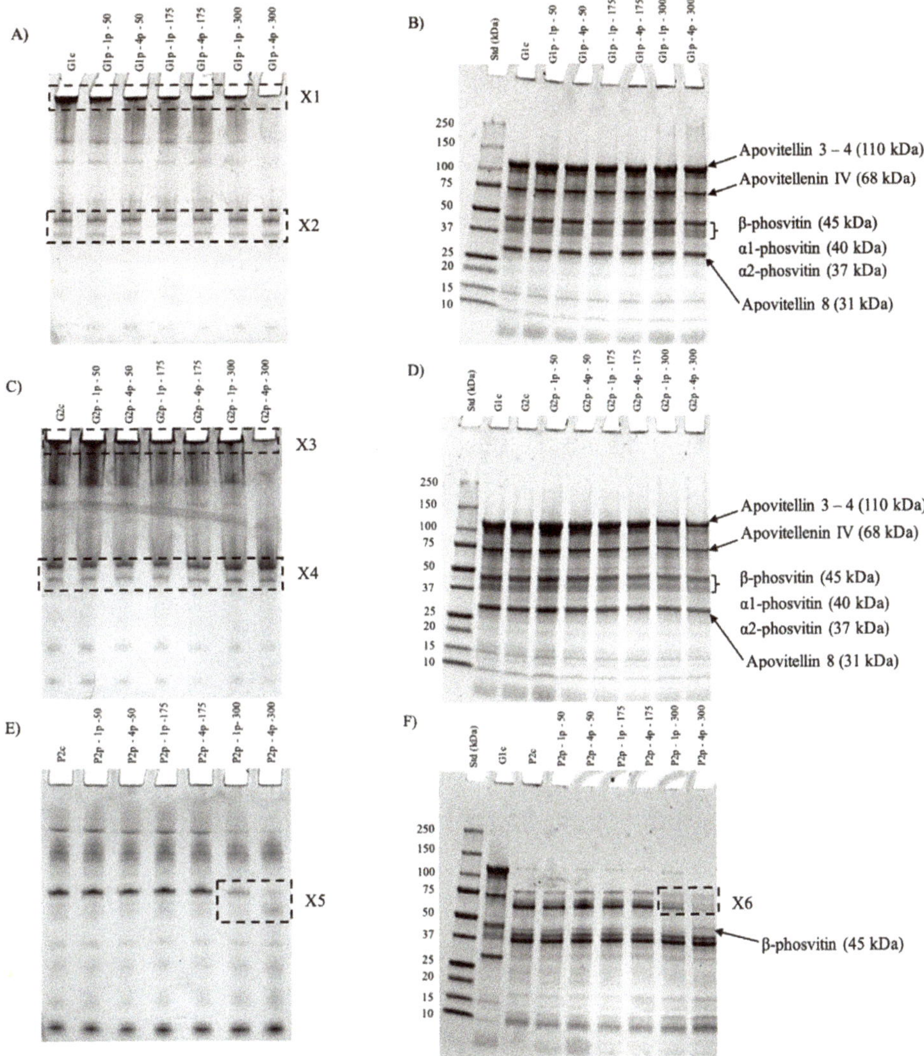


**Figure 5.** Native (**A**,**C**,**E**) and reduced (**B**,**D**,**F**) SDS PAGE of initial granule (G1c) and pressure-treated granule (G1p) (**A**,**B**); control granule from initial granule (G2c) and granule from pressure-treated granule (G2p) (**C**,**D**); and control plasma from initial granule (P2c) and plasma from pressure-treated granule (P2p) (**E**,**F**).

### *3.5. Characterization of Egg Yolk Proteins by Proteomics Analysis*

The band observed in the wells of native PAGE of initial granule (G1c) and pressure-treated granule (G1p) at 300 MPa and 1 pass and 4 passes were analyzed by LC-MS/MS, and the results are presented in Table 2. The total spectrum count (TSC) value corresponds to the total number of spectra identified for a protein and is a semi-quantitative measure for a given protein abundance in proteomic studies [36], and the coverage percentage corresponds to the percentage of the protein sequence represented by the detected peptides in the dataset [37]. Globally, the composition of all G1c and G1p fractions at 300 MPa and 1 pass and G1p at 300 MPa and 4 passes were identical. A total of eight specific proteins from egg yolk granule were identified for all samples, with five predominant proteins with molecular weight ranging from 12 to 523 kDa. These proteins were identified as phosvitin from vitellogenin (VTG)-2 (#1—205 kDa), VTG-1 (#2—211 kDa) and VTG-3 (#3—187 kDa), apolipoprotein-B (#4—523 kDa), and apovitellenin-1 (#5—12 kDa). For these proteins, the coverage percentage and TSC ranged from 41% to 73% and from 17 to 1235, respectively, for all conditions (Table 2).

**Table 2.** Proteomic analysis of native PAGE wells of initial granule (G1c), pressure-treated granule (G1p) at 300 MPa—1 pass and pressure-treated granule (G1p) at 300 MPa—4 passes.

| # Protein | Identified Proteins | Molecular Weight (kDa) | UniProt ID | % Coverage | | | Total Spectrum Count (TSC) | | |
|---|---|---|---|---|---|---|---|---|---|
| | | | | G1c | 300 MPa 1 Pass | 300 MPa 4 Passes | G1c | 300 MPa 1 Pass | 300 MPa 4 Passes |
| 1 | Phosvitin OS = *Gallus gallus* OX = 9031 GN = VTG2 PE = 4 SV = 1 | 205 | F1NFL6_CHICK | 63 | 59 | 73 | 1235 | 660 | 1227 |
| 2 | Phosvitin OS = *Gallus gallus* OX = 9031 GN = VTG1 PE = 4 SV = 1 | 211 | A0A1D5NUW2_CHICK | 56 | 48 | 60 | 523 | 341 | 608 |
| 3 | Phosvitin OS = *Gallus gallus* OX = 9031 GN = VTG3 PE = 4 SV = 1 | 187 | A0A3Q2U347_CHICK | 55 | 49 | 63 | 267 | 197 | 309 |
| 4 | Apolipoprotein B OS = *Gallus gallus* OX = 9031 GN = APOB PE = 4 SV = 2 | 523 | F1NV02_CHICK | 51 | 48 | 59 | 374 | 401 | 571 |
| 5 | Apovitellenin-1 OS = *Gallus gallus* OX = 9031 PE = 1 SV = 1 | 12 | APOV1_CHICK | 65 | 41 | 63 | 34 | 17 | 36 |
| 6 | Albumin OS = *Gallus gallus* OX = 9031 GN = ALB PE = 4 SV = 2 | 64 | A0A1D5NW68_CHICK | 9 | 68 | 82 | 8 | 50 | 86 |
| 7 | Ovotransferrin OS = *Gallus gallus* OX = 9031 GN = TF PE = 3 SV = 4 | 83 | E1BQC2_CHICK | 0 | 34 | 43 | 0 | 30 | 37 |
| 8 | Ovalbumin OS = *Gallus gallus* OX = 9031 GN = SERPINB14 PE = 1 SV = 2 | 43 | OVAL_CHICK (+1) | 0 | 9 | 17 | 0 | 3 | 6 |

## 4. Discussion

The aim of the current work was to investigate the destabilization of egg yolk granule through UHPH treatment. Overall, the approximate compositions of granule and the resulting granule and plasma fractions were not modified following UHPH treatments. However, the application of UHPH decreased the average particle size of the initial granule (G1c). Additionally, the mean diameter of particles expressed as volume ($D_{[4,3]}$) increased for pressures above 300 MPa, and to a greater extent, when treated with 4 passes. These observations correlated with a change in the microstructure of the granule through the formation of a diffuse and thin network. Regardless of the treatment (number of passes or pressure level), phosvitin, apovitellenin and apolipoprotein-B were involved in the formation of these structures.

### *4.1. Impact of UHPH Treatments on Particle Size Distribution of Initial (G1c) and Pressure-Treated Granule (G1p)*

In the granule, HDL, LDL and phosvitin are associated with phosphocalcic bridges. Depending on the conditions (pH and ionic strength), the HDL–phosvitin complex can form soluble micelles with particle sizes ranging from 100 to 200 nm [1]. These structures could correspond to the large population observed in granule samples without application of UHPH treatments (Figure 2A—Population 2). However, after pressurization treatments, a reduction in particle size distribution as well as an increase in the volume density and a decrease in the mean diameter $D_{[4,3]}$ values were observed, compared to the control. These observations were noted regardless of the pressure level and number of passes, indicating that the global UHPH treatment-related mechanical force and high temperatures encountered in the valve could disperse granule aggregates into small particles. These results agree with those obtained by numerous authors using model oil-in-water emulsions [23] and protein-stabilized emulsions [19,38–40]. Our observations were also similar to those of Naderi et al. [11] using ultrasound treatment, which were explained by the cavitation and high shear forces inducing disintegration of the granule into small particles of HDL-phosvitin [11]. In addition, and contrary to the control granule sample, bimodal and polydisperse distributions were observed after UHPH treatments (Figure 2A), as observed by Sirvente et al. [10] after mechanical treatment of yolk granule by conventional high-pressure homogenization (0.3, 10 and 20 MPa). According to Chang et al. [41] and Causeret et al. [8], the smallest particle size population could correspond to a spherical complex of non-soluble HDL-phosvitin linked by phosphocalcic bridges where the largest particle size population was probably granule aggregates [1].

For the smallest population of particles (Figure 2B—hatched), we observed a strong interaction between both pressure and passes on particle size reduction, with the largest decrease observed at 300 MPa and 4 passes. Increasing the homogenization pressure progressively reduced the size particle, while distributions became narrower, potentially due to the reduction in size of HDL caused by homogenization partially disrupting the granules [42], as observed for oil-in water emulsions prepared with egg yolk. Similar results were obtained after application of UHPH to milk fat globule membrane (MFGM), which is mainly composed of phospholipids (like yolk granule), in the presence of milk protein [43]. Simultaneously, increasing the number of passes reduced the sizes of proteins due to the unfolding phenomenon [44]. Increasing the number of passes could also induce greater disruption of the HDL–phosvitin complex and decrease its particle size (Figure 2B), as observed in oil-in-water emulsions composed of whey protein isolate and flaxseed oil [45]. On the other hand, the volume density of the second population observed in Figure 2A was greater than that of the first population. UHPH treatments are known to modify the conformation of globular proteins, which tend to unfold and aggregate, leading to protein–protein interactions [19,21,46]. Consequently, population 2 could be made up of yolk protein aggregates generated by the combination of mechanical homogenization and shear-induced temperature effects throughout UHPH treatment. A detailed analysis of population 2 (Figure 2B—filled) showed that only pressure level had an impact on particle size distribution. More specifically, the particle size decreased between 50 and 175 MPa to eventually increase at 300 MPa. The same tendency was shown by the $D_{[4,3]}$ values (Figure 3). These observations could be the result of increased interaction and subsequent re-coalescence of smaller fat and protein particles, as observed after microfluidization treatment of egg yolk [47]. As indicated previously, the higher the pressure, the smaller the particle size due to a drastic disruption of granule structure, specifically HDL–phosvitin complex. As a result, the protein–protein aggregation phenomenon could also occur at higher pressure levels.

### 4.2. *Impact of UHPH Treatments on the Protein Profile, Composition and Microstructure of Egg Yolk Fractions*

Overall, the results showed a drastic reorganization of the granule's microstructure with the severity of UHPH treatment (up to 300 MPa—4 passes) but little impact on its protein composition. The application of UHPH treatment up to 175 MPa induced the large particles present in the granule to be disrupted into smaller particles, as observed by TEM analysis (Figure 4A,B). The plasma fraction, however, showed irregular leaf-like structures comparable to the one noticed by Staneva et al. [48] with mixture of egg phosphatidylcholine, sphingomyelin and ceramide (Figure 4C). Observations on granule fractions correlated with the particle size measurements discussed previously and were supported by the literature on conventional high-pressure homogenization [10,49,50]. The mechanical forces involved during high pressure homogenization disrupt the quaternary structure of proteins and eventually lead to structural changes through protein unfolding [51]. In addition, when samples were pressurized at 300 MPa, TEM analysis showed the formation of a thin and diffuse network (Figure 4A,B). These results agree with the findings of Floury et al. [19], who observed a gel-like network structure in UHPH-treated soy protein-stabilized oil-in-water emulsions at pressures above 250 MPa. The increase in pressure causes the temperature within the valve to increase, thus increasing the strength of the hydrophobic bonds. The resulting protein–protein interactions can be enhanced to eventually form a gel-like network [19]. In addition, as observed by Marco-Molés et al. [24] with egg/dairy oil-in-water emulsions, the use of pressure up to a critical level (300 MPa in our study) could induce destabilization of an emulsion through the loss of the natural emulsifier barrier, thus potentially increasing protein–protein interactions.

Proteomic analysis (Table 2) and gel electrophoresis (Figure 5B,D) showed that the protein composition of pressure-induced structures was similar regardless of the treatment, and specifically composed of apovitellenin, apolipoprotein and phosvitin. This also indicates the high resistance of granule proteins to UHPH treatment compared to HHP treatment. These observations agree with the literature on heat treatment of egg yolk, which demonstrates very little impact of the treatment on the granule proteins vs. plasma proteins [52,53]. The high resistance of phosphocalcic bridges toward heat could explain granule stability, especially the HDL–phosvitin complex, against UHPH treatment [1]. In this work, the application of UHPH up to 300 MPa—1 pass induced the disintegration of the initial granule structure into smaller particles composed of HDL–phosvitin complexes. However, when treated at 300 MPa—4 passes, the combination of pressure and high temperature, as well as mechanical forces, enhanced protein–protein interactions, and the formation of a protein network. It also caused a very slight dissociation of the β-phosvitin from HDL complexes, leading to the release of a small portion of the β-phosvitin into the soluble fractions (plasma) (Figure 5F). These observations differ from the literature on the application of HHP to egg yolk granule. Indeed, Naderi et al. [11] showed HHP treatment (600 MPa, 5 min) disrupted the egg yolk granule structure and induced a specific and major transfer of phosvitin and folate into the plasma fraction.

## 5. Conclusions

Ultra-high pressure homogenization (50–300 MPa for 1–4 passes) was used as a pretreatment technology to destabilize the native compact structure of egg yolk granule. This study demonstrates that increasing the pressure to 300 MPa and the number of passes to 4 induces structural modification of the granule by decreasing the particle size and increasing protein–protein interactions to eventually form a protein network. Despite the harsh conditions generated during the most severe treatment (300 MPa at 4 passes), the phosvitin–HDL complex showed very high resistance, thus not significantly impacting the composition of the network consisting of apovitellenin, apolipoprotein and phosvitin. However, as this work was preliminary to further investigations, the extent of protein unfolding and the nature of the interactions involved during formation of a protein network by UHPH treatments of egg yolk granule remain to be investigated. Finally, the next step will be to understand the impact of structural modification on the functional properties of egg yolk granules treated by UHPH.

**Author Contributions:** Conceptualization, R.G. and A.D.; Methodology, R.G., G.B., V.P. and A.D.; Software, R.G.; Validation, R.G., A.M., G.B. and A.D.; Formal Analysis, R.G.; Investigation, R.G. and A.M.; Resources, R.G., A.M., V.P., G.B., Y.P. and A.D.; Data Curation, R.G.; Writing—Original Draft Preparation, R.G. and A.M.; Writing—Review and Editing, R.G., A.M., G.B., V.P., J.D.H., Y.P. and A.D.; Visualization, R.G., A.M. and A.D.; Supervision, A.D.; Project Administration, A.D.; Funding Acquisition, A.D. All authors have read and agreed to the published version of the manuscript.

**Funding:** This research was funded by Natural Sciences and Engineering Research Council of Canada (NSERC), grant number RDCPJ-523130-17 and the Consortium de recherche et innovations en bioprocédés industriels au Québec (CRIBIQ), grant number 2017-040-C39. Les Producteurs d'oeufs du Québec are thanked for their financial support.

**Institutional Review Board Statement:** Not applicable.

**Informed Consent Statement:** Not applicable.

**Data Availability Statement:** Not applicable.

**Acknowledgments:** The authors thank Diane Gagnon (Department of Food Science, Laval University) for technical support for the physico-chemical analyses and Sylvie Bourassa and Victor Fourcassié (Proteomics Platform of the Research Center of the Centre Hospitalier Universitaire of Quebec) for proteomics analysis. The authors also thank Richard Janvier (Institute for Integrative Systems Biology (IBIS), Laval University) for technical assistance with TEM image analysis.

**Conflicts of Interest:** The authors declare no conflict of interest.

## Abbreviations

| | |
|---|---|
| AOAC | Association of Official Agricultural Chemists |
| ES MS/MS | Electrospray mass spectrometry |
| G1c | Granule fraction generated after the centrifugation of egg yolk |
| G1p | Granule fraction generated after pressurization of G1c |
| G2c | Granule fraction generated after the centrifugation of G1c |
| G2p | Granule fraction generated after the centrifugation of G1p |
| HDL | high-density lipoproteins |
| ICP-OES | Inductively coupled plasma optical emission spectroscopy |
| LDL | low-density lipoproteins |
| NanoLC | Nanoscale capillary liquid chromatography |
| PAGE | Polyacrylamide gel electrophoresis |
| P2c | Plasma fraction generated after the centrifugation of G1c |
| P2p | Plasma fraction generated after the centrifugation of G1p |
| RP | Reversed-phase |
| SDS | Sodium dodecyl sulphate |
| TEM | Transmission electron microscopy |
| TSC | Total spectrum count |
| UHPH | Ultra-high pressure homogenization |
| VTG | Vitellogenin |

## References

1. Anton, M. Egg yolk: Structures, functionalities and processes. *J. Sci. Food Agric.* **2013**, *93*, 2871–2880. [CrossRef] [PubMed]
2. Powrie, W.D.; Nakai, S. The chemistry of eggs and egg products. In *Egg Science and Technology*, 4th ed.; CRC Press: Boca Raton, FL, USA, 1986; pp. 97–139.
3. Anton. Composition and structure of hen egg yolk. In *Bioactive Egg Compounds*; Springer: Berlin/Heidelberg, Germany, 2007; pp. 1–6.
4. Laca, A.; Sáenz, M.; Paredes, B.; Diaz, M. Rheological properties, stability and sensory evaluation of low-cholesterol mayonnaises prepared using egg yolk granules as emulsifying agent. *J. Food Eng.* **2010**, *97*, 243–252. [CrossRef]
5. Naderi, N.; House, J.D.; Pouliot, Y. Effect of selected pre-treatments on folate recovery of granule suspensions prepared from hen egg yolk. *LWT Food Sci. Technol.* **2016**, *68*, 341–348. [CrossRef]
6. Naderi, N.; House, J.D.; Pouliot, Y. Scaling-up a process for the preparation of folate-enriched protein extracts from hen egg yolks. *J. Food Eng.* **2014**, *141*, 85–92. [CrossRef]
7. Anton, M.; Gandemer, G. Composition, solubility and emulsifying properties of granules and plasma of egg yolk. *J. Food Sci.* **1997**, *62*, 484–487. [CrossRef]
8. Causeret, D.; Matringe, E.; Lorient, D. Ionic strength and pH effects on composition and microstructure of yolk granules. *J. Food Sci.* **1991**, *56*, 1532–1536. [CrossRef]
9. Anton, M.; Le Denmat, M.; Gandemer, G. Thermostability of hen egg yolk granules: Contribution of native structure of granules. *J. Food Sci.* **2000**, *65*, 581–584. [CrossRef]
10. Sirvente, H.; Beaumal, V.; Gaillard, C.; Bialek, L.; Hamm, D.; Anton, M. Structuring and functionalization of dispersions containing egg yolk, plasma and granules induced by mechanical treatments. *J. Agric. Food Chem.* **2007**, *55*, 9537–9544. [CrossRef]
11. Naderi, N.; Doyen, A.; House, J.D.; Pouliot, Y. The use of high hydrostatic pressure to generate folate-enriched extracts from the granule fraction of hen's egg yolk. *Food Chem.* **2017**, *232*, 253–262. [CrossRef]
12. Naderi, N.; Pouliot, Y.; House, J.D.; Doyen, A. High hydrostatic pressure effect in extraction of 5-methyltetrahydrofolate (5-MTHF) from egg yolk and granule fractions. *Innov. Food Sci. Emerg. Technol.* **2017**, *43*, 191–200. [CrossRef]

13. Duffuler, P.; Giarratano, M.; Naderi, N.; Suwal, S.; Marciniak, A.; Perreault, V.; Offret, C.; Brisson, G.; House, J.D.; Pouliot, Y.; et al. High hydrostatic pressure induced extraction and selective transfer of β-phosvitin from the egg yolk granule to plasma fractions. *Food Chem.* **2020**, *321*, 126696. [CrossRef] [PubMed]
14. Giarratano, M.; Duffuler, P.; Chamberland, J.; Brisson, G.; House, J.D.; Pouliot, Y.; Doyen, A. Combination of high hydrostatic pressure and ultrafiltration to generate a new emulsifying ingredient from egg yolk. *Molecules* **2020**, *25*, 1184. [CrossRef] [PubMed]
15. Motta-Romero, H.; Zhang, Z.; Nguyen, A.T.; Schlegel, V.; Zhang, Y. Isolation of egg yolk granules as low-cholesterol emulsifying agent in mayonnaise. *J. Food Sci.* **2017**, *82*, 1588–1593. [CrossRef] [PubMed]
16. Dumay, E.; Chevalier-Lucia, D.; Picart-Palmade, L.; Benzaria, A.; Gràcia-Julià, A.; Blayo, C. Technological aspects and potential applications of (ultra) high-pressure homogenisation. *Trends Food Sci. Technol.* **2013**, *31*, 13–26. [CrossRef]
17. Serra, M.; Trujillo, A.; Guamis, B.; Ferragut, V. Proteolysis of yogurts made from ultra-high-pressure homogenized milk during cold storage. *J. Dairy Sci.* **2009**, *92*, 71–78. [CrossRef]
18. Briviba, K.; Gräf, V.; Walz, E.; Guamis, B.; Butz, P. Ultra high pressure homogenization of almond milk: Physico-chemical and physiological effects. *Food Chem.* **2016**, *192*, 82–89. [CrossRef]
19. Floury, J.; Desrumaux, A.; Legrand, J. Effect of Ultra-high-pressure homogenization on structure and on rheological properties of soy protein-stabilized emulsions. *J. Food Sci.* **2002**, *67*, 3388–3395. [CrossRef]
20. Paquin, P.; Lacasse, J.; Subirade, M.; Turgeon, S. Continuous process of dynamic high-pressure homogenization for the denaturation of proteins. U.S. Patent 6511695B1, 28 January 2003.
21. Keerati-U-Rai, M.; Corredig, M. Effect of dynamic high pressure homogenization on the aggregation state of soy protein. *J. Agric. Food Chem.* **2009**, *57*, 3556–3562. [CrossRef]
22. Paquin, P. Technological properties of high pressure homogenizers: The effect of fat globules, milk proteins, and polysaccharides. *Int. Dairy J.* **1999**, *9*, 329–335. [CrossRef]
23. Floury, J.; Desrumaux, A.; Lardières, J. Effect of high-pressure homogenization on droplet size distributions and rheological properties of model oil-in-water emulsions. *Innov. Food Sci. Emerg. Technol.* **2000**, *1*, 127–134. [CrossRef]
24. Marco-Molés, R.; Hernando, I.; Llorca, E.; Pérez-Munuera, I. Influence of high pressure homogenization (HPH) on the structural stability of an egg/dairy emulsion. *J. Food Eng.* **2012**, *109*, 652–658. [CrossRef]
25. Sørensen, H.; Mortensen, K.; Sørland, G.H.; Larsen, F.H.; Paulsson, M.; Ipsen, R. Dynamic ultra-high pressure homogenisation of whey protein-depleted milk concentrate. *Int. Dairy J.* **2015**, *46*, 12–21. [CrossRef]
26. Speroni, F.; Puppo, M.C.; Chapleau, N.; de Lamballerie, M.; Castellani, O.; Añón, A.M.C.; Anton, M. High-pressure induced physicochemical and functional modifications of low-density lipoproteins from hen egg yolk. *J. Agric. Food Chem.* **2005**, *53*, 5719–5725. [CrossRef] [PubMed]
27. Bouaouina, H.; Desrumaux, A.; Loisel, C.; Legrand, J. Functional properties of whey proteins as affected by dynamic high-pressure treatment. *Int. Dairy J.* **2006**, *16*, 275–284. [CrossRef]
28. Chevalier-Lucia, D.; Blayo, C.; Gràcia-Julià, A.; Picart-Palmade, L.; Dumay, E. Processing of phosphocasein dispersions by dynamic high pressure: Effects on the dispersion physico-chemical characteristics and the binding of α-tocopherol acetate to casein micelles. *Innov. Food Sci. Emerg. Technol.* **2011**, *12*, 416–425. [CrossRef]
29. Grácia-Juliá, A.; René, M.; Cortés-Muñoz, M.; Picart, L.; López-Pedemonte, T.; Chevalier, D.; Dumay, E. Effect of dynamic high pressure on whey protein aggregation: A comparison with the effect of continuous short-time thermal treatments. *Food Hydrocoll.* **2008**, *22*, 1014–1032. [CrossRef]
30. Panozzo, A.; Manzocco, L.; Calligaris, S.; Bartolomeoli, I.; Maifreni, M.; Lippe, G.; Nicoli, M.C. Effect of high pressure homogenisation on microbial inactivation, protein structure and functionality of egg white. *Food Res. Int.* **2014**, *62*, 718–725. [CrossRef]
31. Roach, A.; Harte, F. Disruption and sedimentation of casein micelles and casein micelle isolates under high-pressure homogenization. *Innov. Food Sci. Emerg. Technol.* **2008**, *9*, 1–8. [CrossRef]
32. Wang, C.; Ma, Y.; Liu, B.; Kang, Z.; Geng, S.; Wang, J.; Wei, L.; Ma, H. Effects of dynamic ultra-high pressure homogenization on the structure and functional properties of casein. *Int. J. Agric. Biol. Eng.* **2019**, *12*, 229–234. [CrossRef]
33. Hayes, M.G.; Kelly, A.L. High pressure homogenisation of milk (b) effects on indigenous enzymatic activity. *J. Dairy Res.* **2003**, *70*, 307–313. [CrossRef]
34. Patrignani, F.; Lanciotti, R.J.F.I.M. Applications of high and ultra high pressure homogenization for food safety. *Front. Microbiol.* **2016**, *7*, 1132. [CrossRef] [PubMed]
35. Guilmineau, F.; Krause, A.I.; Kulozik, U. Efficient Analysis of egg yolk proteins and their thermal sensitivity using sodium dodecyl sulfate polyacrylamide gel electrophoresis under reducing and nonreducing conditions. *J. Agric. Food Chem.* **2005**, *53*, 9329–9336. [CrossRef] [PubMed]
36. Lundgren, D.H.; Hwang, S.; Wu, L.; Han, D.K. Role of spectral counting in quantitative proteomics. *Expert Rev. Proteom.* **2010**, *7*, 39–53. [CrossRef] [PubMed]
37. Arenella, M.; Giagnoni, L.; Masciandaro, G.; Ceccanti, B.; Nannipieri, P.; Renella, G. Interactions between proteins and humic substances affect protein identification by mass spectrometry. *Biol. Fertil. Soils* **2014**, *50*, 447–454. [CrossRef]
38. Cortés-Muñoz, M.; Chevalier-Lucia, D.; Dumay, E. Characteristics of submicron emulsions prepared by ultra-high pressure homogenisation: Effect of chilled or frozen storage. *Food Hydrocoll.* **2009**, *23*, 640–654. [CrossRef]

39. Hayes, M.G.; Fox, P.F.; Kelly, A.L. Potential applications of high pressure homogenisation in processing of liquid milk. *J. Dairy Res.* **2005**, *72*, 25–33. [CrossRef]
40. Zamora, A.; Guamis, B.; Trujillo, A. Protein composition of caprine milk fat globule membrane. *Small Rumin. Res.* **2009**, *82*, 122–129. [CrossRef]
41. Chang, C.M.; Powrie, W.D.; Fennema, O. Microstructure of egg yolk. *J. Food Sci.* **1977**, *42*, 1193–1200. [CrossRef]
42. Anton, M.; Gandemer, G. Effect of pH on interface composition and on quality of oil-in-water emulsions made with hen egg yolk. *Colloids Surf. B Biointerfaces* **1999**, *12*, 351–358. [CrossRef]
43. Kiełczewska, K.; Ambroziak, K.; Krzykowska, D.; Aljewicz, M. The effect of high-pressure homogenisation on the size of milk fat globules and MFGM composition in sweet buttermilk and milk. *Int. Dairy J.* **2021**, *113*, 104898. [CrossRef]
44. Song, X.; Zhou, C.; Fu, F.; Chen, Z.; Wu, Q. Effect of high-pressure homogenization on particle size and film properties of soy protein isolate. *Ind. Crop. Prod.* **2013**, *43*, 538–544. [CrossRef]
45. Kuhn, K.R.; Cunha, R. Flaxseed oil—Whey protein isolate emulsions: Effect of high pressure homogenization. *J. Food Eng.* **2012**, *111*, 449–457. [CrossRef]
46. Fernández-Ávila, C.; Escriu, R.; Trujillo, A. Ultra-high pressure homogenization enhances physicochemical properties of soy protein isolate-stabilized emulsions. *Food Res. Int.* **2015**, *75*, 357–366. [CrossRef] [PubMed]
47. Suhag, R.; Dhiman, A.; Thakur, D.; Kumar, A.; Upadhyay, A. Physico-chemical and functional properties of microfluidized egg yolk. *J. Food Eng.* **2021**, *294*, 110416. [CrossRef]
48. Staneva, G.; Momchilova, A.; Wolf, C.; Quinn, P.J.; Koumanov, K. Membrane microdomains: Role of ceramides in the maintenance of their structure and functions. *Biochim. Biophys. Acta Biomembr.* **2009**, *1788*, 666–675. [CrossRef] [PubMed]
49. Aluko, R.E.; Mine, Y. Characterization of oil-in-water emulsions stabilized by hen's egg yolk granule. *Food Hydrocoll.* **1998**, *12*, 203–210. [CrossRef]
50. Anton, M.; Beaumal, V.; Gandemer, G. Adsorption at the oil–water interface and emulsifying properties of native granules from egg yolk: Effect of aggregated state. *Food Hydrocoll.* **2000**, *14*, 327–335. [CrossRef]
51. Harte, F. Food processing by high-pressure homogenization. In *High Pressure Processing of Food*; Springer: New York, NY, USA, 2016; pp. 123–141.
52. Le Denmat, M.; Anton, M.; Gandemer, G. Protein denaturation and emulsifying properties of plasma and granules of egg yolk as related to heat treatment. *J. Food Sci.* **1999**, *64*, 194–197. [CrossRef]
53. Zambrowicz, A.; Dąbrowska, A.; Bobak, Ł.; Szołtysik, M. Egg yolk proteins and peptides with biological activity. *Adv. Hyg.* **2014**, *68*, 1524–1529. [CrossRef]

 *foods*

*Article*

# Evaluating the Anti-Inflammatory and Antioxidant Effects of Broccoli Treated with High Hydrostatic Pressure in Cell Models

**Yi-Yuan Ke, Yuan-Tay Shyu and Sz-Jie Wu ***

Department of Horticulture and Landscape Architecture, National Taiwan University, No. 1, Section 4, Roosevelt Road, Taipei 10617, Taiwan; r07628208@ntu.edu.tw (Y.-Y.K.); tedshyu@ntu.edu.tw (Y.-T.S.)
* Correspondence: chiehwu@ntu.edu.tw; Tel.:+886-233-664-850; Fax: +886-223-696-278

**Abstract:** Isothiocyanates (ITCs) are important functional components of cruciferous vegetables. The principal isothiocyanate molecule in broccoli is sulforaphane (SFN), followed by erucin (ERN). They are sensitive to changes in temperature, especially high temperature environments where they are prone to degradation. The present study investigates the effects of high hydrostatic pressure on isothiocyanate content, myrosinase activity, and other functional components of broccoli, and evaluates its anti-inflammatory and antioxidant effects. Broccoli samples were treated with different pressures and for varying treatment times; 15 min at 400 MPa generated the highest amounts of isothiocyanates. The content of flavonoids and vitamin C were not affected by the high-pressure processing strategy, whereas total phenolic content (TPC) exhibited an increasing tendency with increasing pressure, indicating that high-pressure processing effectively prevents the loss of the heat-sensitive components and enhances the nutritional content. The activity of myrosinase (MYR) increased after high-pressure processing, indicating that the increase in isothiocyanate content is related to the stimulation of myrosinase activity by high-pressure processing. In other key enzymes, the ascorbate peroxidase (APX) activity was unaffected by high pressure, whereas peroxidase (POD) and polyphenol oxidase (PPO) activity exhibited a 1.54-fold increase after high-pressure processing, indicating that high pressures can effectively destroy oxidases and maintain food quality. With regards to efficacy evaluation, NO production was inhibited and the expression levels of inducible nitric oxide synthase (iNOS) and Cyclooxygenase-2 (COX-2) were decreased in broccoli treated with high pressures, whereas the cell viability remained unaffected. The efficacy was more significant when the concentration of SFN was 60 $mg \cdot mL^{-1}$. In addition, at 10 $mg \cdot mL^{-1}$ SFN, the reduced/oxidized glutathione (GSH/GSSG) ratio in inflammatory macrophages increased from 5.99 to 9.41. In conclusion, high-pressure processing can increase the isothiocyanate content in broccoli, and has anti-inflammatory and anti-oxidant effects in cell-based evaluation strategies, providing a potential treatment strategy for raw materials or additives used in healthy foods.

**Keywords:** high hydrostatic pressure; broccoli; Isothiocyanates; sulforaphane; erucin; anti-inflammatory; antioxidant

**Citation:** Ke, Y.-Y.; Shyu, Y.-T.; Wu, S.-J. Evaluating the Anti-Inflammatory and Antioxidant Effects of Broccoli Treated with High Hydrostatic Pressure in Cell Models. *Foods* **2021**, *10*, 167. https://doi.org/10.3390/foods10010167

Received: 18 December 2020
Accepted: 13 January 2021
Published: 15 January 2021



## 1. Introduction

*Brassica oleracea* var. *italica* (broccoli) is a vegetable widely consumed worldwide. It is derived from genetic mutations and the evolution of wild cabbage, and is a cultivar of *Brassica oleracea*, which belongs to the family Brassicaceae together with cabbage, gai lan, and cauliflower. Broccoli is rich in a variety of nutrients, including vitamin A, vitamin C, dietary fibers, and isothiocyanates. Among these, isothiocyanates are formed primarily through hydrolysis of glucosinolates by the enzyme myrosinase, and are the most representative functional component in cruciferous vegetables. They can inhibit the proliferation, development, and metastasis of cancer cells, regulate the production of inflammation-related factors, and enhance the expression of antioxidant-related proteins [1–4]. The most abundant isothiocyanate molecule found in broccoli is sulforaphane, followed by erucic acid [5].

Myrosinases and glucosinolates not only comprise the isothiocyanate production system in plants but also act as a chemical defense mechanism. Under normal circumstances, the myrosinase enzyme is in an inactive state and the glucosinolates are stored in plants in a precursor form since these two are spatially separated. When attacked by herbivores, insects, or microorganisms, the cells are destroyed, releasing these enzymes and glucosinolate molecules, which then interact to produce the biologically active isothiocyanates, nitriles, and thiocyanates, such as, sulforaphanes [5,6]. The degree of myrosinase hydrolysis and the types of products formed are strongly affected by environmental changes, including the substrates and cofactors of myrosinase, the presence of specific proteins, pH, stress, carbon dioxide concentration, and temperature. When the pH of the hydrolysis environment is slightly acidic or neutral, the principal product produced are isothiocyanates, whereas under acidic conditions (pH $< 3$) or in the presence of iron or epithiospecifier protein (ESP), nitriles are produced instead, which, unlike the isothiocyanates, have no physiological effects [5,7,8].

The inflammatory response is the natural defense mechanism activated by the body when subjected to noxious stimuli. A proper inflammatory response not only protects the human body from injuries to tissues and microbial invasion, but also increases the ability of tissues and cells to restore stability and enhance the immune system [9,10]. During the inflammatory response processes, inflammation-related factors are activated to promote the production of inflammatory mediators (e.g., TNF-$\alpha$ and IL-6). Some inflammatory mediators may drive blood vessels to be remodeled during inflammation [11]. The oxidative stress leads to large amounts of reactive oxygen species (ROS) being produced [12]. In addition, cells and tissues absorb more oxygen due to swelling, leading to gradual accumulation of ROS. Inflammation and oxidative stress increase the likelihood of injury or pathology to tissues and cells, which can lead to immune-related diseases such as cancers and multiple sclerosis (neurodegenerative diseases) [13].

High-pressure processing (HPP), which is also known as high hydrostatic pressure (HHP) processing or ultra-high pressure (UHP) processing, is a non-thermal processing technology that has undergone rapid development in recent years. Its mechanism of action is primarily based on the use of liquids as a pressure transmission medium. Food processing under high-pressure environments at appropriate temperatures and times can inactivate microorganisms and enzymes, thereby achieving sterilization, prolonging shelf life, and reducing the use of chemical preservatives. When compared to traditional thermal processing strategies, this technique can do a much better job in preserving nutrients, flavor, appearance, and texture without any heat treatment [14–16]. Since isothiocyanates and their production systems are easily damaged in high-temperature environments, many studies have investigated this problem by using high-pressure techniques to treat cruciferous crops, and have noticed that not only was the isothiocyanate content retained, but there was also a tendency of improvement in the content [8,17–19]. Current research in this topic is focused on investigating cruciferous crops, such as samples of cabbage or broccoli sprouts, and the efficacy evaluation is based primarily on purified isothiocyanates. In this study, the No. 42 broccoli grown commercially in Taiwan was selected as the sample for processing under different high-pressure conditions to investigate the effects of high-pressure processing techniques on the changes in isothiocyanate content, myrosinase activity, and other functional components in broccoli, and also to develop the optimal processing conditions, to analyze the mechanisms behind these changes, and to evaluate efficacy through cell-based experiments.

## 2. Materials and Methods

### 2.1. Materials/Processing

Fresh broccoli was purchased from Erlun Produce Cooperative, Yunlin County, Taiwan. The broccoli was cleaned and was cut into 2-cm pieces under the florets, and equal amounts of broccoli was divided into several clean airtight bags and vacuum-sealed. The bags were then randomly divided into two groups: control group and high-pressure group. The high-pressure group was further subdivided into 9 batches based on the applied pressure and the incubation time, namely 200 MPa (3, 10, 15 min), 400 MPa (3, 10, 15 min), and 600 MPa (3, 10, 15 min). The high-pressure groups were placed into an HPP600 MPa/6.2 L (Bao Tou KeFa High Pressure Technology Co. Ltd., Inner Mongolia, Baotou, China) high-pressure apparatus for high-pressure treatment, and then allowed to stand at room temperature for 1 h to allow the myrosinase enzyme and glucosinolates to react with each other [18]. The broccoli samples were blanched in boiling water for 1 min to inactivate the polyphenol oxidase (PPO) and Peroxidase (POD) activities to reduce the browning reaction, to retain the highest commercial values. After cooling, the broccoli was freeze-dried and ground using a food processor. This was followed by filtration through a mesh and the powder was stored at −20 °C for use in subsequent analyses.

### *2.2. Extraction and High Performance Liquid Chromatography (HPLC) Analysis of Isothiocyanates*

#### 2.2.1. Preparation of the Extract

The isothiocyanate extraction process was performed as described previously by Hwang and Lim (2014) [12]. A total of 0.5 g of broccoli powder was added to a 6 mL solution of 80% methanol (Macron Fine Chemicals, Center Valley, PA, USA) and the extraction process was performed with constant shaking at 1260 rpm for 1 h, followed by centrifugation at 10,000× *g* and 4 °C for 20 min. The supernatant was collected and the precipitate was resuspended with the same volume of 80% methanol, followed by extraction and centrifugation under the same conditions. The two supernatants were combined as a crude isothiocyanate extract and stored at −20 °C.

#### 2.2.2. HPLC

The analysis was performed as described by You et al. (2008) [20], with modifications. The analytical instrument used for this process was a Waters$^{TM}$ 600 series Controller pump with a Waters 717 Plus Autosampler and SPD-20AV UV-VIS detector (Shimazu Co., Kyoto, Japan). A C18 reversed-phase chromatography column (Zorbax Eclipse XDB C-18, 4.6 × 150 mm) was used for the separation. The detection wavelength was set to 241 nm, and the sample injection volume was set to 20 μL. Water and methanol were used as the HPLC mobile phases. Initial conditions consisted of 10% methanol, followed by a linear increase to 90% methanol at 40 min, reduced to 10% at the end of the 50 min analysis time, and the column was then equilibrated at 10% methanol for 10 min. The mobile phase flow rate was kept at 1 mL·min$^{-1}$.

### *2.3. Chemical Characterization*

#### 2.3.1. Polyphenol Determination

The Folin–Ciocalteu reagent (Sigma Chemical Inc., St. Louis, MO, USA) was used to determine the content of the phenolic compounds in broccoli. A total of 0.1 g of broccoli powder was taken and suspended in 3 mL of an 80% methanol solution and the extraction process was performed for 1 h, followed by centrifugation of the homogenate at 10,000× *g* and 4 °C for 20 min. The supernatant was collected, and the precipitate underwent an extraction process for a second time in the same manner as described above. The two supernatants were combined and used as the crude extract. A total of 60 μL of extract and 60 μL of Folin–Ciocalteu reagent were mixed with 480 μL of water and was placed in the dark and allowed to react for 90 min. The change in absorbance was measured at a wavelength of 760 nm. The concentration of the TPC is expressed in GAE mg·g$^{-1}$. Gallic acid (Sigma Chemical Inc., St. Louis, MO, USA) was used as a standard for plotting the standard curve [21].

#### 2.3.2. Flavonoid Determination

The sample extraction was performed, as described above, for the flavonoid content determination. The supernatants of the two extractions were combined and used as the crude extract. A total of 150 μL of the extract was dissolved in 450 μL 95% ethanol, a

30 μL 10% aluminum chloride solution ($AlCl_3$) (Thermo Fisher Scientific Inc., Waltham, MA, USA), 30 μL potassium acetate (Merck KGaA, Darmstadt, DA, Germany), and 840 μL water, and allowed to react at 25 °C for 30 min. Then, the change in absorbance was measured at a wavelength of 415 nm. The standard curve was plotted using quercetin as a standard, and the concentration is expressed as QuE mg·g$^{-1}$ [22].

2.3.3. Vitamin C Determination

A total of 0.05 g broccoli powder was dissolved in 0.95 μL dd$H_2O$ (double-distilled $H_2O$) to yield a 20-fold diluted solution. A reflectometer (Merck KGaA, Darmstadt, Germany) along with its Ascorbinsaure-test (Reflectoquant®116981) was used to determine the content of vitamin C in the solution. The concentration is expressed as mg·g$^{-1}$.

### 2.4. *Enzyme Activity Assays*

2.4.1. Myrosinase Activity

The experimental method followed is performed as described by Yuan et al. (2010) [23], Li et al. (2008) [24], and Zhao et al. (2008) [25], with slight modifications. A total of 0.05 g of broccoli powder was dissolved in 0.45 g of water, and then 1.8 mL of MES buffer (50 mM, pH 6.0) was added and the mixture was extracted with shaking for 5 min and centrifuged at 10,000× *g* and 4 °C for 10 min, and the supernatant was collected. A total of 100 μL of 1 mM glucoraphanin (USBiological Inc., Salem, MA, USA) was mixed with 20 μL of supernatant and allowed to react at 40–60 °C for 15 min. Then, 240 μL of DNS reagent (Sigma Aldrich Inc., St. Louis, MO, USA) was added and the reaction was carried out at 100 °C for 5 min, and then immediately cooled by placing it in an ice bath. After cooling to room temperature, 720 μL of water was added and mixed well, and a spectrophotometer was used to measure absorbance change at a wavelength of 540 nm. A standard curve was made using glucose solution as standard and the concentration was expressed as μmol glucose (Sigma Aldrich Inc., St. Louis, MO, USA) produced per min (μmol·g$^{-1}$·min$^{-1}$).

2.4.2. Ascorbate Peroxidase (APX) Activity

The APX activity measurements were performed as described by Chen and Liu (2012) [26], with slight modifications. A total of 10 mL extraction solvent (containing 100 mM $KH_2PO_4$ (J.T. Baker Chemical Inc., Oklahoma, PA, USA), pH 7.8; 1% Triton X-100 (Sigma Chemical Inc., St. Louis, MO, USA); 1 mM EDTA-$Na_2$ (Sigma Chemical Inc., St. Louis, MO, USA) was added to 0.5 g broccoli powder and the mixture was centrifuged at 10,000× *g* and 4 °C for 20 min. Then, the supernatant was collected, with the enzyme extract at 0.047–0.077 mg/g protein, and the protein content was determined by the Bradford method, with the standard curves prepared using BSA. A 0.5 mL $KH_2PO_4$ (250 mM, pH 7), 0.05 mL EDTA-$Na_2$ (0.5 mM), 0.2 mL $H_2O_2$ (10 mM), 0.2 mL ascorbic acid (Honeywell Riedel-de Haen, Seelze, Germany), and 0.05 mL enzyme extract was sequentially added to the quartz tube, and the changes in absorbance were measured immediately after mixing at a wavelength of 290 nm within 5 min. The APX activity was calculated using the extinction coefficient of $H_2O_2$ (2.8 mM$^{-1}$ cm$^{-1}$) and the enzyme activity was expressed in units of mmol ascorbate min$^{-1}$ mg$^{-1}$ protein.

2.4.3. Peroxidase (POD) and Polyphenol Oxidase (PPO) Activities

The enzymatic activities of these two enzymes were determined as described by Yang (2016) [27], with slight modifications. A total of 0.5 g of broccoli powder was added to 10 mL of a 0.2 M sodium phosphate buffer solution (pH 6.5) containing 1% PVPP (Sigma Chemical Inc., St. Louis, MO, USA), mixed well; the extraction was conducted by continuous shaking for 5 min, followed by centrifugation at 4 °C and 10,000× *g* for 20 min. The supernatant was collected and used as the enzyme extract. Peroxidase activity was determined by mixing 25 μL of the extract with 2.7 mL of a sodium phosphate buffer (pH 6.5), 200 μL of 1% p-phenylenediamine (Sigma Chemical Inc., St. Louis, MO, USA), and 100 μL of 1.5% hydrogen peroxide (Honeywell Riedel-de Haen, Seelze, Germany), measuring the absorbance at a wavelength of 485 nm every min for 10 min. Polyphenol oxidase activity was determined by mixing 100 μL of the enzyme extract and 3 mL of 0.15 M catechin (Sigma Chemical Inc., St. Louis, MO, USA), measuring the change in absorbance at a wavelength of 420 nm every min for 10 min. The activity of the two enzymes is expressed in the units of absorbance change per min ($\Delta A \cdot min^{-1}$).

### 2.5. Cell Culture

RAW264.7 mouse macrophages were purchased from the Bioresources Collection and Research Center (BCRC) of the Food Industry Research and Development Institute (Hsinchu, Taiwan). The cells were cultured in Dulbecco's Modified Eagle Medium (DMEM) (Thermo Fisher Scientific Inc., Waltham, MA, USA), containing 10% fetal bovine serum (FBS) (Thermo Fisher Scientific Inc., Waltham, MA, USA) and $NaHCO_3$ (Merck KGaA, Darmstadt, Germany), and placed in a 37 °C incubator containing 5% carbon dioxide ($CO_2$) for growth.

### 2.6. Cell Viability Assay

The 3-(4,5-dimethylthiazol-2-yl)-2,5-diphenyltetrazolium bromide (MTS) assay is a method for evaluating the toxicity of target substances to cells. Cells were seeded in a 96-well plate at $1 \times 10^4$ cells per well ($cells \cdot mL^{-1}$), allowed to grow for a day, and then were treated with LPS (2 $\mu g \cdot mL^{-1}$) and the broccoli extract and allowed to react for a further 24 h. The medium was then changed and CellTiter 96® AQueous (Promega Co., Madison, WI, USA) One Solution was added, followed by the incubation of the cells for 1 h. Then, the change in absorbance was measured at a wavelength of 490 nm.

### 2.7. Measurement of Nitric Oxide (NO) Production

NO production was indirectly assessed by measuring the nitrite levels in the cultured media and serum determined by a colorimetric method based on the Griess reagent (BioVision, Milpitas, CA, USA). Briefly, the cells were seeded in a 24-well plate at $1 \times 10^5$ cells per well ($cells \cdot mL^{-1}$). After culturing for 24 h, the positive control group and the treatment group were first activated with LPS and treated with DMEM (Thermo Fisher Scientific Inc., Waltham, MA, USA) or the broccoli extract. The cells that were not activated with LPS for 24 h served as the control group. A total of 100 μL of cell culture medium was added to a 96-well plate and a similar volume of Griess reagent was added. The mixture was allowed to react at room temperature for 10 min, and the absorbance was measured at a wavelength of 550 nm. $NaNO_2$ (Sigma Chemical Inc., St. Louis, Louis, MO, USA) was used as a standard to plot a standard curve, and the inhibitory effect was expressed as a percentage.

### 2.8. Measurement of $PGE_2$ Content

Cells were seeded in a 24-well plate at $1 \times 10^5$ ($cells \cdot mL^{-1}$) cells per well, and allowed to grow for 24 h. The medium was changed and LPS and the broccoli extract (5, 10, 20, 40, and 60 ppm) were added and allowed to react for a further 24 h. The Prostaglandin $E_2$ ELISA Kit (Cayman Co., Ann Arbor, MI, USA) commercial kit was used to determine the $PGE_2$ content of the RAW264.7 macrophage supernatants, as per the manufacturer's instructions.

### 2.9. Expression Levels of iNOS and COX-2

#### 2.9.1. Total Cellular RNA Extraction

RAW264.7 macrophages were seeded onto a 24-well plate and cultured for 24 h. The medium was removed, and the samples were treated with different concentrations (5, 10, 20, 40, and 60 ppm) of broccoli extract for one day, depending on the experimental treatment, and washed twice with 1 × PBS. An appropriate amount of Trypsin-EDTA was added to detach the cells, and the cells were collected in a microcentrifuge tube and centrifuged at 500× *g* for 5 min. The supernatant was removed, and 1-thioglycerol/homogenization solution and lysis buffer were sequentially added to the cell suspension as per the instructions of the Maxwell® RSC simplyRNA Cells Kit (Promega Co., Madison, WI, USA). Total RNA was extracted using the Maxwell® RSC Instrument.

2.9.2. RT-PCR Analysis

Extracted RNA was reverse transcribed into cDNA using GoScriptTM Reverse Transcription Mix (Promega Co.,, Madison, WI, USA) and oligo (dT) primers, and then PCR was performed using GoTaq® Green Master Mix (Promega Co., Madison, WI, USA). The resulting product was subjected to gel electrophoresis in 1% agarose (Amresco Inc., Solon Ind. Pkwy., Solon, OH, USA) and 0.003% HealthyView nucleic acid stain (Genomics, Taipei, Taiwan) for analyzing the size of the specific fragments, and GeneTools 4.3.7 software was used to quantify and compare the fragments in the gel. The following primers were used: iNOS forward, 5′-AAT GGC AAC ATC AGG TCG GCC ATC ACT-3′, reverse, 5′-GCT GTG TGT CAC AGA AGT CTC GAA CTC-3′; COX-2 forward, 5′-GGA GAG ACT ATC AAG ATA GT-3′, reverse, 5′-ATG GTC AGT AGA CTT TTA CA-3′; β-actin forward, 5′-TCA TGA AGT GTG ACG TTG ACA TCC GT-3′, reverse, 5′-CCT AGA AGC ATT TGC GGT GCA CGA TG-3′ (Mission biotech, Taipei, Taiwan).

2.9.3. GSH/GSSG Ratio

The GSH/GSSG-Glo™ assay (Promega Co., Madison, WI, USA) was used to measure the ratio. RAW264.7 cells were cultured in a 96-well plate containing a medium. After 3 h, the old medium was removed, and either DMEM (control group), 2 $\mu g \cdot mL^{-1}$ LPS (positive control group), or broccoli extract with different concentrations of SFN were added, and the cells were cultured for 20 h. Total glutathione reagent (50 $\mu L \cdot well^{-1}$) or oxidized glutathione reagent (50 $\mu L \cdot well^{-1}$) was added to each well and shaken for 5 min. In addition, the glutathione standard was diluted into 8 different concentrations by a 2-fold serial dilution method, and the total glutathione reagent (50 $\mu L \cdot well^{-1}$) was added. Luciferin generation reagent (50 $\mu L \cdot well^{-1}$) was added to each treatment group and the standard group and mixed well and incubated at room temperature for 30 min. Next, the luciferin detection reagent (100 $\mu L \cdot well^{-1}$) was added and allowed to stand for 15 min, and the luminescence was measured (integration time = 0.3 s). The GSH/GSSG ratio is calculated as follows: ratio GSH/GSSG treated = (μM total glutathione treated − (μM GSSG treated × 2))/μM GSSG treated.

2.9.4. Statistical Analyses

XLSTAT statistical software was used for analysis of variance (ANOVA). Differences in the means between the groups of data were analyzed using Tukey's test (Tukey's Honestly Significant Difference Test, Tukey's HSD). The significance level was kept at $p < 0.05$. Statistical results are expressed in lowercase English letters. Two sets of data marked with completely different letters indicate a statistically significant difference; duplicate or identical letters indicate a lack of a statistically significant difference between the data.

## 3. Results and Discussion

The calibration curves of sulforaphane (SFN) and erucin (ERN) were established by HPLC analysis. The R2 value of SFN and ERN was 0.9994 and 0.9998, respectively, and the elution time was 16 min and 35 min, respectively. Figure 1 displays the SFN and ERN metabolite content of broccoli treated under different conditions. In the untreated broccoli samples, the SFN content was measured to be 35.59 $mg \cdot 100\ g^{-1}$, and the ERN content was 10.30 ± 0.21 $mg \cdot 100\ g^{-1}$. After microwave and hot water treatment, the content is significantly reduced by at least 70%. However, in the high-pressure treatment group, the SFN content increased significantly when the pressure range was kept between 200 and 400 MPa and increased with increasing pressures and treatment times. The highest content was achieved in the group that underwent treatment at 400 MPa for 15 min, obtaining the highest SFN content of 154.79 ± 7.64 $mg \cdot 100\ g^{-1}$. When the pressure was increased to 600 MPa, the SFN content decreased. Similarly, the highest ERN content was achieved in the group with treatment at 400 MPa for 15 min, obtaining the highest ERN content of 109.86 ± 7.45 mg·100 $g^{-1}$. When the pressure was increased to 600 MPa, there was a significant decrease in ERN content as the treatment time was increased.

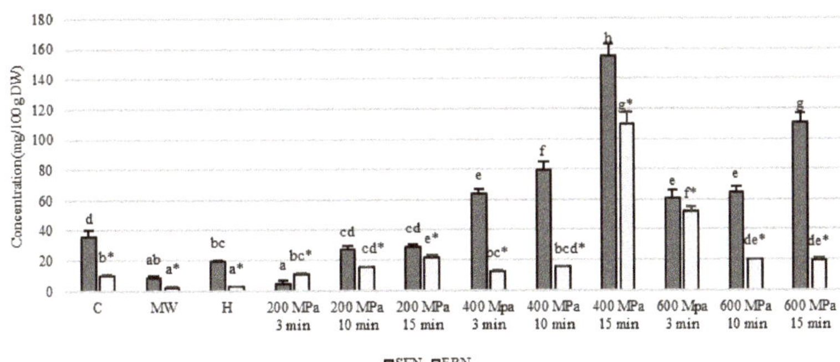


**Figure 1.** The sulforaphane (SFN) and erucin (ERN) contents of broccoli under different conditions. C: control (untreated); MW: microwave treatment (800 W, 3 min); H: hot water treatment (100 °C, 3 min). The same symbols, noted as superscripts after the letter, means that the ANOVA performed is in the same group. Bars carrying different letters are statistically different at $p < 0.05$.

Both SFN and ERN are isothiocyanates that are primarily produced by the glucosinolates–myrosinase system. According to the literature, hot water soaks into cruciferous crops during the cooking process, resulting in a 90% loss in glucosinolate and isothiocyanate content. Microwave treatment also destroys myrosinase activity, which prevents glucosinolates from being converted into isothiocyanate and lowers their content [28,29]. In the present study, broccoli was treated with a non-thermal, high hydrostatic pressure of 400 MPa for 15 min. The contents of both the isothiocyanates, SFN and ERN, in the broccoli were significantly increased, but their content decreased when the pressure reached 600 MPa. A study by Westphal et al. (2017) [5] indicated that an increase in pressure can promote the disintegration of plant cell structures, releasing large quantities of myrosinase, which then interacts with the glucosinolates to increase the isothiocyanate content. A study by Eylen et al. (2009) [17] showed that myrosinase enzyme begins to inactivate with increasing pressures. The rate of myrosinase inactivation increases when the pressure is increased over 500 MPa, which reduces the amount of isothiocyanate produced.

All data are presented as the mean ± SD ($n$ = 3). The same symbols, noted as a superscript after the letter, means that the ANOVA performed is in the same group. Bars carrying different letters are statistically different ($p < 0.05$).

Figure 2 indicates the changes in the functional components of broccoli after high-pressure treatment. TPC is the most significant component before and after high-pressure treatment. The initial TPC of broccoli was 4.73 ± 0.23 GAE mg·$g^{-1}$. After high-pressure treatment, the content increased to 7.28 ± 0.26 GAE mg·$g^{-1}$, an increase of about 1.5-fold. The content of flavonoids and vitamin C were not significantly altered after high-pressure treatment, indicating that high-pressure processing can result in better nutrient retention in foods. Vinicio et al. (2017) [30] reviewed numerous kinetic studies reporting the HPP effects of phytochemicals focused on microstructural changes and found the effects of HPP on the concentration of phenolic compounds are not clear and may either increase, decrease, or not be affected by HPP. This might be due to the large and complex phenolics family that exists in many forms in plants, some found as soluble conjugated glycosides, and some found as insoluble forms typically bound to structural components of the polysaccharides or proteins of the cell wall [31,32]. Different forms of phenolic compounds may react differently under high-pressure treatments. Disruption of the cellular structure causes the compartments to release their contents, and the disassociation of the phenolic compounds from the bound polysaccharides or proteins is generally hypothesized to play a major role.

However, the mechanisms by which phytochemicals are released from plant cells remain mostly unknown [30]. Liu et al. (2020) [33] reviewed the current state of knowledge on the internal factors that influence cell wall polysaccharides and polyphenol interactions. In that article, many advanced instrumental analysis methods (ITC, TSC, DLS, NTA, NIR, NMR, CLSM, etc.) were also introduced for the discovery of the exact interaction mechanism, through studies of their morphology, chemical composition, and molecular architecture.

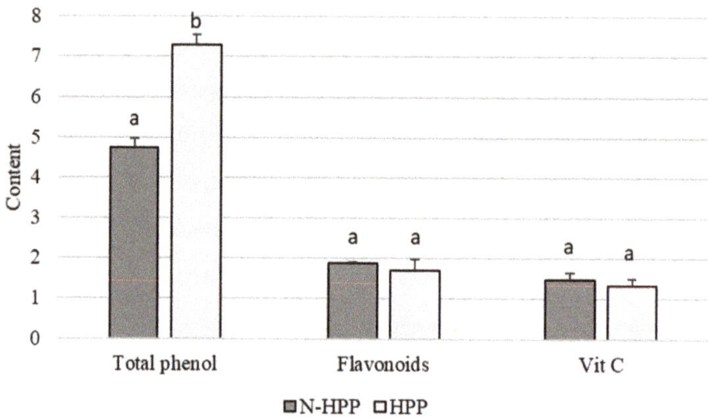


**Figure 2.** Content of the functional components in the untreated group (N-HPP) and high-pressure group. (HPP-400 MPa, 15 min). Total phenol: GAE mg·g$^{-1}$. Flavonoids: QuE mg·g$^{-1}$. Vit C: mg·g$^{-1}$. All data are presented as the mean ± SD ($n$ = 3). Bars carrying different letters on the same parameter are statistically different ($p < 0.05$).

Hence, the content of flavonoids and vitamin C were not significantly altered by the high-pressure treatment of broccoli, whereas TPC increased, indicating that a high hydrostatic pressure is effective in maintaining the heat-sensitive components. Comparisons based on the contents of vitamin C and isothiocyanates showed that these two functional components exhibited different tendencies, suggesting that the increase in isothiocyanate content could be unrelated to vitamin C levels. The results of the present study are similar to those of a study by Prasad et al. (2009) [34], in which high pressures were used to extract these components from longan peels. Since high-pressure treatments can disrupt the hydrophobic bonds in the cellular walls and cell membranes, thereby increasing the rates of substance transfer and facilitating the penetration of solvents into cells, this leads to an increase in the phenolic content. Rodríguez-Roque et al. (2015) [17] and Landl et al. (2010) [35] indicated that vitamin C is a substance that is sensitive to environmental changes, and its stability is easily affected by the presence of oxygen, heat, and heavy metals. High-pressure treatment not only involves no heat, but also inhibits the activity of oxidases, greatly reducing the loss of vitamin C. In addition, Tola and Ramaswamy (2015) [36] also indicated that high-pressure processing does not destroy the covalent, hydrophobic, or ionic bonds present in small molecule components, resulting in better nutrient retention in food.

With respect to the analysis of enzymatic activities, the activities of the MYR, APX, POD, and PPO enzymes were analyzed, and the results are indicated in Table 1. The activity of MYR after high-pressure treatment was significantly higher than in the untreated group, regardless of the blanching process, indicating that high-pressure processing can effectively increase MYR activity. As a comparison with isothiocyanate content, these two values exhibited a similar tendency of change, suggesting that high-pressure stimulation of MYR activity increases isothiocyanate content. There was no significant difference in APX activity after blanching, but after high-pressure treatment, the APX activity exhibited an increasing tendency. This was not consistent with the values obtained for the vitamin C content, suggesting that the increase in isothiocyanate content might not be related to changes in

APX activity. The activities of PPO and POD were significantly reduced after blanching and high-pressure treatment, indicating that both processing methods could inhibit oxidase effectively and preserve food quality.

**Table 1.** Changes in enzyme activity in broccoli before and after high-pressure treatment.

| | Non-Blanched | | Blanched | |
|---|---|---|---|---|
| | **Non Processed** | **HPP Processed** | **Non Processed** | **HPP Processed** |
| MYR ($\mu mol \cdot g^{-1} \cdot min^{-1}$) | $165.75 \pm 3.75^{b}$ | $267.25 \pm 28.75^{c}$ | $123.50 \pm 2.00^{a}$ | $250.75 \pm 15.75^{c}$ |
| APX ($mM \cdot min^{-1} \cdot mg^{-1}$ protein) | $0.142 \pm 0.002^{a}$ | $0.324 \pm 0.004^{b}$ | $0.112 \pm 0.011^{a}$ | $0.361 \pm 0.026^{b}$ |
| PPO ($\Delta A \cdot min^{-1}$) | $0.0067 \pm 0.0003^{d}$ | $0.0049 \pm 0.000^{c}$ | $0.0041 \pm 0.0001^{b}$ | $0.0025 \pm 0.0001^{a}$ |
| POD ($\Delta A \cdot min^{-1}$) | $0.161 \pm 0.007^{d}$ | $0.148 \pm 0.015^{c}$ | $0.043 \pm 0.002^{b}$ | $0.021 \pm 0.004^{a}$ |

All data are presented as the mean ± SD ($n$ = 3). Different letters in the same row indicate significantly different results ($p < 0.05$). MYR: myrosinase; APX: ascorbate peroxidase; PPO: polyphenol oxidase; POD: peroxidase. The activity of the PPO and POD is expressed in the units of absorbance change per min ($\Delta A \cdot min^{-1}$).

The activity of the enzyme myrosinase after high-pressure treatment was significantly higher than in the untreated group, and this increase in activity exhibited a tendency similar to the change in isothiocyanate content, indicating that high-pressure processing can promote myrosinase activation and thereby increase isothiocyanate production. Wang et al. (2016) [37] and Okunade et al. (2015) [38] applied high-pressure treatment to measure the myrosinase activity in Brussels sprouts and mustards, and their results were similar to that of the present study. Furthermore, Wang et al. (2016) [37] indicated that myrosinase activity is altered based on the environmental pH and that high-pressure treatment affects the ionic balance in food, suggesting that environmental pH is more suitable for myrosinase survival and thus improves the overall activity. The effects of high-pressure processing on the enzymes' activities are complex and depends a lot on the matrix composition of the tested samples. Wang et al. (2018) [39] studied the high-pressure effects on myrosinase activity and glucosinolate preservation in seedlings of Brussels sprouts and proposed that the effect depends on myrosinase activity and cell permeabilization. The measurable increase of myrosinase activity and content of isothiocyanates can be explained by the interplay of the increased contact between the glucosinolates and myrosinase via membrane permeabilization, induced by the high pressure and availability of myrosinase.

When compared to other enzymes, there were no significant differences in the APX activity levels before and after blanching. This result is in accordance with a previous study by Vicente et al. (2006) [40] in heat-treated strawberries. Their study indicated that heat and oxidative stress induced an increased expression of the gene *apx1* upstream of APX, causing the activity of APX to be retained after heat treatment. APX activity was significantly increased after high-pressure treatment. This was not consistent with the vitamin C content changes, suggesting that an increase in isothiocyanate content might not be related to changes in APX activity. Although the APX enzyme plays an important role in vitamin C metabolism, the measurable content of vitamin C in HPP processed fruits and vegetables is variable due to many possible mechanisms, such as the enhanced extraction of bioactive compounds and the cells' rupture that releases their cytosol content, caused by the compression effect of the high pressure [41].

POD and PPO are oxidases commonly found in plant cells that primarily use phenolic compounds as substrates. When these two enzymes interact with phenolic compounds, brown colored substances are formed, causing enzymatic browning of plants. In addition, peroxidases are known to oxidize lipids, causing unpleasant odors that affect food quality [42]. After high-pressure treatment of broccoli, the POD and PPO activities were significantly reduced by 51% and 39%, respectively. According to previous studies by Denoya et al. (2015) [43] and Fang et al. (2008) [44], high-pressure techniques can reduce the activity of these enzymes by altering the structures of these proteins, and hence the activities of POD and PPO are significantly reduced after high-pressure processing.

Figure 3 shows the effect of the broccoli extract on the viability of RAW264.7 macrophages. The control group was not induced by LPS or the given sample treatment, so its survival rate was 100%. The LPS induction and sample treatment resulted in a higher cell viability compared to the control group (Figure 3A). When different concentrations of SFN broccoli extract were added to the cells without the effect of LPS, cell viability after treatment at every concentration was significantly higher than that of the control group (Figure 3B). With LPS induction, cell viability was not reduced by the increased extract concentration and was significantly higher than that of the control group (Figure 3C).

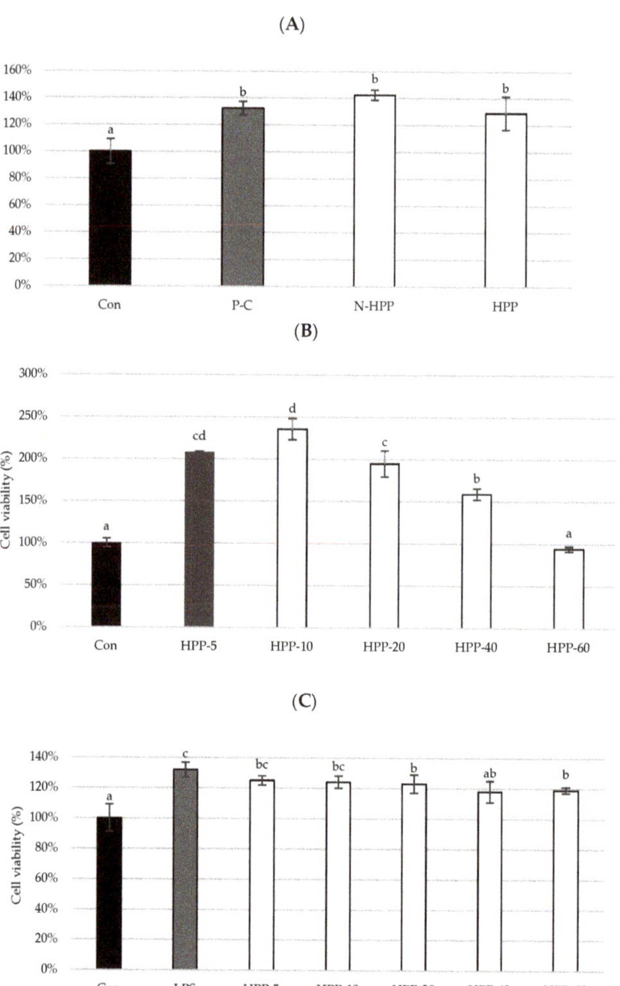


**Figure 3.** Cell survival rate. (**A**) Effects of the untreated group and high-pressure treated group. ■Con: control (not LPS induced); ■P-C: positive control (LPS induced); □N-HPP: LPS induction + Non HPP untreated group; □HPP: LPS induction + high-pressure group, 400 MPa, 15 min. (**B**) Effects of the broccoli extracts of different SFN concentrations without LPS induction. ■Con: Control; □HPP-5 (not LPS induced + 5 ppm broccoli extract). (**C**) Effects of the broccoli extracts of different SFN concentrations. ■Con: control (not LPS induced); ■LPS:(LPS induced); □HPP-5 (LPS induced +5 ppm broccoli extract). All data are presented as the mean ± SD ($n$ = 3). Different letters in the same row indicate significantly different results ($p$ < 0.05).

LPS is a polysaccharide present in the cell wall of Gram-negative bacteria. Its structure contains lipid A, which is a source of endotoxin and can activate macrophages and cause inflammation [45]. After LPS-induced inflammation, the cell viability was not decreased below that of the control group. According to Brandenburg et al. (2010) [46], exposure to low concentrations of LPS did not lead to cell death, but instead stimulated cell viability, so the cell viability of the positive control group was higher than that of the control group. When the broccoli extract was added to the cells, the viability of the treatment group was higher than that of the control group regardless of whether inflammation was induced, indicating that broccoli extract is not toxic to cells. This result is consistent with those of a study by Hwang and Lim (2014) [12] in which different SFN concentrations (7.8–1000 mg·mL$^{-1}$) of the broccoli extract were added to RAW264.7 inflammatory cells. In addition, Guerrero-Beltrán et al. (2010) [47] indicated that SFN could enhance cytoprotection by inducing nuclear translocation of the Nrf2 protein, thereby improving viability.

The effect of broccoli extract addition to RAW264.7 inflammatory macrophages, and specifically its effects on nitric oxide production, are shown in Figure 4. After the inflammatory cells were treated with broccoli extract, the amount of NO produced was significantly lower than in the positive control group, indicating that the extract has anti-inflammatory potential. After high-pressure treatment, the effect of the extract on inhibiting NO production was improved 1.3-fold compared to the untreated group (Figure 4A). Using SFN as an indicator, the broccoli extract was diluted to 5, 10, 20, 40, and 60 mg.mL$^{-1}$ and added to the inflammatory cells. Figure 4B indicates that the inhibitory effect of the extract is at its best when the concentration of SFN is 60 mg.mL$^{-1}$ and the NO content is reduced by 85% compared to the positive control group.

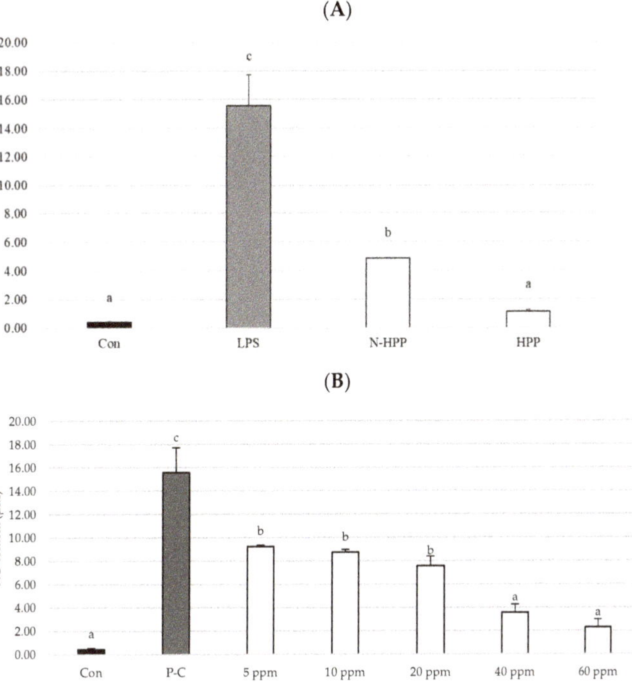


**Figure 4.** Nitric oxide production. (**A**) Untreated group and high-pressure treated group. ■Con: control (not LPS induced); ■LPS: (LPS induced); □N-HPP: LPS induction + Non HPP untreated group; □HPP: LPS induction + high-pressure group, 400 MPa, 15 min. (**B**) Effects of the broccoli extracts of different concentrations. ■Con: control (not LPS induced); ■P-C: positive control (LPS induced); □HPP-5 (LPS induced + 5 ppm broccoli extract). All data are presented as the mean ± SD ($n$ = 3). Different letters in the same row indicate significantly different results ($p$ < 0.05).

NO is an important biological indicator in cells that can regulate the physiological processes associated with inflammation [48]. Treatment of cells with LPS causes an inflammatory response, which promotes upregulation of iNOS in macrophages affected by the inflammatory response, leading to a massive increase in the production of NO [49]. In the present study, the NO content in the cells was significantly reduced when treated with broccoli extracts with different SFN concentrations, with higher SFN concentrations yielding better inhibitory effects. The results of the present study are similar to those of Subedi et al. (2019) [50], in which glial cells were treated with purified SFN. This study indicated that SFN can effectively downregulate iNOS expression, thereby decreasing NO production. Yang et al. (2007) [51] applied SFN to retinal microglia in which inflammation was induced by LPS. The NO content decreased significantly when the concentration of SFN was increased between 1.25 and 10 μM. Furthermore, it was observed that the changes in iNOS expression were proportional to the NO content. These studies suggest that SFN primarily reduces iNOS expression and NO production, thereby delaying inflammation.

All data are presented as the mean ± SD ($n$ = 3). Different letters in the same row indicate significantly different results ($p < 0.05$).

A pre-LPS induction and a post-LPS induction group were used to simulate the treatment and prevention, respectively. Figure 5 shows the effect of the broccoli extract on the prostaglandin $E_2$ ($PGE_2$) content in RAW264.7 inflammatory macrophages. After LPS induction, the $PGE_2$ content of the cells increased significantly, indicating that the cells were in an inflamed state. After the addition of broccoli extract, the high-pressure broccoli extracts inhibited $PGE_2$ production compared to the untreated group (Figure 5A). Regardless of the LPS induction strategy done first or later, once the broccoli extract was diluted, the inhibitory effects on $PGE_2$ was reduced both before and after LPS induction, and the $PGE_2$ content doubled compared to pre-dilution, indicating that the concentration of SFN in the broccoli extract must be higher than 60 mg·mL$^{-1}$ in order for the inhibitory effect to be significant (Figure 5B).

$PGE_2$ is one of the most abundant prostaglandins in the human body and is involved in many physiological and pathological processes, including cancer and inflammation [52]. Upon LPS induction, COX is rapidly activated in cells, prompting the conversion of large amounts of arachidonic acid into prostaglandins (PG), including PGI2, PGE2, and other molecules, which lead to inflammation. A large amount of $PGE_2$ is present after LPS-induced inflammation. The $PGE_2$ content is significantly reduced after addition of high-pressure-treated broccoli extracts, and this effect is greater than in the untreated group. However, when the extract is diluted to different SFN concentrations, the inhibitory effect on $PGE_2$ exhibited a decreasing tendency. Park et al. (2019) [53] treated cells with LPS-induced inflammation with broccoli extracts and found that the $PGE_2$ content was significantly reduced, which is similar to the findings of the present study. Qi et al. (2016) [54] used LPS to induce lung injury in BALB/c mice that were previously treated with SFN and found that the $PGE_2$ content in these mice were significantly reduced, indicating that SFN has the potential to delay inflammation. In addition, these two studies also indicated that the amounts of $PGE_2$ produced are correlated with the COX-2 expression levels, suggesting that SFN primarily inhibits $PGE_2$ by reducing COX-2 expression.

Figure 6 shows the effect of broccoli extracts on iNOS and COX-2 in inflammatory cells. After the cells were treated with broccoli extracts, the expression level of iNOS was lower than in the positive control group, with the post-induction group, treated with a SFN concentration of 60 mg.mL$^{-1}$, exhibiting the best inhibitory effect on iNOS, around 61%. With regards to COX-2 expression, the changes in expression levels were similar to that in iNOS, and the pre-LPS induction group exhibited a significant concentration-dependent effect, with expression decreasing with increasing SFN concentration. In addition, both induction treatment strategies exhibited the best inhibitory effects on COX-2 gene expression at a concentration of 60 mg.mL$^{-1}$, and the inhibitory effects were measured to be 46% and 35%, respectively.

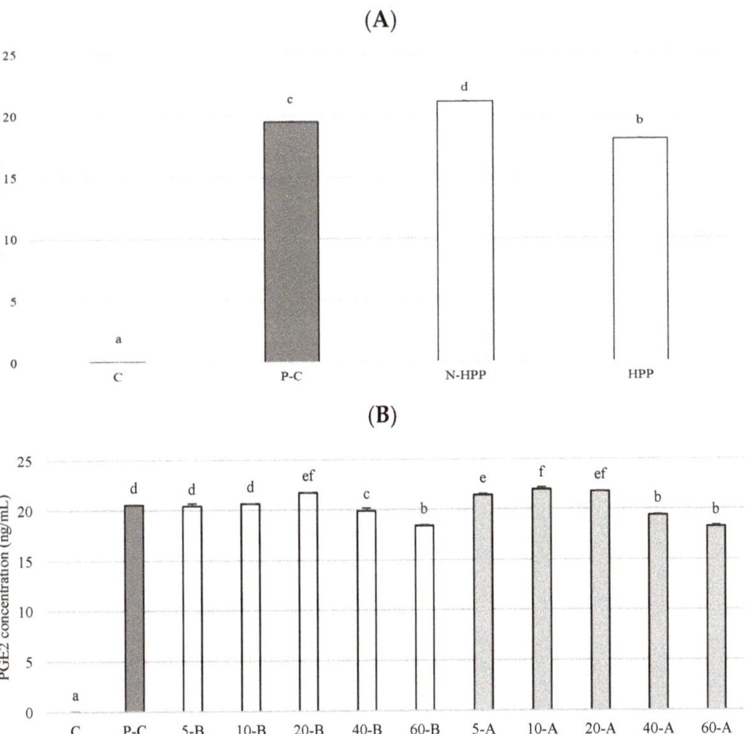


**Figure 5.** $PGE_2$ content. (**A**) Untreated group and high-pressure-treated group. ■C: control, ■P-C: positive control (LPS induced); □N-HPP: LPS induction + Non HPP untreated group; □HPP: LPS induction + high-pressure group, 400 MPa, 15 min. (**B**) Broccoli extracts of different concentrations, pre-/post-LPS treated. ■C: control (not LPS induced); ■P-C: positive control; □5-B broccoli extracts 5 ppm, pre-LPS treated; ■5-A broccoli extracts 5 ppm, post-LPS treated. All data are presented as the mean ± SD ($n$ = 3). Different letters in the same row indicate significantly different results ($p < 0.05$).

iNOS is an upstream enzyme that produces nitric oxide, and COX-2 is a pivotal enzyme for the production of $PGE_2$. Inflammation stimulates increased expression of both enzymes and promotes production of large amounts of inflammatory cytokines [55]. Many studies have found that excessive expression of iNOS and COX-2 causes massive production of inflammatory factors such as NO and $PGE_2$. SFN can effectively inhibit the expression of these two proteins, thereby regulating inflammatory response [56–59]. In the present study, applying high-pressure-treated broccoli extracts to inflammatory macrophages had the best inhibitory effect on iNOS or COX-2 when the SFN concentration was 60 mg.mL$^{-1}$. Comparisons based on the results of iNOS and COX-2 inhibition with NO production and $PGE_2$ content showed that the changes in iNOS and NO production were correlated, which is consistent with the results of the aforementioned studies, but the COX-2 and $PGE_2$ content were different. Cells contain two COX isoenzymes, namely, COX-1 and COX-2. COX-1 functions primarily as a housekeeping gene and can stabilize the physiological functions of cells. COX-2 is generally considered to be activated by inflammation. However, studies related to neurodegenerative diseases and neuroinflammation have found that the expression levels of COX-1 is associated with the production of $PGE_2$ and inflammatory cytokines in microglia, showing that COX-1 not only stabilizes cell physiology, but also promotes inflammation [60,61]. In addition, Qin et al. (2016) [62] and Zhou et al. (2012) [52] found that the mechanism of SFN inhibition of $PGE_2$ content might be achieved through regulating the expression of microsomal prostaglandin E synthase 1 (mPGES-1) downstream of COX-2, rather than by inhibiting COX-2 expression. These studies suggest that, in addition to COX-2, the factors affecting the synthesis of $PGE_2$ are also regulated by COX-1 and mPGES-1, resulting in differences in the COX-2 expression levels and $PGE_2$ content.

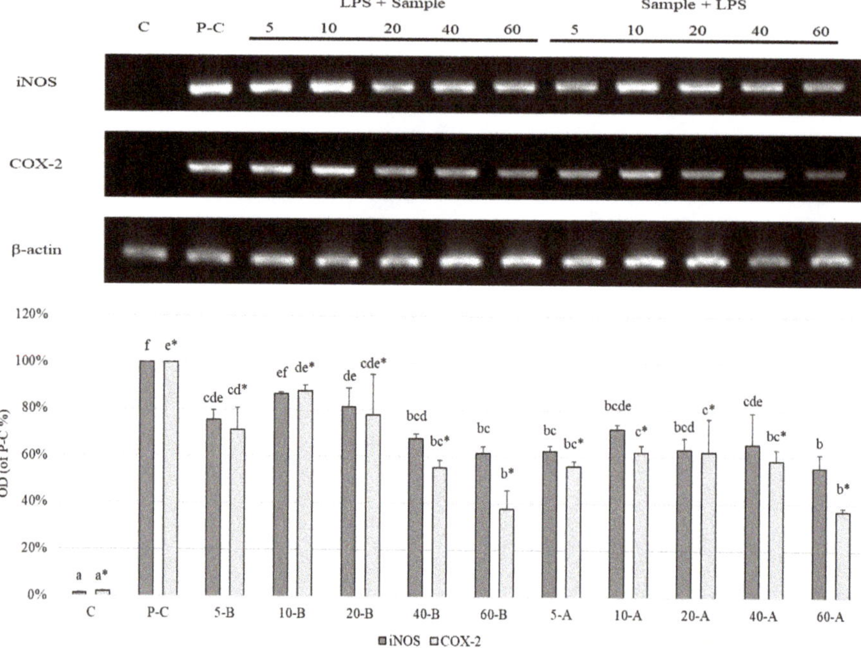


**Figure 6.** Effects of different extracts of broccoli on the iNOS, COX-2, and β-actin gene expression in RAW264.7 macrophage cells. C: control; P-C: positive control; 5-B: broccoli extracts 5 ppm, pre-LPS treated; 5-A: broccoli extracts 5 ppm, post-LPS treated. All data are presented as the mean ± SD ($n$ = 3). The same symbols, noted as superscripts after the letter, means that the ANOVA performed is in the same group. Bars carrying different letters are statistically different at $p < 0.05$.

The antioxidant effect of broccoli was evaluated by the ratio of reduced to oxidized glutathione (GSH/GSSG) in cells. Figure 7 shows that the GSH/GSSG ratio of the control group not induced by LPS was 9.90, whereas this ratio (5.99) was significantly reduced in the positive control group treated with only LPS. In the LPS pre-induction group, the ratio was significantly higher than that of the positive control group when the SFN concentration was 5 $mg.L^{-1}$ and 10 $mg.L^{-1}$. The antioxidant effect was best when the SFN concentration was 10 $mg.L^{-1}$, with a GSH/GSSG ratio of 9.41.

Glutathione (GSH) is an important indicator of oxidative/nitrative stress in organisms. It can metabolize ROS and RNS to clear potentially toxic oxidation products and reduce oxidative and nitrative damage in cells. In addition, GSH is also a coenzyme of glutathione peroxidase, which protects the sulfhydryl group of this enzyme from oxidation and preserves its activity [63]. Oxidation of GSH yields glutathione disulfide (GSSG), and the alterations in the ratio between these two are associated with redox balance in cells. Hence, the GSH/GSSG ratio is often used as an indicator to evaluate the degree of cellular oxidation [64]. In our present study, the GSH/GSSG ratio was significantly reduced, which is consistent with the results of a previous study by Yamada et al. (2006) [65] regarding $n$ dendritic cells induced with LPS. The literature indicates that ROS is generated in large quantities when cells are in an inflamed state, which reduces the GSH/GSSG ratio. When treated with broccoli extracts with different SFN concentrations, the antioxidant effect is best at an SFN concentration of 5 $mg.L^{-1}$ or 10 $mg.L^{-1}$, which is consistent with the results of studies by Kim et al. (2003) [66] and Heiss et al. (2001) [67], who respectively used HepG2-C8 cells and RAW264.7 macrophages for analyzing oxidative stress. After treatment with low concentrations of SFN, the GSH content in the cells increased with increasing time in culture, and the antioxidant capacity of the cells was significantly improved.

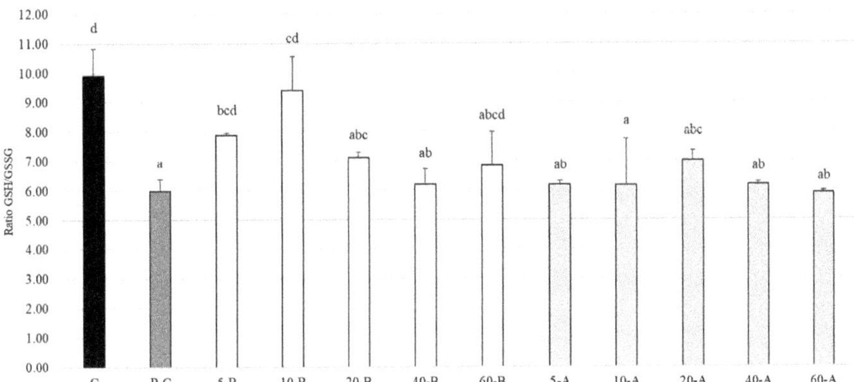


**Figure 7.** Effects of the broccoli extracts with different SFN concentrations on the ratio of GSH/GSSG in RAW264.7 macrophages. All data are presented as the mean ± SD ($n$ = 3). Different letters in the same row indicate significantly different results ($p < 0.05$).

## 4. Conclusions

This study demonstrated that high-pressure treatment could effectively increase the isothiocyanate content in broccoli. Specifically, the best results were achieved with processing conditions of 400 MPa for 15 min. The mechanism for this change is primarily due to the high-pressure processing strategy, which stimulates myrosinase activity, thereby increasing the efficiency of the glucosinolate hydrolysis and increasing the isothiocyanate content. With regards to the functional components, high-pressure processing did not affect the vitamin C or flavonoid content in broccoli, and increased the TPC. When compared to traditional thermal processing, high-pressure processing can prevent the loss of heat-sensitive components. In addition, the activity of PPO and POD in broccoli tended to decrease after being subjected to high-pressure treatment, indicating that this processing technique can help inhibit oxidase activity and maintain food quality. With respect to cellular experiments, the varying concentrations of the broccoli extracts applied to the cells did not affect the viability of the RAW264.7 macrophages. Further evaluation of its anti-inflammatory and antioxidant effects showed that broccoli extracts could effectively inhibit NO production, $PGE_2$ content, and iNOS and COX-2 protein expression levels. At low concentrations, SFN significantly increased the GSH content and reduced GSSG production, indicating that broccoli has the potential to delay inflammation and reduce oxidative stress. In conclusion, high-pressure processing of broccoli not only adds value to fresh food and provides increased nutritional value, but also allows it to be used as a raw material or additive for the development of healthy foods, in so doing maximizing the utilization value of broccoli.

**Author Contributions:** Conceptualization, Y.-T.S., S.-J.W.; methodology,., Y.-Y.K., Y.-T.S. and S.-J.W; software, Y.-Y.K., S.-J.W.; validation, Y.-T.S., S.-J.W.; formal analysis, S.-J.W.; investigation, Y.-T.S., Y.-Y.K., and S.-J.W.; resources, Y.-T.S., S.-J.W.; data curation, Y.-T.S., S.-J.W.; Writing—Original draft preparation, S.-J.W.; Writing—Review and editing, S.-J.W., Y.-T.S.; visualization, S.-J.W., Y.-Y.K. and Y.-T.S.; supervision, S.-J.W., Y.-T.S.; project administration, S.-J.W.; funding acquisition, S.-J.W., Y.-T.S. All authors have read and agreed to the published version of the manuscript.

**Funding:** This research was funded by the Council of Agriculture, Taiwan, ROC, grant number 109AS-22.1.1-ST-aA.

**Acknowledgments:** The authors would like to thank Chen Yung Memorial Foundation for their support on parts of the funding for this research.

**Conflicts of Interest:** This manuscript has not been published or presented elsewhere in part or entirety and is not under consideration by another journal. All the authors have approved the manuscript and agree with submission to your esteemed journal. There are no conflicts of interest to declare.

## References


1. Lafarga, T.; Bobo, G.; Viñas, I.; Collazo, C.; Aguiló-Aguayo, I. Effects of thermal and non-thermal processing of cruciferous vegetables on glucosinolates and its derived forms. *J. Food Sci. Technol.* **2018**, *5*, 1973–1981. [CrossRef] [PubMed]
2. Lin, Y.S. Construction and Transformation of Sulforaphane Biosynthetic Related Genes of Broccoli. Master's Thesis, National Chung Hsing University, Taichung, Taiwan, 2017.
3. Saw, C.L.; Huang, M.T.; Liu, Y.; Khor, T.O.; Conney, A.H.; Kong, A.N. Impact of Nrf2 on UVB-induced skin inflammation/photoprotection and photoprotective effect of sulforaphane. *Mol. Carcinogen.* **2010**, *50*, 479–486. [CrossRef] [PubMed]
4. Piao, C.S.; Gao, S.; Lee, G.H.; Kim, D.S.; Park, B.H.; Chae, S.W.; Chae, H.J.; Kim, S.H. Sulforaphane protects ischemic injury of hearts through antioxidant pathway and mitochondrial $K_{ATP}$ channels. *Pharmacol. Res.* **2010**, *61*, 342–348. [CrossRef] [PubMed]
5. Westphal, A.; Riedl, K.M.; Cooperstone, J.L.; Kamat, S.; Balasubramaniam, V.M.; Schwartz, S.J.; Böhm, V. High-pressure processing of broccoli sprouts: Influence on bioactivation of glucosinolates to isothiocyanates. *J. Agric. Food. Chem.* **2017**, *65*, 8578–8585. [CrossRef]
6. Shirakawa, M.; Ueda, H.; Shimada, T.; Hara-Nishimura, I. Myrosin cells are differentiated directly from ground meristem cells and are developmentally independent of the vasculature in *Arabidopsis* leaves. *Plant Signal. Behav.* **2016**, *11*, e1150403. [CrossRef]
7. Halkier, B.A.; Gershenzon, J. Biology and biochemistry of glucosinolates. *Annu. Rev. Plant Biol.* **2006**, *57*, 303–333. [CrossRef]
8. Eylen, D.V.; Oey, I.; Hendrickx, M.; Loey, A.V. Effects of pressure/temperature treatments on stability and activity of endogenous broccoli (*Brassica oleracea* L. cv. *Italica*) myrosinase and on cell permeability. *J. Food Eng.* **2008**, *89*, 178–186.
9. Medzhitov, R. Origin and physiological roles of inflammation. *Nature* **2008**, *454*, 428–435. [CrossRef]
10. Tung, Y.T.; Chua, M.T.; Wang, S.Y.; Chang, S.T. Anti-inflammation activities of essential oil and its constituents from indigenous cinnamon (*Cinnamomum osmophloeum*) twigs. *Bioresour. Technol.* **2008**, *99*, 3908–3913. [CrossRef]
11. Baluk, P.; Yao, L.C.; Feng, J.; Romano, T.; Jung, S.S.; Schreiter, J.L.; Yan, L.; Shealy, D.J.; McDonald, D.M. TNF-α drives remodeling of blood vessels and lymphatics in sustained airway inflammation in mice. *J. Clin. Investig.* **2009**, *119*, 2954–2964. [CrossRef]
12. Hwang, J.H.; Lim, S.B. Antioxidant and anti-inflammatory activities of broccoli florets in LPS-stimulated RAW264.7 cells. *Prev. Nutr. Food Sci.* **2014**, *19*, 89–97. [CrossRef] [PubMed]
13. Rakariyatham, K.; Wu, X.; Tang, Z.; Han, Y.; Wang, Q.; Xiao, H. Synergism between luteolin and sulforaphane in anti-inflammation. *Food Funct.* **2018**, *9*, 5115–5123. [CrossRef] [PubMed]
14. Muntean, M.V.; Marian, O.; Barbieru, V.; Cătunescu, G.M.; Ranta, O.; Drocas, I.; Terhes, S. High pressure processing in food industry-characteristics and applications. *Agric. Agric. Sci. Procedia* **2016**, *10*, 377–383. [CrossRef]
15. Maitland, J.E.; Boyer, R.R.; Eifert, J.D.; Williams, R.C. High hydrostatic pressure processing reduces *Salmonella enterica* serovars in diced and whole tomatoes. *Int. J. Food Microbiol.* **2011**, *149*, 113–117. [CrossRef]
16. Rodríguez-Roque, M.J.; de Ancos, B.; Sánchez-Moreno, C.; Cano, M.P.; Elez-Martínez, P.; Martín-Belloso, O. Impact of food matrix and processing on the in vitro bioaccessibility of vitamin C, phenolic compounds, and hydrophilic antioxidant activity from fruit juice-based beverages. *J. Funct. Foods* **2015**, *14*, 33–43. [CrossRef]
17. Van Eylen, D.; Bellostas, N.; Strobel, B.W.; Oey, I.; Hendrickx, M.; Van Loey, A.; Sørensen, H.; Sørensen, J.C. Influence of pressure/temperature treatments on glucosinolate conversion in broccoli (*Brassica oleraceae* L. *cv Italica*) heads. *Food Chem.* **2009**, *112*, 646–653. [CrossRef]
18. Koo, S.Y.; Cha, K.H.; Song, D.G.; Chung, D.; Pan, C.H. Amplification of sulforaphane content in red cabbage by pressure and temperature treatments. *J. Korean Soc. Appl. Biol. Chem.* **2011**, *54*, 183–187. [CrossRef]
19. Alvarez-Jubete, L.; Valverde, J.; Patras, A.; Mullen, A.M.; Marcos, B. Assessing the impact of high-pressure processing on selected physical and biochemical attributes of white cabbage (*Brassica oleracea* L. var. *capitata alba*). *Food Bioprocess Technol.* **2014**, *7*, 682–692. [CrossRef]
20. You, Y.; Wu, Y.; Mao, J.; Zou, L.; Liu, S. Screening of Chinese brassica species for anti-cancer sulforaphane and erucin. *Afr. J. Biotechnol.* **2008**, *7*, 147–152.
21. Yu, G.; Bei, J.; Zhao, J.; Li, Q.; Cheng, C. Modification of carrot (*Daucus carota* Linn. var. Sativa Hoffm.) pomace insoluble dietary fiber with complex enzyme method, ultrafine comminution, and high hydrostatic pressure. *Food Chem.* **2018**, *257*, 333–340. [CrossRef]
22. Chang, C.C.; Yang, M.H.; Wen, H.M.; Chern, J.C. Estimation of total flavonoid content in propolis by two complementary colorimetric methods. *J. Food Drug Anal.* **2002**, *10*, 178–182.
23. Yuan, G.F.; Wang, X.P.; Guo, R.F.; Wang, Q.M. Effect of salt stress on phenolic compounds, glucosinolates, myrosinase and antioxidant activity in radish sprouts. *Food Chem.* **2010**, *121*, 1014–1019. [CrossRef]
24. Li, J.; Wang, Q.Z.; Zhou, Q.; Sun, Y.; Xie, B.J. Study on properties of myrosinase from radish. *Food Sci.* **2008**, *29*, 451–455.
25. Zhao, K.; Xue, P.J.; Gu, G.Y. Study on determination of reducing sugar content using 3,5-dinitrosalicylic acid method. *Food Sci.* **2008**, *29*, 534–536.
26. Chen, K.M.; Liu, C.P. Effects of pretreating calcium on *Acacia confusa* seedlings under simulated acid rain stress. *Q. J. For. Res.* **2012**, *34*, 237–250.

27. Yang, H.A. Effect of High Pressure Processing on Functional Ingredient and Quality of Heat-Sensitive Juice. Master's Thesis, National Taiwan University, Taipei, Taiwan, 2016.
28. Rungapamestry, V.; Duncan, A.J.; Fuller, Z.; Ratcliffe, B. Effect of cooking brassica vegetables on the subsequent hydrolysis and metabolic fate of glucosinolates. *Proc. Nutr. Soc.* **2007**, *66*, 69–81. [CrossRef]
29. Jones, R.B.; Frisina, C.L.; Winkler, S.; Imsic, M.; Tomkins, R.B. Cooking method significantly effects glucosinolate content and sulforaphane production in broccoli florets. *Food Chem.* **2010**, *123*, 237–242. [CrossRef]
30. Vinicio, S.M.; Daniel, A.J.V.; Jose, A.T.; Jorge, W.C. Microstructural and physiological changes in plant cell induced by pressure: Their role on the availability and pressure-temperature stability of phytochemicals. *Food Eng. Rev.* **2017**, *9*, 314–334.
31. Acosta-Estrada BA, Gutiérrez-Uribe JA, Serna-Saldívar SO (2014) Bound phenolics in foods, a review. *Food Chem.* **2014**, *152*, 46–55.
32. Jakobek, L. Interactions of polyphenols with carbohydrates, lipids and proteins. *Food Chem.* **2015**, *175*, 556–567. [CrossRef]
33. Liu, X.W.; Bourvellec, C.L.; Renard, C.M.G.C. Interactions between cell wall polysaccharides and polyphenols: Effect of molecular internal structure. Compr. *Rev. Food Sci. Food Saf.* **2020**, *19*, 3574–3617. [CrossRef] [PubMed]
34. Prasad, K.N.; Yang, E.; Yi, C.; Zhao, M.; Jiang, Y. Effects of high pressure extraction on the extraction yield, total phenolic content and antioxidant activity of longan fruit pericarp. *Innov. Food Sci. Emerg. Technol.* **2009**, *10*, 155–159. [CrossRef]
35. Landl, A.; Abadias, M.; Sárraga, C.; Viñas, I.; Picouet, P.A. Effect of high pressure processing on the quality of acidified Granny Smith apple purée product. *Innov. Food Sci. Emerg. Technol.* **2010**, *11*, 557–564. [CrossRef]
36. Tola, Y.B.; Ramaswamy, H.S. Temperature and high pressure stability of lycopene and vitamin C of watermelon juice. *Afr. J. Food Sci.* **2015**, *9*, 351–358.
37. Wang, J.; Barba, F.J.; Frandsen, H.B.; Sørensen, S.; Olsen, K.; Sørensen, J.C.; Orlien, V. The impact of high pressure on glucosinolate profile and myrosinase activity in seedlings from Brussels sprouts. *Innov. Food Sci. Emerg. Technol.* **2016**, *38*, 342–348. [CrossRef]
38. Okunade, O.A.; Ghawi, S.K.; Methven, L.; Niranjan, K. Thermal and pressure stability of myrosinase enzymes from black mustard (*Brassica nigra* L. W.D.J. Koch. var. *nigra*), brown mustard (*Brassica juncea* L. Czern. var. *juncea*) and yellow mustard (*Sinapsis alba* L. subsp. *maire*) seeds. *Food Chem.* **2015**, *187*, 485–490. [CrossRef]
39. Wang, J.; Brba, F.J.; Sorensen, J.; Frandsen, H.B.; Sorensen, S.; Olsen, K.; Orlien, V. High pressure effects on myrosinase activity and glucosinolate preservation in seedliings of brussels sprouts. *Food Chem.* **2018**, *15*, 1212–1217. [CrossRef]
40. Vicente, A.R.; Martínez, G.A.; Chaves, A.R.; Civello, P.M. Effect of heat treatment on strawberry fruit damage and oxidative metabolism during storage. *Postharvest Biol. Technol.* **2006**, *40*, 116–122. [CrossRef]
41. Tewari, S.; Sehrawat, R.; Nema, P.K.; Kaur, B.P. Preservation effect of high pressure processing on ascorbic acid of fruits and vegetables: A review. *J. Food Biochem.* **2017**, *41*, e12319. [CrossRef]
42. Woolf, A.B.; Wibisono, R.; Farr, J.; Hallett, I.; Richter, L.; Oey, I.; Wohlers, M.; Zhou, J.; Fletcher, G.C.; Requejo-Jackman, C. Effect of high pressure processing on avocado slices. *Innov. Food Sci. Emerg. Technol.* **2013**, *18*, 65–73. [CrossRef]
43. Denoya, G.I.; Vaudagna, S.R.; Polenta, G. Effect of high pressure processing and vacuum packaging on the preservation of fresh-cut peaches. *LWT-Food Sci. Technol.* **2015**, *62*, 801–806. [CrossRef]
44. Fang, L.; Jiang, B.; Zhang, T. Effect of combined high pressure and thermal treatment on kiwifruit peroxidase. *Food Chem.* **2008**, *109*, 802–807. [CrossRef] [PubMed]
45. Kim, H.G.; Shrestha, B.; Lim, S.Y.; Yoon, D.H.; Chang, W.C.; Shin, D.J.; Han, S.K.; Park, S.M.; Park, J.H.; Park, H.I.; et al. Cordycepin inhibits lipopolysaccharide-induced inflammation by the suppression of NF-κB through Akt and p38 inhibition in RAW 264.7 macrophage cells. *Eur. J. Pharmacol.* **2006**, *545*, 192–199. [CrossRef]
46. Brandenburg, L.O.; Kipp, M.; Lucius, R.; Pufe, T.; Wruck, C.J. Sulforaphane suppresses LPS-induced inflammation in primary rat microglia. *Inflamm. Res.* **2010**, *59*, 443–450. [CrossRef] [PubMed]
47. Guerrero-Beltrán, C.E.; Calderón-Oliver, M.; Tapia, E.; Medina-Campos, O.N.; Sánchez-González, D.J.; Martínez-Martínez, C.M.; Ortiz-Vega, K.M.; Franco, M.; Pedraza-Chaverri, J. Sulforaphane protects against cisplatin-induced nephrotoxicity. *Toxicol. Lett.* **2010**, *192*, 278–285. [CrossRef] [PubMed]
48. Banskota, A.H.; Stefanova, R.; Sperker, S.; Lall, S.P.; Craigie, J.S.; Hafting, J.T.; Critchley, A.T. Polar lipids from the marine macroalga *Palmaria palmata* inhibit lipopolysaccharide-induced nitric oxide production in RAW264.7 macrophage cells. *Phytochemistry* **2014**, *101*, 101–108. [CrossRef]
49. Sawle, P.; Foresti, R.; Mann, B.E.; Johnson, T.R.; Green, C.J.; Motterlini, R. Carbon monoxide-releasing molecules (CO-RMs) attenuate theinflammatory response elicited by lipopolysaccharide in RAW264.7 murine macrophages. *Br. J. Pharmacol.* **2009**, *145*, 800–810. [CrossRef] [PubMed]
50. Subedi, L.; Lee, J.H.; Yumnam, S.; Ji, E.; Kim, S.Y. Anti-inflammatory effect of sulforaphane on LPS-activated microglia potentially through JNK/AP-1/NF-κB inhibition and Nrf2/HO-1 activation. *Cells* **2019**, *8*, 194. [CrossRef]
51. Yang, L.P.; Zhu, X.A.; Tso, M.O.M. Minocycline and sulforaphane inhibited lipopolysaccharide-mediated retinal microglial activation. *Mol. Vis.* **2007**, *13*, 1083–1093.
52. Zhou, J.; Joplin, D.G.; Cross, J.V.; Templeton, D.J. Sulforaphane inhibits prostaglandin E2 synthesis by suppressing microsomal prostaglandin E synthase 1. *PLoS ONE.* **2012**, *7*, e49744. [CrossRef]
53. Park, S.J.; An, I.; Noh, G.P.; Yoo, B.H.; Lee, J.R. Inhibitory effect of broccoli leaf extract on $PGE_2$ production by NF-κB inhibition. *Korea J. Herbol.* **2019**, *34*, 117–124.

54. Qi, T.; Xu, F.; Yan, X.; Li, S.; Li, H. Sulforaphane exerts anti-inflammatory effects against lipopolysaccharide-induced acute lung injury in mice through the Nrf2/ARE pathway. *Int. J. Mol. Med.* **2016**, *37*, 182–188. [CrossRef] [PubMed]
55. Cho, H.; Yun, C.W.; Park, W.K.; Kong, J.Y.; Kim, K.S.; Park, Y.; Lee, S.; Kim, B.K. Modulation of the activity of pro-inflammatory enzymes, COX-2 and iNOS, by chrysin derivatives. *Pharmacol. Res.* **2004**, *49*, 37–43. [CrossRef]
56. Ruhee, R.T.; Ma, S.; Suzuki, K. Sulforaphane protects cells against lipopolysaccharide-stimulated inflammation in murine macrophages. *Antioxidants* **2019**, *8*, 577. [CrossRef] [PubMed]
57. Ma, L.L.; Xing, G.P.; Yu, Y.; Liang, H.; Yu, T.X.; Zheng, W.H.; Lai, T.B. Sulforaphane exerts neuroprotective effects via suppression of the inflammatory response in a rat model of focal cerebral ischemia. *Int. J. Clin. Exp. Med.* **2015**, *8*, 17811–17817.
58. Guo, S.; Qiu, P.; Xu, G.; Wu, X.; Dong, P.; Yang, G.; Zheng, J.; McClements, D.J.; Xiao, H. Synergistic anti-inflammatory effects of nobiletin and sulforaphane in lipopolysaccharide-stimulated RAW 264.7 cells. *J. Agric. Food Chem.* **2012**, *60*, 2157–2164. [CrossRef]
59. Lin, W.; Wu, R.T.; Wu, T.; Khor, T.O.; Wang, H.; Kong, A.N. Sulforaphane suppressed LPS-induced inflammation in mouse peritoneal macrophages through Nrf2 dependent pathway. *Biochem. Pharmacol.* **2008**, *76*, 967–973. [CrossRef]
60. Choi, S.H.; Aid, S.; Bosetti, F. The distinct roles of cyclooxygenase-1 and -2 in neuroinflammation: Implications for translational research. *Trends Pharmacol. Sci.* **2009**, *30*, 174–181. [CrossRef]
61. Smith, C.J.; Zhang, Y.; Koboldt, C.M.; Muhammad, J.; Zweifel, B.S.; Shaffer, A.; Talley, J.J.; Masferrer, J.L.; Seibert, K.; Isakson, P.C. Pharmacological analysis of cyclooxygenase-1 in inflammation. *Proc. Natl. Acad. Sci. USA* **1998**, *95*, 13313–13318. [CrossRef]
62. Qin, W.S.; Deng, Y.H.; Cui, F.C. Sulforaphane protects against acrolein-induced oxidative stress and inflammatory responses: Modulation of Nrf-2 and COX-2 expression. *Arch. Med. Sci.* **2016**, *12*, 871–880. [CrossRef]
63. Giustarini, D.; Dalle-Donne, I.; Colombo, R.; Milzani, A.; Rossi, R. An improved HPLC measurement for GSH and GSSG in human blood. *Free Radic. Biol. Med.* **2003**, *35*, 1365–1372. [CrossRef] [PubMed]
64. Afzal, M.; Afzal, A.; Jones, A.; Armstrong, D. A rapid method for the quantification of GSH and GSSG in biological samples. In *Oxidative Stress Biomarkers and Antioxidant Protocols*; Armstrong, D., Ed.; Humana Press: Totowa, NJ, USA, 2002; Volume 186, pp. 117–122. ISBN 978-0-89603-850-9.
65. Yamada, H.; Arai, T.; Endo, N.; Yamashita, K.; Fukuda, K.; Sasada, M.; Uchiyama, T. LPS-induced ROS generation and changes in glutathione level and their relation to the maturation of human monocyte-derived dendritic cells. *Life Sci.* **2006**, *78*, 926–933. [CrossRef] [PubMed]
66. Kim, B.R.; Hu, R.; Keum, Y.S.; Hebbar, V.; Shen, G.; Nair, S.S.; Kong, A.T. Effects of glutathione on antioxidant response element-mediated gene expression and apoptosis elicited by sulforaphane. *Cancer Res.* **2003**, *63*, 7520–7525. [PubMed]
67. Heiss, E.; Herhaus, C.; Klimo, K.; Bartsch, H.; Gerhäuser, C. Nuclear factor κB is a molecular target for sulforaphane-mediated anti-inflammatory mechanisms. *J. Biol. Chem.* **2001**, *276*, 32008–32015. [CrossRef] [PubMed]

 *foods*

*Article*

# The Impact of $N_2$-Assisted High-Pressure Processing on the Microorganisms and Quality Indices of Fresh-Cut Bell Peppers

**Fan Zhang [1,2], Jingjing Chai [1,2], Liang Zhao [1,2], Yongtao Wang [1,2,*] and Xiaojun Liao [1,2]**

1 College of Food Science and Nutritional Engineering, China Agricultural University, Beijing 100085, China; bs20183060546@cau.edu.cn (F.Z.); s20183060850@cau.edu.cn (J.C.); zhaoliang1987@cau.edu.cn (L.Z.); liaoxjun@cau.edu.cn (X.L.)

2 National Engineering Research Center for Fruit & Vegetable Processing, Beijing Key Laboratory for Food Nonthermal Processing, Key Laboratory of Fruit and Vegetable Processing, Ministry of Agriculture, Beijing 100085, China

* Correspondence: wangyongtao102@cau.edu.cn; Tel.: +86-010-62737434

**Abstract:** This work aimed to evaluate the effects of $N_2$-assisted high-pressure processing (HPP, 400 MPa/7.5 min and 500 MPa/7.5 min) on the microorganisms and physicochemical, nutritional, and sensory characteristics of fresh-cut bell peppers (FCBP) during 25 days of storage at 4 °C. Yeasts and molds were not detected, and the counts of total aerobic bacteria were less than 4 $log_{10}$ CFU/g during storage at 4 °C. The total soluble solids and L* values were maintained in HPP-treated FCBP during storage. After the HPP treatment, an 18.7–21.9% weight loss ratio and 54–60% loss of hardness were found, and the polyphenol oxidase (PPO) activity was significantly inactivated (33.87–55.91% of its original activity). During storage, the weight loss ratio and PPO activity of the samples increased significantly, but the hardness of 500 MPa/7.5 min for treated FCBP showed no significant change (9.79–11.54 N). HPP also effectively improved the total phenol content and antioxidant capacity of FCBP to 106.69–108.79 mg GAE/100 g and 5.76–6.55 mmol Trolox/L; however, a non-negligible reduction in total phenols, ascorbic acid, and antioxidant capacity was found during storage. Overall, HPP treatments did not negatively impact the acceptability of all sensory attributes during storage, especially after the 500 MPa/7.5 min treatment. Therefore, $N_2$-assisted HPP processing is a good choice for the preservation of FCBP.

**Keywords:** bell peppers; high-pressure processing; fresh-cut bell peppers; nitrogen-assisted; storage quality; structural damage

**Citation:** Zhang, F.; Chai, J.; Zhao, L.; Wang, Y.; Liao, X. The Impact of $N_2$-Assisted High-Pressure Processing on the Microorganisms and Quality Indices of Fresh-Cut Bell Peppers. *Foods* **2021**, *10*, 508. https://doi.org/10.3390/foods10030508

Academic Editors: Asgar Farahnaky, Mahsa Majzoobi and Mohsen Gavahian

Received: 4 January 2021
Accepted: 22 February 2021
Published: 28 February 2021



## 1. Introduction

Bell peppers (or sweet peppers), a cultivar group of the species *Capsicum annuum* L., are characterized by their blocky shape, attractive color, and mild taste. Bell peppers contain extremely high levels of the antioxidants vitamin C and E. Furthermore, they contain moderate to high amounts of phenolics, flavonoids, and carotenoids, and capsanthin, lutein, and cryptoflavin have been found in peppers [1,2]. The consumption of these bioactive compounds provides the human body with protection against oxidative damage, thus reducing the incidence of degenerative diseases. Currently, the fresh market consumption of bell peppers is becoming increasingly popular, mainly attributed to their availability in a wide variety of shapes, sizes, colors, and distinctive flavors [3].

For the convenience of food preparation processes in restaurants and fast-food industries worldwide and personal consumption, numerous fruits and vegetables are packaged according to the process of fresh cutting [4]. Compared with fresh vegetables, fresh-cut vegetables are ready to eat without further treatment and are still in their fresh state, keeping all the advantages of fresh vegetables (e.g., color, shape, and nutrition). The market demand for fresh-cut vegetables has grown rapidly because of their health benefits and convenience. However, the short shelf life of fresh-cut vegetables is an undesirable problem, which needs to be addressed [5]. Respiration is well known as an important metabolic process, which causes the deterioration of fruits and vegetables both after harvest and during food processing, and throughout the storage time. Hence, in order to reduce the respiration rate and maintain the quality of vegetables, lowing the temperature of storage (i.e., cold storage and cold-chain) and decreasing the $O_2$ level in packaged vegetables are recommended [6]. Additionally, $N_2$ is widely used in packaged food processing, thereby helping to isolate $O_2$ and retard oxidative processes and the growth of aerobic spoilage microorganisms [7]. Overall, the modification of the gaseous composition in food packaging has been demonstrated to reduce the physical–chemical deterioration and prolong the shelf life of packaged food.

As a successfully commercialized non-thermal technology, high-pressure processing (HPP) achieves a remarkably good sterilization effect and ensures that the original flavor and nutritional value of packaged products are retained [8,9]. Compared with traditional sterilization, HPP technology is less time-consuming and can maintain the nutritional value and delicate sensory properties of fruits and vegetables, owing to its restricted effect on the covalent bonds of low-molecular-mass compounds [10]. When it comes to the applications of HPP, the improved physiochemical and storage qualities of frozen albacore tuna [11], broccoli hummus [12], and multi-fruit smoothies [13] were reported. However, similar to other agricultural produce, bell peppers are highly susceptible to spoilage, especially after a series of fresh-cutting processes, and their quality tends to deteriorate. [14]. A previous study found that the storage of peppers under 5 kPa $O_2$ + 5 kPa $CO_2$ greatly reduced the ion leakage, controlled soft rotting, delayed softening, and maintained a lower metabolic activity of fresh-cut peppers [15]. Meanwhile, there has been little discussion about the application of HPP combined with $N_2$ on the shelf-life of fresh-cut bell peppers.

Therefore, this study was conducted to evaluate the impact of the $N_2$-assisted HPP treatment of fresh-cut bell peppers (FCBP) on the microorganisms, physicochemical properties, antioxidant capacity, and sensory quality after processing and during storage, to preserve the quality and enhance the shelf life of FCBP.

## 2. Materials and Methods

### 2.1. Chemicals

Folin–Ciocalteu, 2,2-diphenyl-1-picrylhydrazyl (DPPH), 2,4,6-tri-2-pyridyl-1,3,5-triazine (TPTZ), 7,8-tetramethyl-chroman-2-carboxylicacid (Trolox), and ascorbic acid were purchased from Sigma Aldrich (St. Louis, MO, USA). Guaiacol was supplied by Sinopharm Chemical Reagent (Shanghai, China). Acetonitrile of HPLC grade was purchased from Merck(Darmstadt, Germany). $N_2$ (99.999%) gases were purchased from Beijing Beiwen Gas Co., Ltd. (Beijing, China). Other analytical-grade chemicals were provided by Beijing Chemicals Co., Ltd. (Beijing, China).

### 2.2. Preparation of FCBP

Red bell peppers *(Capsicum annuum* L.) at commercial maturity were purchased from a local market in Beijing (China) and washed for 3 min in distilled water. The cleaned peppers had their stems and seeds removed then were sliced into 3 cm × 3 cm with a ceramic knife and drained off for 3 min. A total of 200 g of pepper was packed in square plastic boxes (16 cm × 10 cm × 4 cm), covered with polyethylene (PE) film, then filled with 99.999% $N_2$ using an atmosphere packaging machine (Dajiang Mechanical Equipment Co., Ltd., Wenzhou, China). The packed samples were temporarily stored at 4 °C until they were treated by HPP within 3 h. In the present study, both the untreated and HPP-treated samples were packed with $N_2$ and the subsequent discussion will focus on the impact of different HPP treatments on the qualities of FCBP with the assistance of $N_2$.

### 2.3. HPP Treatments

High-pressure processing was carried out with HPP equipment (30.0 L, Baotou Kefa Co., Ltd., Inner Mongolia, Baotou, China) at room temperature (25 ± 2 °C). The pressurization rate was about 120 MPa/min and the pressure was immediately released to 0.1 MPa (<3 s) after treatment. The pressure-transmitting fluid was distilled water. The treatment time did not include the pressure increase and the releasing time.

According to our pre-experiment, yeasts and molds (Y&M) were not detected in FCBP treated by 400 MPa/7.5 min (HPP-400) and 500 MPa/7.5 min (HPP-500) and the counts of total aerobic bacteria (TAB) were less than 3.00 $\log_{10}$ CFU/g, which also had a better performance in sensory evaluation compared to other HPP processing methods (300/400/500 MPa with 1/2.5/5/7.5 min, Table 1). Therefore, the samples were processed by HPP-400 and HPP-500 treatments in this study.

**Table 1.** Sensory evaluation of fresh-cut peppers by high-pressure processing (HPP) treatment.

| HPP Process | | Color | Flavor | Texture | Appearance | Overall Acceptability |
|---|---|---|---|---|---|---|
| Pressure (MPa) | Holding Time (min) | | | | | |
| Untreated | - | 4.20 ± 0.62a | 3.90 ± 0.99a | 4.20 ± 0.63d | 4.40 ± 0.52d | 4.50 ± 0.53c |
| 300 | 1 | 3.78 ± 0.67a | 4.11 ± 1.17a | 3.44 ± 0.73cd | 4.33 ± 0.71d | 3.67 ± 0.71b |
| | 2.5 | 3.78 ± 0.83a | 3.22 ± 1.39 | 3.78 ± 0.67abc | 3.56 ± 1.24abcd | 3.22 ± 0.67ab |
| | 5 | 3.44 ± 1.33a | 3.00 ± 1.00a | 2.56 ± 0.53ab | 2.78 ± 1.20a | 3.11 ± 0.78ab |
| | 7.5 | 3.78 ± 0.67a | 3.44 ± 0.88a | 3.78 ± 0.67cd | 4.00 ± 1.12bcd | 3.50 ± 0.50ab |
| 400 | 1 | 3.78 ± 0.83a | 3. 33 ± 1.00a | 3.22 ± 0.67bcd | 4.11 ± 0.78cd | 3.50 ± 0.50ab |
| | 2.5 | 3.33 ± 1.12a | 3.11 ± 1.17a | 2.56 ± 0.53ab | 3.11 ± 1.17abc | 3.00 ± 0.50ab |
| | 5 | 3.33 ± 1.23a | 3.00 ± 1.32a | 2.33 ± 0.50a | 3.11 ± 1.17abc | 2.78 ± 0.97a |
| | 7.5 | 4.22 ± 0.67a | 3.44 ± 0.88a | 3.78 ± 0.97cd | 4.33 ± 0.87d | 3.72 ± 0.67b |
| 500 | 1 | 4.11 ± 1.05a | 3.22 ± 1.20a | 3.33 ± 0.71cd | 4.33 ± 0.87d | 3.67 ± 0.50b |
| | 2.5 | 3.33 ± 1.12a | 3.00 ± 1.58a | 2.33 ± 0.50a | 3.00 ± 1.12ab | 3.00 ± 0.71ab |
| | 5 | 3.56 ± 1.10a | 2.89 ± 1.05a | 2.33 ± 0.87a | 2.89 ± 1.17a | 2.78 ± 0.67a |
| | 7.5 | 3.44 ± 0.88a | 2.89 ± 1.17a | 3.00 ± 1.00abc | 4.11 ± 0.78cd | 2.83 ± 1.00a |

All data were the mean ± S.D., $n = 3$. Values with different letters within one column are significantly different ($p < 0.05$). -, represents the pressure holding time of untreated groups was zero.

### 2.4. Storage Conditions

The untreated and HPP-treated samples were stored at 4 °C in the dark. Sample analyses were carried out at 0, 4, 8, 12, 16, 20, and 25 days of storage. After 8 days, the untreated samples were badly spoilt and not able to be analyzed for quality parameters.

### 2.5. Microbiological Analysis

When counting the viable microorganisms in samples, the total plate count method was applied [16]. A homogenizer bagmixer 400 (Interscience, Mourjou, France) was used to homogenize 25 g of samples and 225 mL of sterile normal saline (0.85% sodium chloride). The obtained pepper suspensions were serially diluted in sterile saline solution and plated in triplicate on nutritional agar for TAB counts at 37 °C for 48 ± 2 h and on the rose bengal medium for Y&M at 28 °C for 72 ± 2 h. After proper incubation, all the colonies were counted.

### 2.6. Physicochemical Characteristics Analysis

A GT6G7 juice extractor (Zhejiang Light Industry Machinery Co., Ltd., Zhejiang, China) was used to prepare FCBP pulp for pH, total soluble solids (TSSs), and weight loss ratio measurements.

The pH value was measured in triplicate at 25 ± 2 °C with a Thermo Orion 868 pH meter (Thermo Fisher Scientific, Inc., Waltham, MA, USA).

An Abbe refractometer (WAY-2S, Shanghai Precision and Scientific Instrument Co., Shanghai, China) was used to detect the total soluble solids (TSSs) at 25 ± 2 °C. The final results were reported as °Brix [16].

To analyze the weight loss ratio, the weights of untreated and HPP-treated samples were measured, respectively [17]. The weight loss ratio was calculated using Equation (1) as follows:

$$\text{Weight loss ratio } (\%) = \frac{m_0 - m_1}{m_0} \times 100\%, \quad (1)$$

where $m_0$ indicates the weight of untreated FCBP and $m_1$ indicates the weight of HPP-treated samples during storage.

### 2.7. Color Assessment

The color assessment was evaluated at 25 ± 2 °C using a color measurement spectrophotometer (Hunter Lab Color Quest XE, Hunter Associates Laboratory, Inc., Reston, VA, USA) in the reflectance mode. The standard illuminant source inside the instrument (type of light source: D65) was used. Samples of crushed FCBP were loaded into a quartz cuvette (50 mm diameter), carefully removing air bubbles, and placed under the measuring aperture of the spectrophotometer [18]. The color of FCBP was expressed in L*, a*, and b* values. The total color difference ΔE is a parameter that describes the overall color difference of HPP-treated samples compared to the reference sample. It was calculated using Equation (2) as follows:

$$\Delta E = \sqrt[2]{\left[(\Delta L^*)^2 + (\Delta a^*)^2 + (\Delta b^*)^2\right]}, \quad (2)$$

where $\Delta L^* = (L^*_1 - L^*_0)$; $\Delta a^* = (a^*_1 - a^*_0)$; and $\Delta b^* = (b^*_1 - b^*_0)$. Subscript "0" indicates the color value for the reference sample (untreated FCBP at day 0) and subscript "1" indicates the color value for the sample being analyzed. All the measurements were conducted ten times, and the results were averaged.

### 2.8. Texture Profile Analysis

The texture profile analysis was valued by a texture analyzer (TX-XT Plus, Stable Micro System, Scarsdale, NY, USA) to evaluate the hardness of FCBP, following the method of Tangwongchai et al. [19] with some modifications. Each sample was cut into a 10 mm × 10 mm × 3 mm size for testing and was axially compressed two times to 30% of the original height with a 38 mm cylinder probe at a pretest speed of 1 mm/s, a test speed of 1 mm/s, and a post-test speed of 2 mm/s. Ten determinations were performed for each treatment. From the resulting force–time curves, the hardness (N) parameters were obtained and the average hardness values were calculated.

### 2.9. Determination of Total Phenols

The FCBP was ground and filtered, then the pepper pulp was centrifuged (GR21G, Hitachi Koki Co., Ltd., Tokyo, Japan) at 8000× *g* for 15 min at 4 °C. The supernatant was gathered and diluted 10 times with distilled water for further analysis.

Total phenols were evaluated according to Ryu and Koh [20] with slight modifications. A total of 0.1 mL of diluted sample was mixed with 2 mL of the Folin–Ciocalteu reagent (previously diluted 10-fold with distilled water). After being incubated for 5 min, 1.8 mL of 7.5% $Na_2CO_3$ ($m/v$) solution was added. After 1 h, the absorbance of the mixture was measured at 765 nm. (UV-726, Shimadzu, Shanghai, China). The results were expressed as the milligrams of gallic acid equivalent per 100 g of FCBP (mg GAE/100 g).

### 2.10. HPLC Analysis of Ascorbic Acid

For the extraction and analysis of ascorbic acid in FCBP, the method was proposed by Cao et al. [21] with modifications. A total of 2 g of the ground FCBP flesh was mixed with 4 mL of metaphosphoric acid (2.5%) and diluted to 10 mL using distilled water after filtration. After passing through a 0.45 μm cellulose nitrate membrane, the FCBP was ready for testing.

Ascorbic acid was separated and detected by a liquid chromatograph (LC-20AT) equipped with a UV-Vis detector (SPD-20AV) from Shimadzu Corporation (Kyoto, Japan). The separation was performed using Sunfire TM C18 from Waters (Milford, Massachusetts, USA). The mobile phase was an isocratic solvent system consisting of 95.00% monopotassium phosphate (50 mM, pH = 3.0) and 5.00% acetonitrile. The flow rate was 1.0 mL/min and aliquots of 20 μm were injected. The detection was performed in absorbance mode at 245 nm. Whole analyses were conducted at room temperature (25 ± 2 °C). A calibration curve was calculated using an external standard and used for quantification. Results were expressed as milligrams of ascorbic acid per 100 g of FCBP.

### 2.11. PPO Activity Assay

The polyphenol oxidase activity of the samples was analyzed according to Cao et al. [12], with some modifications. To extract the crude enzymes, 5 mL of FCBP flesh was blended with 25 mL of phosphate buffer solution (0.2 M, pH 6.5) and then centrifuged at 4000× *g* for 10 min at 4 °C. Crude enzymes were obtained from the supernatant after centrifugation. Subsequently, 0.5 mL of crude enzymes was thoroughly mixed with 3 mL of 1.0% o-methoxyphenol (diluted with 0.2 M, pH 6.5 phosphate buffer solution) and 10 μL of 1.5% hydrogen peroxide. The absorbance of the mixed solution was recorded every 10 s for 5 min at 470 nm (UV-726, Shimadzu, Shanghai, China). The enzyme activity unit (U) was defined as the change in absorbance of 0.001 units caused by 1 mL of enzyme extraction in 1 min.

### 2.12. Determination of Antioxidant Capacity

To study the antioxidant activity of FCBP, the free-radical scavenging effect on the ·DPPH radical and ferric reducing/antioxidant power (FRAP) was evaluated, following the method described by Gao et al. [22].

#### 2.12.1. DPPH Assay

At the beginning of the reaction, 100 μL of 10-fold diluted FCBP flesh was added to a cuvette containing 4 mL of a methanol solution (0.14 mol/L) of the methanolic ·DPPH solution. The mixture was kept in the dark for 50 min at room temperature and then its absorption was measured at 517 nm. Determinations were made using a UV-726 spectrophotometer (Shimadzu, Shanghai, China). Trolox solutions within the range of 100–1000 μM were used for calibration and the baseline was corrected by methanol.

#### 2.12.2. FRAP Assay

Freshly prepared FRAP solution contained 25 mL of 0.3 M acetate buffer (pH 3.6), 2.5 mL of 10 mM TPTZ (dissolved in 40 mM HCl), and 2.5 mL of 20 mM ferric chloride. A total of 4 mL of FRAP solution was mixed with 10-fold diluted pepper flesh in the dark at 37 °C for 10 min. The ferric reducing ability of the samples was measured by detecting the increase in absorbance at 593 nm with a UV-726 spectrophotometer (Shimadzu, Shanghai, China). Trolox solutions within the range of 100–1000 μM were used for calibration. The results were expressed as the radical scavenging activity of ·DPPH and FRAP, and were calculated by Equation (3) as follows:

$$\text{Radical scavenging activity} = \frac{A_1 - A_2}{A_1} \times 100, \tag{3}$$

where $A_1$ is the absorbance of the untreated sample at 517 or 593 nm and $A_2$ is the absorbance in the presence of FCBP extract.

Antioxidant activity is expressed as millimoles of Trolox equivalents per kilogram of sample.

### 2.13. Sensory Evaluation

Ten volunteers (College of Food Science and Nutritional Engineering at the China Agricultural University, six woman and four man, aged 22–25) participated on the sensory test. HPP-treated and untreated samples were given to the participants, for a sensory evaluation on storage days 0, 4, 8, 12, 16, 20, and 25. To understand the characteristics of good-quality FCBP and the meaning of the different terminologies used in the sensory evaluation, all 10 participants had been trained for a sensory test at least once previously. Samples were evaluated for their sensory characteristics (taste, flavor, texture, and appearance) and overall acceptability on a 5-point scale (Table 2).

**Table 2.** The standard score sheet for the sensory evaluation of the fresh-cut bell pepper.

| Score | Taste | Flavor | Texture | Appearance | Overall Acceptability |
|---|---|---|---|---|---|
| 5 | Refreshing, juicy and sweet; appropriate brittleness | Special pepper aroma; favorable soft and comfortable | Complete fruit tissue; stiff and springy | Full flesh; no drip loss | Excellent |
| 4 | Less sweet or juicy; a certain degree of brittleness | Special pepper aroma; relatively soft and comfortable | Certain springy | Full flesh; a little drip loss | Good |
| 3 | Lighter sweetness; general brittleness | Special pepper aroma | Slightly soft | Partly wrinkled; a little drip loss | General |
| 2 | No sweetness; tender | A little special pepper aroma | Soft | Partly wrinkled; serious drip loss | Bad |
| 1 | No sweetness; soft rotten | Pungent odor | Rotten | Sever wrinkled; serious drip loss | Unacceptable |

### 2.14. Statistical Analysis

All the experiments were carried out in triplicate and the average values were reported. The data were analyzed using statistical software (SPSS 17.0, Chicago, IL, USA). The results were expressed as mean ± S.D. Data were analyzed with one-way analysis of variance and Tukey multiple comparison tests (significance level $p < 0.05$) to verify whether mean values were significantly different.

## 3. Results and Discussion

### 3.1. Microbiological Analysis

The number of surviving cells after HPP treatments was determined by monitoring the TAB and Y&M counts. As shown in Table 3, the initial counts of TAB and Y&M in the untreated sample were 4.18 and 2.23 $\log_{10}$ CFU/g, respectively.

Both HPP-400 and HPP-500 treatments resulted in the reduction in Y&M to a level below the detection limit (10 CFU/g) during 25 days of storage at 4 °C. Similarly, several studies showed that Y&M were not detected immediately after HPP treatments, and survivors were kept below the detection limit in cupped strawberry [22] and banana puree [16].

Meanwhile, the counts of TAB in samples were significantly reduced to 2.69 and 2.15 $\log_{10}$ CFU/g after HPP-400 and HPP-500 treatments, respectively. The counts of TAB increased both in HPP-treated samples and untreated ones during storage. Meanwhile, the counts of TAB were consistently lower than 4 $\log_{10}$ CFU/g after the HPP-400 and HPP-500 treatments during storage, which showed that there was better microbiological stability after HPP treatment in comparison with untreated ones. This demonstrated that HPP is an effective technique for inactivating the microorganisms in FCBP. Similar results were found in fresh-cut cucumber slices [23] and precut lettuce [24]. With adiabatic compression and rapid expansion during HPP treatment, structural damage to the cell membranes of microorganisms occurs, and the inactivation of enzymes and the denaturation of active compounds, both of which are lethal to bacteria [10]. Furthermore, the presence of $N_2$ in packed FCBP could also inhibit the proliferation of microorganisms during storage in the present study.

**Table 3.** Changes in microorganisms, pH, total soluble solids (TSSs), weight loss ratio, and hardness of fresh-cut bell peppers during storage at 4 °C.

| | Treatment | Storage Time (Days) | | | | | | |
|---|---|---|---|---|---|---|---|---|
| | | 0 | 4 | 8 | 12 | 16 | 20 | 25 |
| TAB ($log_{10}$ CFU/g) | Untreated | 4.18 ± 0.36Ca | 5.21 ± 0.11b | 6.91 ± 0.37c | — | — | — | — |
| | HPP-400 | 2.69 ± 0.07Ba | 4.05 ± 0.17c | 2.84 ± 0.50a | 2.79 ± 0.13a | 3.64 ± 0.18b | 3.85 ± 0.35b | 3.92 ± 0.11b |
| | HPP-500 | 2.15 ± 0.07Aa | 2.35 ± 0.12a | 3.80 ± 0.28b | 1.69 ± 0.21c | 2.73 ± 0.15c | 3.61 ± 0.20b | 3.83 ± 0.13b |
| Y&M ($log_{10}$ CFU/g) | Untreated | 2.33 ± 0.14a | 2.52 ± 0.13a | 3.18 ± 0.22b | — | — | — | — |
| | HPP-400 | ND | ND | ND | ND | ND | ND | ND |
| | HPP-500 | ND | ND | ND | ND | ND | ND | ND |
| pH | Untreated | 5.09 ± 0.01Ba | 5.08 ± 0.03b | 5.05 ± 0.03b | — | — | — | — |
| | HPP-400 | 4.97 ± 0.02Aa | 4.99 ± 0.01a | 4.99 ± 0.03a | 5.04 ± 0.27b | 4.53 ± 0.33c | 4.18± 0.03d | 4.17± 0.04d |
| | HPP-500 | 4.99 ± 0.02Aa | 4.99 ± 0.01a | 4.98 ± 0.01a | 5.00 ± 0.01a | 5.03 ± 0.02b | 4.99 ± 0.02a | 4.45 ± 0.02c |
| TSS (°Brix) | Untreated | 6.80 ± 0.17Aa | 6.50 ± 0.17a | 6.60 ± 0.10a | — | — | — | — |
| | HPP-400 | 7.27 ± 0.12ABab | 7.83 ± 0.12c | 7.70± 0.10bc | 7.73± 0.15bc | 7.80 ± 0.26c | 7.77 ± 0.25c | 7.00 ± 0.17a |
| | HPP-500 | 7.33 ± 0.12Babc | 7.83 ± 0.21c | 7.47± 0.12abc | 7.73± 0.31bc | 7.27 ± 0.15ab | 7.60 ± 0.20bc | 7.07 ± 0.12a |
| Weight loss ratio (%) | Untreated | 0 | 1.1 | 4.3 | — | — | — | — |
| | HPP-400 | 21.9 | 22.7 | 22.8 | 25.7 | 25.9 | 26.3 | 27.2 |
| | HPP-500 | 18.7 | 23.1 | 23.5 | 24.4 | 24.2 | 24.4 | 24.5 |
| Hardness (N) | Untreated | 28.92 ± 1.22Ba | 28.05 ± 1.28ab | 21.31± 1.17b | — | — | — | — |
| | HPP-400 | 13.44 ± 1.45ABa | 10.59 ± 0.80bc | 11.09 ± 0.37bc | 11.76 ± 0.22bc | 10.88 ± 0.57bc | 10.51 ± 0.33bc | 9.65 ± 0.26c |
| | HPP-500 | 11.54 ± 0.93Aa | 10.99 ± 0.45a | 10.58 ± 0.78a | 11.11 ± 0.28a | 9.79 ± 0.24a | 10.18 ± 0.29a | 10.72 ± 1.95a |

—, not tested; ND, not detected (detection limit <10 CFU/g); TAB, total aerobic bacteria; Y&M, yeasts and molds. All data were the mean ± S.D., $n = 3$. Values with different letters within one row are significantly different ($p < 0.05$). The capital letters within one column are significantly different at day 0 ($p < 0.05$).

### 3.2. Chemical and Physical Analysis

The changes in the pH, TSS, and weight loss ratio of FCBP during storage at 4 °C are shown in Table 3. There was a significant change in the pH values after both HPP-400 and HPP-500 treatments. This might have been caused by the instantaneous leaching of acidic components after HPP treatment (e.g., phenolic compounds). Furthermore, the pH values were reduced generally in all FCBP during the 25-day refrigeration period, which may be attributed to the formation of organic acids produced by the microbial proliferation of FCBP.

The increase in TSS in FCBP was found after HPP treatment. Partially due to the textural damage of FCBP, the ingredients were concentrated relatively. Meanwhile, fluctuating TSS values were observed in both HPP-400 and HPP-500 treatments, which could have been caused by the dynamic balance between the loss of water molecules and the leaking of ruptured FCBP cells after HPP treatment. Whereas, Gallotta et al. [25] found that the TSS content of fresh-cut nectarines rose during 15-day storage with a reduced sample weight owning to the fruit ripening and the increasing production of ethylene. However, these effects may have been suppressed in the present study, as the packages contained 99.99% $N_2$ and the after-ripening effect of fruit requires oxygen [26].

Due to the textural damage, the weight loss ratio of the FCBP increased substantially after the HPP treatment. Throughout the storage period, the rates of increase in the weight loss ratio after HPP-400 and HPP-500 treatments were 0.21% and 0.23%, respectively, and the weight loss ratio of HPP-500 tended to be stabilized for 12–25 days. There is no doubt that a series of processing operations, including HPP, could lead to the mechanical wounding of fresh-cut vegetable tissues [27], providing physical conditions for FCBP to lose weight after HPP-400 and HPP-500 treatments. Meanwhile, peppers are highly prone to lose water and corrupt naturally during long-term storage [28]. Presently, the HPP ensured the microbiological safety of FCBP, while a higher weight loss ratio was inevitably found after HPP treatment. Further studies need to focus on this undesirable result of FCBP.

### 3.3. Hardness Analysis

As shown in Table 3, the hardness of FCBP was 11.54–13.44 N after the HPP treatments, whereas the hardness of the untreated samples was 28.92 N. The texture loss observed in FCBP after HPP-400 and HPP-500 treatments could be defined as an instantaneous pressure softening. The leaching and non-enzymatic depolymerization of cell wall pectin in FCBP after the HPP application could be another reason for this effect [29,30]. For cherry tomatoes, Tangwongchai, Ledward, and Ames [19] found that the firmness was reduced by 90% after 400 MPa/20 min treatment. Similarly, strawberry halves were processed at 400–550 MPa for 0.1–20 min, reaching a maximum loss of 80% in hardness [31]. Moreover, the hardness of untreated samples decreased rapidly during 0–8 days of storage at 4 °C. However, none of the HPP-treated FCBP experienced a significant reduction in hardness over 25 days.

### 3.4. Color Analysis

The color of vegetables has a remarkable impact on consumer appreciation and acceptance. Processing had a significant effect on the color variables of FCBP in the present study (Table 4). There was no significant difference in the L* value between the HPP-treated and untreated samples at day 0. This result was in agreement with previous studies with fresh-cut peaches [32] and nectarines [33]. It has also been found that, regardless of the HPP treatment conditions, the chromatic parameters remained almost unaffected [32]. A lower L* value is an indicator of darkening and enzymatic browning, which is one of the factors most limiting the shelf life of fresh-cut products. Since HPP treatment could induce damage to the structure and the breaking of cell walls in fruits, it would facilitate enzyme (mainly PPO) and substrate contact, thus affecting the color of fruits. Meanwhile, $N_2$-assisted HPP treatment, a technology that ensures the absence of oxygen inside packaging and inactivates PPO activity effectively, contributes to the preservation of L* by inhibiting enzymatic browning for 25 days.

**Table 4.** Changes in the color parameters of fresh-cut bell pepper during storage at 4 °C.

| Process | Storage (Days) | L* | a* | b* | ΔE |
|---|---|---|---|---|---|
| Untreated | 0 | 32.90 ± 2.53Aa | 28.84 ± 4.47Aa | 17.45 ± 3.89Aa | 0 |
| | 4 | 30.93 ± 3.79a | 28.67 ± 3.99a | 20.04 ± 3.57ab | 3.26 |
| | 8 | 31.52 ± 3.65a | 33.67 ± 5.41b | 26.56 ± 10.93b | 10.40 |
| HPP-400 | 0 | 31.06 ± 2.31Aa | 36.38 ± 4.27Ba | 27.94 ± 7.85Ba | 13.05 |
| | 4 | 29.04 ± 3.13a | 35.49 ± 6.52ab | 30.30 ± 11.63a | 14.97 |
| | 8 | 29.68 ± 2.01a | 30.10 ± 4.33cd | 22.15 ± 5.05ab | 5.83 |
| | 12 | 32.00 ± 1.90a | 27.70 ± 2.15cd | 15.26 ± 2.35b | 2.63 |
| | 16 | 30.01 ± 4.50a | 30.50 ± 5.05bc | 20.50 ± 6.36ab | 4.52 |
| | 20 | 30.65 ± 8.613a | 31.22 ± 8.79abc | 21.78 ± 12.18ab | 5.43 |
| | 25 | 30.94 ± 2.90a | 25.13 ± 3.33d | 14.53 ± 3.03ab | 5.11 |
| HPP-500 | 0 | 33.49 ± 1.37ABa | 34.18 ± 6.54Ba | 28.31 ± 14.93Bab | 12.12 |
| | 4 | 30.08 ± 2.53ab | 36.31 ± 6.69a | 30.89 ± 8.42a | 15.63 |
| | 8 | 27.83 ± 3.45b | 30.75 ± 5.36b | 24.16 ± 10.26bc | 8.62 |
| | 12 | 28.01 ± 6.70b | 26.82 ± 3.80b | 16.76 ± 2.21c | 5.34 |
| | 16 | 27.52 ± 3.57b | 29.61 ± 4.12b | 21.71 ± 5.19bc | 6.91 |
| | 20 | 27.90 ± 3.76b | 26.33 ± 2.06b | 17.29 ± 3.85c | 5.60 |
| | 25 | 31.62 ± 3.52ab | 25.91 ± 4.31b | 16.91 ± 6.15c | 3.24 |

All the data were the mean ± S.D., $n = 3$. The capital letters within one column show significant differences at day 0 ($p < 0.05$). Values with different letters within one column are significantly different ($p < 0.05$).

There were significant ($p < 0.05$) effects of HPP treatment on the a* and b* values of FCBP. The HPP-treated samples became redder (higher a* values) and more yellow (higher b* values) at day 0. This result can probably be attributed to the cell disruption and release of pigment compounds after HPP treatment. Oliveira et al. [29] found a significant decrease in the a* and b* values in Peruvian carrot after both 600 MPa/5 min and 600 MPa/30 min. Meanwhile, significant increases in the a* and b* values of pawpaw pulp were found after 600 MPa/76 s [34]. The disparity between our observations and those reported from other studies may be attributed to differences in the HPP treatment conditions and the types of fruit and vegetables used. During storage for 25 days, HPP-treated samples showed a decrease in their a* and b* values. Thus, a color shift toward negative a* and negative b* directions indicated less red and less yellow in the samples, which was probably due to the significant degradation of chromogenic compounds, such as decreases in the amounts of carotenoids, flavonoids, and anthocyanins during storage.

The ΔE value, which is an indicator of total color difference, also showed that there were significant differences between untreated and treated samples. It has been considered that a ΔE of two would be a noticeable visual difference for a number of situations [35]. Thus, in this study noticeable changes were observed in the color of HPP-treated samples in comparison to that of untreated ones, which was probably due to the significant differences in the L*, a*, and b* values between the untreated and treated samples. During whole storage, the decrease in the ΔE of HPP-treated FCBP was related to the gradual degradation of chromogenic compounds, which diminished the color difference from the untreated sample. In contrast, the color behaviors of HPP-400-treated FCBP were more stable than those of HPP-500, as manifested by the lower ΔE and smaller changes in color values.

### 3.5. PPO Activity

Figure 1 shows the change in the residual PPO activity of FCBP during storage. After HPP-400 and HPP-500 treatments, the residual PPO activity in the FCBP was 44.09% and 66.13%, indicating that the PPO activity was passivated by the HHP treatment. Similarly, the application of HPP treatment has also been found to inhibit the PPO activity in fresh-cut potato [36] and fresh-cut peach [37].

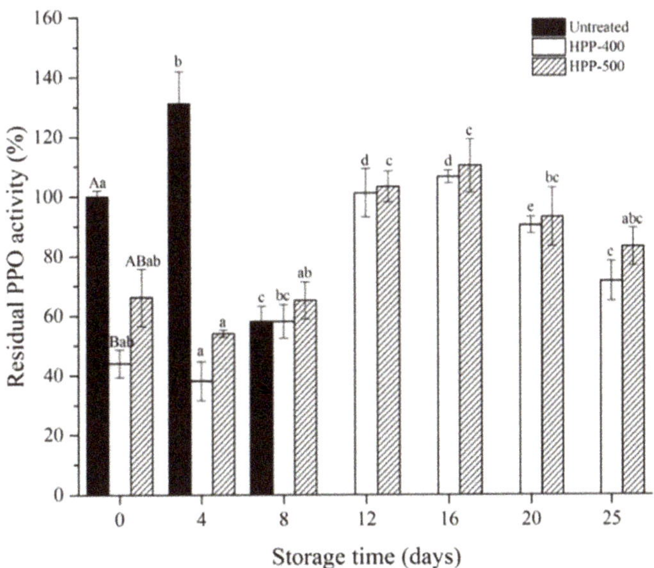


**Figure 1.** Changes in residual PPO activities of fresh-cut bell pepper during storage at 4 °C. The capital letters within one column indicate significant differences at day 0 ($p < 0.05$). Values with different letters within one column are significantly different during storage ($p < 0.05$).

During the first 8 days, HPP-treated FCBP showed lower PPO activity, whereas the PPO activity increased significantly after 8 days ($p < 0.05$). There was no significant difference in PPO activity between the HPP-400- and HPP-500-treated samples at day 25. The inactivation of PPO is highly dependent on several factors, especially the acidity of the medium, so the decrease in pH value in FCBP at the later stage of storage (8–25 days) was helpful for the recovery of PPO residual activity [38]. To maintain the color stability of FCBP during storage, additional hurdles should be considered to completely inactivate the residual activity of PPO in future studies.

### 3.6. Total Phenols and Ascorbic Acid

Changes in the total phenols and ascorbic acid contents of FCBP after HPP treatment during storage are shown in Figure 2. Compared to the total phenols content in an untreated sample (99.09 mg GAE/100 g), the total phenols in FCBP increased to 106.69 mg GAE/100 g and 108.80 mg GAE/100 g after treatments with HPP-400 and HPP-500, respectively. The previous study showed that HPP could cause a great compression of fruit microstructure and disruptions of the cell walls and membranes, leading to an increase in permeability and the leaking of total phenols in cells [38].

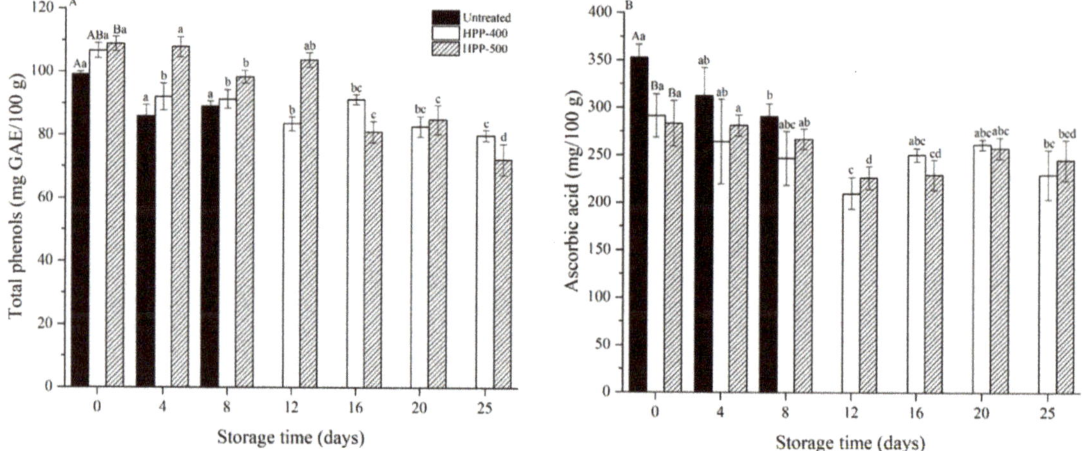


**Figure 2.** Changes in total phenols (**A**) and ascorbic acid contents (**B**) of fresh-cut bell pepper during storage at 4 °C. The capital letters within one column indicated significant differences at day 0 ($p < 0.05$). Values with different letters within one column are significantly different during storage ($p < 0.05$).

All the samples showed a significant reduction in their content of total phenols during storage (Figure 2A). After 25 days of storage, the losses of total phenols in the HPP-400- and HPP-500-treated FCBP were 19.44% and 27.19%, respectively. The leaching of phenolic compounds from flesh to juice was the main reason for this [22]. Although there was a high residual PPO activity, which was related to the degradation of phenolic compounds during storage [39], the presence of $N_2$ inside the packaging did not provide proper conditions for the degradation of total phenols by PPO.

Figure 2B shows the changes in the ascorbic acid content of FCBP during storage. In all samples, the content of ascorbic acid in HPP-treated FCBP was significantly lower than that in untreated ($p < 0.05$). HPP treatment has a good performance in accelerating the extraction of ascorbic acid and other soluble compounds [40]. Therefore, in the present study due to the freely soluble characteristics of ascorbic acid, the decrease in ascorbic acid may be related to the drip loss of FCBP caused by HPP treatments. No difference in ascorbic acid content was observed between two HPP-treated samples.

The content of ascorbic acid decreased significantly in the untreated sample over 8 days ($p < 0.05$), yet no significant losses of ascorbic acid were found in the HPP-treated ones. At the end of storage, the ascorbic acid content decreased by 21.16% and 13.57% in the HPP-400- and HPP-500-treated FCBP, respectively. Drip loss, evidenced by the increasing weight loss ratio during storage, could be the main reason for this decrease. Furthermore, although FCBP was completely packaged with $N_2$ instead of oxygen in this study, ascorbic acid can degrade anaerobically. Its influence factors include the presence or lack of oxygen, the amount of light, the presence of cupric ions, the temperature of processing, the storage time, and especially the pH [41]. The maximum degradation rate was reported at pH 4.0 and the minimum at pH 2.5–3.0 [42]. Therefore, it was reasonable to infer that after 8 days, with the pH decreasing to around 4.0 and the extension of storage time, the ascorbic acid degraded gradually in the HPP-treated samples.

### 3.7. Antioxidant Capacity Analysis

The antioxidant capacity in FCBP significantly increased after HPP-500 ($p < 0.05$), regardless of whether the DPPH or FRAP method was used (Figure 3). A similar result for antioxidant capacity was also found in pressurized swedes [43] and carrots [44].

In particular, $N_2$-assisted HPP either increased or maintained the antioxidant capacity compared with untreated samples for 0–8 days, indicating that the $N_2$-assisted HPP treatment was useful in preserving the antioxidant capacity of FCBP. Furthermore, the antioxidant capacity in HPP-500-treated samples was higher than that in HPP-400-treated ones for 0–8 days. However, HPP-400 treatment had a better performance in terms of the stability and persistence of antioxidant capacity in the second half of the storage period (12–25 days). After 25 days of storage, the DPPH antioxidant capacity decreased by 30.32% and 57.35% after treatments with HPP-400 and HPP-500, respectively. The FRAP antioxidant capacity decreased by 10.6% and 58.6% after 25 days, respectively. These results were probably due to the fact that a higher HPP treatment level (HPP-500) induced more severe structural damage to FCBP, leading to a better extracting effect of the antioxidant. Antioxidant capacity is closely related to the content of antioxidant substances (e.g., polyphenols and ascorbic acid) in samples. Therefore, the change in antioxidant capacity during storage can be explained by the change in antioxidant substances. During storage, the antioxidant capacity of the samples decreased significantly, which may be due to two reasons. One is that HPP treatment destroys the texture of the sample and that the weight loss ratio increases with the extension of storage time, resulting in the loss of polyphenols and other antioxidant substances; the other is that the rupture of the cells in samples allows the polyphenols to contact PPO, causing an enzymatic reaction, thus the polyphenols of the sample are consumed.

### 3.8. Sensory Evaluation

A sensory evaluation of the FCBP was carried out during 25 days of storage. The sensory scores were given by trained panelists ($N = 10$) according to a standard score sheet (Table 2) and they are shown in Table 5. The results showed that the sensory attributes of taste in the HPP-400-treated samples were similar to those for the untreated ones at day 0. Compared to HPP-400 treatment, the FCBP processed by HPP-500 received a higher score for the attributes of texture and appearance, probably because a better brightness ($L^*$ value) and lower weight loss ratio were found. Interestingly, a higher score for the attributes of taste and flavor were also found in the HPP-500-treated FCBP. Obviously, the HPP-500-treated samples had higher consumer acceptability. These results demonstrated that although there were significant changes in color, hardness, and other quality parameters, HPP processing did not negatively affect the sensory quality of FCBP.

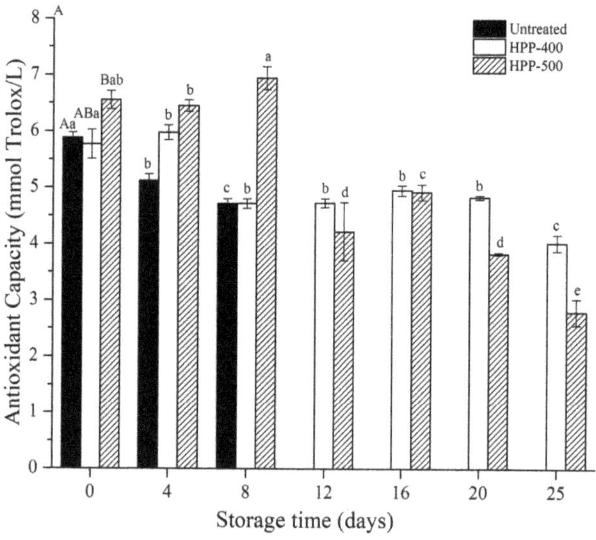


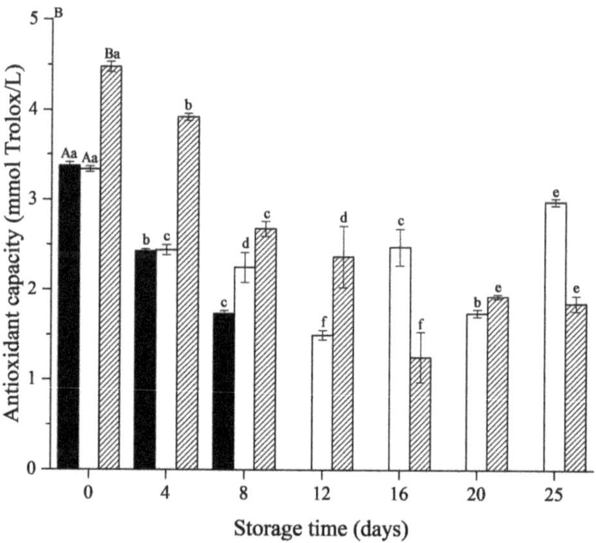


**Figure 3.** Changes in the antioxidant capacity (DPPH (**A**) and FRAP (**B**)) of fresh-cut bell pepper during storage at 4 °C. The capital letters within one column show significant differences at day 0 ($p < 0.05$). Values with different letters within one column are significantly different during storage ($p < 0.05$).

The sensory scores of HPP-treated samples also decreased during storage, but the rate of decline was lower than in untreated ones, and it was not until day 25 that the samples began to corrupt. FCBP treated by HPP-500 performed better in terms of sensory quality than those treated by HPP-400 in the whole storage. The conclusion can be drawn that the sensory qualities of FCBP can be guaranteed after the application of $N_2$-assisted HPP treatment. Furthermore, different processing conditions affected the sensory qualities of FCBP, both in terms of the storage stability and consumer acceptance of all sensory qualities.

**Table 5.** Variations in the sensory scores of fresh-cut bell pepper during storage at 4 °C.

| Process | Storage (Days) | Taste | Flavor | Texture | Appearance | Overall Acceptability |
|---|---|---|---|---|---|---|
| Untreated | 0 | 3.78 ± 0.67Aa | 4.33 ± 0.86Aa | 3.44 ± 0.73Aa | 4.33 ± 0.50Aa | 3.67 ± 0.61Aa |
| | 4 | 3.78 ± 0.83a | 3.67 ± 1.22a | 3.33 ± 0.56a | 3.55 ± 1.24a | 3.22 ± 0.64ab |
| | 8 | 2.00 ± 0.71b | 1.67 ± 0.71b | 3.40 ± 0.73a | 3.44 ± 0.72a | 2.78 ± 0.76b |
| HPP-400 | 0 | 3.75 ± 0.89Aa | 3.50 ± 1.20BCa | 2.75 ± 0.70Bab | 3.88 ± 0.83Ba | 3.38 ± 0.52Ba |
| | 4 | 3.67 ± 1.12ab | 3.22 ± 0.83a | 3.03 ± 0.86ab | 3.33 ± 0.70a | 3.11 ± 0.78a |
| | 8 | 3.78 ± 0.67a | 3.44 ± 0.88a | 2.77 ± 0.66a | 4.00 ± 1.11a | 3.50 ± 0.56a |
| | 12 | 3.33 ± 1.12ab | 3.11 ± 1.17a | 2.55 ± 0.53ab | 3.11 ± 1.17a | 3.00 ± 0.51a |
| | 16 | 2.78 ± 0.97ab | 2.44 ± 0.73a | 2.30 ± 0.50a | 3.10 ± 1.17a | 2.78 ± 0.97a |
| | 20 | 2.78 ± 0.67ab | 2.33 ± 0.50a | 2.33 ± 0.50a | 3.00 ± 1.12a | 3.00 ± 0.71a |
| | 25 | 2.33 ± 1.00b | 2.44 ± 0.73a | 2.33 ± 0.50a | 3.11 ± 1.17a | 2.78 ± 0.91a |
| HPP-500 | 0 | 4.20 ± 0.63Ba | 3.90 ± 0.99ABa | 3.40 ± 0.97Aab | 4.40 ± 0.51Aa | 3.50 ± 0.53Aab |
| | 4 | 4.22 ± 0.83a | 3.22 ± 1.20a | 3.30 ± 0.70ab | 4.33 ± 0.85a | 3.67 ± 0.40ab |
| | 8 | 4.22 ± 0.67a | 3.40 ± 0.88a | 3.70 ± 0.90a | 4.33 ± 0.87a | 3.72 ± 0.70ab |
| | 12 | 3.56 ± 0.73abc | 3.33 ± 1.00a | 3.22 ± 0.67abc | 4.11 ± 0.78ab | 3.50 ± 0.50bc |
| | 16 | 3.78 ± 0.67ab | 3.22 ± 0.97a | 3.50 ± 0.52ab | 4.00 ± 1.12ab | 3.50 ± 0.53bc |
| | 20 | 2.78 ± 0.83bc | 2.56 ± 0.88a | 2.56 ± 0.50bc | 3.00 ± 1.00b | 3.11 ± 0.78bc |
| | 25 | 2.63 ± 0.74c | 2.50 ± 0.93a | 2.25 ± 0.88c | 3.00 ± 1.20b | 2.75 ± 0.71c |

All the data were the mean ± S.D., $n = 3$. The capital letters within one column show significant differences at day 0 ($p < 0.05$). Values with different letters within one column are significantly different ($p < 0.05$).

## 4. Conclusions

$N_2$-assisted HPP FCBP exhibited a high microbial reduction after processing and a better microbiological stability during storage. Color changes were noticeable between the untreated and treated samples after processing, and the ΔE values significantly decreased during storage. HPP effectively improved the total phenols content and antioxidant capacity of FCBP and significantly inactivated the PPO activity when compared to untreated samples. Besides this, HPP processing did not negatively impact the acceptability of all sensory attributes in contrast to untreated samples. However, due to the instantaneous pressure, softening was found in FCBP after HPP treatment, and significant weight loss ratios and losses of hardness were also found.

Therefore, $N_2$-assisted HPP processing may be a good choice for the preservation of FCBP. Moreover, to further improve the quality and prolong the shelf-life of fresh-cut vegetables and decrease the damage to the natural structure after HPP, further research is required.

**Author Contributions:** Conceptualization, Y.W. and X.L.; methodology, J.C. and L.Z.; writing—original draft preparation, F.Z.; writing—review and editing, Y.W.; visualization, F.Z.; supervision, X.L. All authors have read and agreed to the published version of the manuscript.

**Funding:** This study was financially supported by Project No. 2016YFD0400700 and 2016YFD0400704 of the National Key Research and Development Program of China.

**Conflicts of Interest:** The authors declare no conflict of interest.

## References

1. Bae, H.; Jayaprakasha, G.K.; Jifon, J.; Patil, B.S. Variation of antioxidant activity and the levels of bioactive compounds in lipophilic and hydrophilic extracts from hot pepper (*Capsicum* spp.) cultivars. *Food Chem.* **2012**, *134*, 1912–1918. [CrossRef]
2. Hallmann, E.; Marszalek, K.; Lipowski, J.; Jasinska, U.; Kazimierczak, R.; Srednicka-Tober, D.; Rembialkowska, E. Polyphenols and carotenoids in pickled bell pepper from organic and conventional production. *Food Chem.* **2019**, *278*, 254–260. [CrossRef]
3. Devgan, K.; Kaur, P.; Kumar, N.; Kaur, A. Active modified atmosphere packaging of yellow bell pepper for retention of physico-chemical quality attributes. *J. Food Sci. Technol. Mys.* **2019**, *56*, 878–888. [CrossRef] [PubMed]
4. Bahram-Parvar, M.; Lim, L.T. Fresh-Cut Onion: A Review on Processing, Health Benefits, and Shelf-Life. *Compr. Rev. Food Sci.* **2018**, *17*, 290–308. [CrossRef]
5. Yousuf, B.; Qadri, O.S.; Srivastava, A.K. Recent developments in shelf-life extension of fresh-cut fruits and vegetables by application of different edible coatings: A review. *LWT Food Sci. Technol.* **2018**, *89*, 198–209. [CrossRef]

6. Saenmuang, S.; Al-Haq, M.I.; Samarakoon, H.C.; Makino, Y.; Kawagoe, Y.; Oshita, S. Evaluation of Models for Spinach Respiratory Metabolism Under Low Oxygen Atmospheres. *Food Bioprocess Technol.* **2012**, *5*, 1950–1962. [CrossRef]
7. Caleb, O.J.; Mahajan, P.V.; Al-Said, F.A.-J.; Opara, U.L. Modified Atmosphere Packaging Technology of Fresh and Fresh-cut Produce and the Microbial Consequences—A Review. *Food Bioprocess Technol.* **2013**, *6*, 303–329. [CrossRef] [PubMed]
8. Aaby, K.; Grimsbo, I.H.; Hovda, M.B.; Rode, T.M. Effect of high pressure and thermal processing on shelf life and quality of strawberry purée and juice. *Food Chem.* **2018**, *260*, 115–123. [CrossRef]
9. Pega, J.; Denoya, G.I.; Castells, M.L.; Sarquis, S.; Aranibar, G.F.; Vaudagna, S.R.; Nanni, M. Effect of High-Pressure Processing on Quality and Microbiological Properties of a Fermented Beverage Manufactured from Sweet Whey Throughout Refrigerated Storage. *Food Bioprocess Technol.* **2018**, *11*, 1101–1110. [CrossRef]
10. Bhattacharjee, C.; Saxena, V.K.; Dutta, S. Novel thermal and non-thermal processing of watermelon juice. *Trends Food Sci. Technol.* **2019**, *93*, 234–243. [CrossRef]
11. Cartagena, L.; Puertolas, E.; Martinez de Maranon, I. Application of High Pressure Processing After Freezing (Before Frozen Storage) or Before Thawing in Frozen Albacore Tuna (*Thunnus alalunga*). *Food Bioprocess Technol.* **2020**, *13*, 1791–1800. [CrossRef]
12. Venzke Klug, T.; Benito Martinez-Hernandez, G.; Collado, E.; Artes, F.; Artes-Hernandez, F. Effect of Microwave and High-Pressure Processing on Quality of an Innovative Broccoli Hummus. *Food Bioprocess Technol.* **2018**, *11*, 1464–1477. [CrossRef]
13. Picouet, P.A.; Hurtado, A.; Jofre, A.; Banon, S.; Ros, J.-M.; Dolors Guardia, M. Effects of Thermal and High-pressure Treatments on the Microbiological, Nutritional and Sensory Quality of a Multi-fruit Smoothie. *Food Bioprocess Technol.* **2016**, *9*, 1219–1232. [CrossRef]
14. Kabir MS, N.; Chowdhury, M.; Lee, W.-H.; Hwang, Y.-S.; Cho, S.-I.; Chung, S.-O. Influence of delayed cooling on quality of bell pepper (*Capsicum annuum* L.) stored in a controlled chamber. *Emir. J. Food. Agric.* **2019**, *31*, 271–280.
15. Rodoni, L.; Vicente, A.; Azevedo, S.; Concellon, A.; Cunha, L.M. Quality retention of fresh-cut pepper as affected by atmosphere gas composition and ripening stage. *LWT Food Sci. Technol.* **2015**, *60*, 109–114. [CrossRef]
16. Xu, Z.; Wang, Y.; Ren, P.; Ni, Y.; Liao, X. Quality of Banana Puree During Storage: A Comparison of High Pressure Processing and Thermal Pasteurization Methods. *Food Bioprocess Technol.* **2016**, *9*, 407–420. [CrossRef]
17. Wang, J.; Yang, X.H.; Mujumdar, A.S.; Wang, D.; Zhao, J.H.; Fang, X.M.; Zhang, Q.; Xie, L.; Gao, Z.J.; Xiao, H.W. Effects of various blanching methods on weight loss, enzymes inactivation, phytochemical contents, antioxidant capacity, ultrastructure and drying kinetics of red bell pepper (*Capsicum annuum* L.). *LWT Food Sci. Technol.* **2017**, *77*, 337–347. [CrossRef]
18. Patras, A.; Brunton, N.; Da Pieve, S.; Butler, F.; Downey, G. Effect of thermal and high pressure processing on antioxidant activity and instrumental colour of tomato and carrot purees. *Innov. Food Sci. Emerg.* **2009**, *10*, 16–22. [CrossRef]
19. Tangwongchai, R.; Ledward, D.A.; Ames, J.M. Effect of high-pressure treatment on the texture of cherry tomato. *J. Agric. Food Chem.* **2000**, *48*, 1434–1441. [CrossRef] [PubMed]
20. Ryu, D.; Koh, E. Application of response surface methodology to acidified water extraction of black soybeans for improving anthocyanin content, total phenols content and antioxidant activity. *Food Chem.* **2018**, *261*, 260–266. [CrossRef]
21. Cao, X.; Bi, X.; Huang, W.; Wu, J.; Hu, X.; Liao, X. Changes of quality of high hydrostatic pressure processed cloudy and clear strawberry juices during storage. *Innov. Food Sci. Emerg.* **2012**, *16*, 181–190. [CrossRef]
22. Gao, G.; Ren, P.; Cao, X.; Yan, B.; Liao, X.; Sun, Z.; Wang, Y. Comparing quality changes of cupped strawberry treated by high hydrostatic pressure and thermal processing during storage. *Food Bioprocess Technol.* **2016**, *100*, 221–229. [CrossRef]
23. Guo, Y.; Li, M.; Han, H.; Cai, J. Salmonella enterica serovar Choleraesuis on fresh-cut cucumber slices after reduction treatments. *Food Control* **2016**, *70*, 20–25. [CrossRef]
24. Ou, Y.; Liu, Q.; Zhou, B.; Hu, X.; Zhang, Y. Influence of gas and High Hydrostatic Pressure on quality of prefabricated lettuce during shelf life. *Food Res. Dev.* **2017**, *38*, 203–207.
25. Gallotta, A.; Allegra, A.; Inglese, P.; Sortino, G. Fresh-cut storage of fruit and fresh-cuts affects the behaviour of minimally processed Big Bang nectarines (*Prunus persica* L. Batsch) during shelf life. *Food Packag. Shelf.* **2017**, *15*, 62–68. [CrossRef]
26. Arpaia, M.L.; Collin, S.; Sievert, J.; Obenland, D. Influence of cold storage prior to and after ripening on quality factors and sensory attributes of 'Hass' avocados. *Postharvest Biol. Tec.* **2015**, *110*, 149–157. [CrossRef]
27. Liu, C.; Chen, C.; Jiang, A.; Sun, X.; Guan, Q.; Hu, W. Effects of plasma-activated water on microbial growth and storage quality of fresh-cut apple. *Innov. Food Sci. Emerg.* **2020**, *59*, 102256. [CrossRef]
28. Gil, M.I.; Tudela, J.A. Fresh and fresh-cut fruit vegetables: Peppers. In *D and Modified Atmospheres for Fresh and Fresh-Cut Produce*; Gil, M.I., Beaudry, R., Eds.; Elsevier Inc.: Amsterdam, The Netherlands, 2020; pp. 521–525.
29. De Oliveira, M.M.; Tribst, A.A.L.; Leite, B.R.D.; de Oliveira, R.A.; Cristianini, M. Effects of high pressure processing on cocoyam, Peruvian carrot, and sweet potato: Changes in microstructure, physical characteristics, starch, and drying rate. *Innov. Food Sci. Emerg.* **2015**, *31*, 45–53. [CrossRef]
30. Sila, D.N.; Doungla, E.; Smout, C.; Van, L.A.; Hendrickx, M. Pectin fraction interconversions: Insight into understanding texture evolution of thermally processed carrots. *J. Agric. Food Chem.* **2006**, *54*, 8471–8479. [CrossRef] [PubMed]
31. Duvetter, T.; Fraeye, I.; Hoang, T.V.; Buggenhout, S.V.; Verlent, I.; Smout, C.; Loey, A.V.; Hendrickx, M. Effect of Pectin-methylesterase Infusion Methods and Processing Techniques on Strawberry Firmness. *J. Food Sci.* **2005**, *70*, s383–s388. [CrossRef]
32. Denoya, G.I.; Polenta, G.A.; Apóstolo, N.M.; Budde, C.O.; Sancho, A.M.; Vaudagna, S.R. Optimization of high hydrostatic pressure processing for the preservation of minimally processed peach pieces. *Innov. Food Sci. Emerg.* **2016**, *33*, 84–93. [CrossRef]

33. Miguel-Pintado, C.; Nogales, S.; Fernández-León, A.M.; Delgado-Adámez, J.; Hernández, T.; Lozano, M.; Cañada-Cañada, F.; Ramírez, R. Effect of hydrostatic high pressure processing on nectarine halves pretreated with ascorbic acid and calcium during refrigerated storage. *LWT Food Sci. Technol.* **2013**, *54*, 278–284. [CrossRef]
34. Zhang, L.; Dai, S.; Brannan, R.G. Effect of high pressure processing, browning treatments, and refrigerated storage on sensory analysis, color, and polyphenol oxidase activity in pawpaw (*Asimina triloba* L.) pulp. *LWT Food Sci. Technol.* **2017**, *86*, 49–54. [CrossRef]
35. Francis, F.J.; Clydesdale, F.M. *Food Colorimetry: Theory and Applications*; The AVI Publishing Co., Inc.: Westport, CT, USA, 1975; pp. 477–478.
36. Amaral, R.D.A.; Benedetti, B.C.; Pujola, M.; Achaerandio, I.; Bachelli, M.L.B. Effect of Ultrasound on Quality of Fresh-Cut Potatoes During Refrigerated Storage. *Food Eng. Rev.* **2015**, *7*, 176–184. [CrossRef]
37. Chang, Y.H.; Wu, S.J.; Chen, B.Y.; Huang, H.W.; Wang, C.Y. Effect of high pressure processing and thermal pasteurization on overall quality parameters of white grape juice. *J. Sci. Food Agric.* **2017**, *97*, 3166–3172. [CrossRef]
38. Denoya, G.I.; Vaudagna, S.R.; Polenta, G. Effect of high pressure processing and vacuum packaging on the preservation of fresh-cut peaches. *LWT Food Sci. Technol.* **2015**, *62*, 801–806. [CrossRef]
39. Isabel, O.S.; Robert, S.F.; Teresa, H.; Olga, M. Carotenoid and phenolic profile of tomato juices processed by high intensity pulsed electric fields compared with conventional thermal treatments. *Food Chem.* **2009**, *112*, 258–266.
40. Vega-Gálvez, A.; López, J.; Galotto, M.J.; Puente-Díaz, L.; Quispe-Fuentes, I.; Scala, K.D. High hydrostatic pressure effect on chemical composition, color, phenolic acids and antioxidant capacity of Cape gooseberry pulp (*Physalis peruviana* L.). *LWT Food Sci. Technol.* **2014**, *58*, 519–526. [CrossRef]
41. Tewari, S.; Sehrawat, R.; Nema, P.K.; Kaur, B.P. Preservation effect of high pressure processing on ascorbic acid of fruits and vegetables: A review. *J. Food Biochem.* **2017**, *41*, e12319. [CrossRef]
42. Moura, T.; Gaudy, D.; Jacob, M.; Cassanas, G. PH influence on the stability of ascorbic acid spray-drying solutions. *Pharm. Acta Helv.* **1994**, *69*, 77–80. [CrossRef]
43. Clariana, M.; Valverde, J.; Wijngaard, H.; Mullen, A.M.; Marcos, B. High pressure processing of swede (*Brassica napus*): Impact on quality properties. *Innov. Food Sci. Emerg.* **2011**, *12*, 85–92. [CrossRef]
44. Evrendilek, G.A.; Ozdemir, P. Effect of various forms of non-thermal treatment of the quality and safety in carrots. *LWT Food Sci. Technol.* **2019**, *105*, 344–354. [CrossRef]

 *foods*

*Article*

# Effect of Freeze Crystallization on Quality Properties of Two Endemic Patagonian Berries Juices: Murta (*Ugni molinae*) and Arrayan (*Luma apiculata*)

**María Guerra-Valle [1,2], Siegried Lillo-Perez [1,3], Guillermo Petzold [1,*] and Patricio Orellana-Palma [4,*]**

[1] Laboratory of Cryoconcentration, Department of Food Engineering, Universidad del Bío-Bío, Av. Andrés Bello 720, 3780000 Chillán, Chile; maria.guerra1601@egresados.ubiobio.cl (M.G.-V.); silillo@egresados.ubiobio.cl (S.L.-P.)
[2] Doctorado en Ingeniería de Alimentos, Universidad del Bío-Bío, Av. Andrés Bello 720, 3780000 Chillán, Chile
[3] Magíster en Ciencias e Ingeniería en Alimentos, Universidad del Bío-Bío, Av. Andrés Bello 720, 3780000 Chillán, Chile
[4] Department of Biotechnology, Universidad Tecnológica Metropolitana, Las Palmeras 3360, P.O. Box, 7800003 Ñuñoa, Santiago, Chile
* Correspondence: gpetzold@ubiobio.cl (G.P.); p.orellanap@utem.cl (P.O.-P.); Tel.: +56-42-2463173 (G.P.); +56-2-27877032 (P.O.-P.)

**Abstract:** This work studied the effects of centrifugal block freeze crystallization (CBFC) on physicochemical parameters, total phenolic compound content (TPCC), antioxidant activity (AA), and process parameters applied to fresh murta and arrayan juices. In the last cycle, for fresh murta and arrayan juices, the total soluble solids (TSS) showed values close to 48 and 54 Brix, and TPCC exhibited values of approximately 20 and 66 mg gallic acid equivalents/100 grams dry matter (d.m.) for total polyphenol content, 13 and 25 mg cyanidin-3-glucoside equivalents/100 grams d.m. for total anthocyanin content, and 9 and 17 mg quercetin equivalents/100 grams d.m. for total flavonoid content, respectively. Moreover, the TPCC retention indicated values over 78% for murta juice, and 82% for arrayan juice. Similarly, the AA presented an increase over 2.1 times in relation to the correspondent initial AA value. Thus, the process parameters values were between 69% and 85% for efficiency, 70% and 88% for percentage of concentrate, and 0.72% and 0.88 (kg solutes/kg initial solutes) for solute yield. Therefore, this work provides insight about CBFC on valuable properties in fresh Patagonian berries juices, for future applications in health and industrial scale.

**Keywords:** freeze crystallization; murta; arrayan; physicochemical properties; bioactive compounds; antioxidant activity; process parameters

**Citation:** Guerra-Valle, M.; Lillo-Perez, S.; Petzold, G.; Orellana-Palma, P. Effect of Freeze Crystallization on Quality Properties of Two Endemic Patagonian Berries Juices: Murta (*Ugni molinae*) and Arrayan (*Luma apiculata*). *Foods* **2021**, *10*, 466. https://doi.org/10.3390/foods10020466

Academic Editors: Asgar Farahnaky, Mohsen Gavahian and Mahsa Majzoobi

Received: 25 January 2021
Accepted: 19 February 2021
Published: 20 February 2021



## 1. Introduction

In the last decades, berries have gained a lot of attention due to their attractive colors, interesting physicochemical properties, and excellent nutritional and organoleptic characteristics. Thus, berries are not only consumed for their physical appearance but also for their countless and positive effects on the consumers' health, since these fruits have a significant source of micronutrients, phenolic compounds, and antioxidant activity that provide important beneficial effects for human health [1,2].

In this context, Chile has an important diversity of wild and endemic berries due to the different ecosystems of each region, since Chile presents from dry desert climate (North) to high-intensity rainfall rate and low temperatures (South) [3]. Hence, in southern-central Chile, there are various endemic berries such as maqui (*Aristotelia chilensis*), calafate (*Berberis microphylla*), Chilean strawberry (*Fragaria chiloensis ssp. chiloensis*), murta (*Ugni molinae*), and arrayan (*Luma apiculata*), and till date, murta and arrayan have been poorly studied. However, these berries have been used since ancient times as food, ingredients, colorants, and/or traditional medicines [4,5], and it opens the possibility of future scientific analysis in the fresh fruits and/or the development of studies for commercial exploitation through new processing technologies.

Specifically, murta and arrayan are two interesting and exotic berry fruits recognized as "superfruits," since these fruits present high levels in terms of fiber, vitamins, minerals, nutrients, and phytochemical composition [6]. Murta (also called murtilla, myrtle berry, or mutilla) is a wild Myrtaceae bush, with black/purple color and an intense sweet taste, and it grows from Talca (35°25′36″ S, 71°40′18″ W, Maule Region, central Chile) to the Palena River (41°28′18″ S, 72°56′12″ W, Los Lagos Region, southern Chile) [7]. Similarly, arrayan (also called cauchao) is other wild Myrtaceae endemic plant that grows in forests from Valparaíso (33°03′47″ S, 71°38′22″ W, Valparaiso Region, central Chile) to Aysen (45°34′12″ S, 72°03′58″ W, Aysen del General Carlos Ibáñez del Campo region, southern Chile). The fruits have nearly globular shape with characteristic aroma [8]. Moreover, these fruits present an important content of phenolic compounds (phenolic acids, anthocyanins, flavonoids, flavonols, and tannins), and thus, it contributes significantly to the high value of antioxidant activity measured for these fruits [9–14]. Therefore, murta and arrayan have excellent potential to be incorporated into the diet for daily consumption, either as fresh fruits or derived products such as yogurt, jams, jellies, purees, and fruit juices [15].

Additionally, different innovations have been adopted in the food industry due to the demands of consumers, every day more interested and aware, for healthy, safe, tasty, and ideal functional foods with low impact on the environment [16]. Furthermore, these innovations can be made at any stage during the food processing, and in recent years, the changes have been observed through the implementation of emerging thermal and non-thermal technologies with the intention to increase the extraction of components, efficiency, percentage of recovery, stability, and preservation of nutritional and organoleptic characteristics in each food product [17–19].

Accordingly, freeze crystallization (FC) has attracted increasing attention as emerging non-thermal technology due to their important implications for the concentration of food solutions. Specifically, FC uses low temperatures to concentrate liquid solutions and it is based on the freezing of the water, and then, the unfrozen solution (cryoconcentrated) is separated from the ice crystals (frozen fraction). Thereby, FC allows increasing the solutes and bioactive compounds in the cryoconcentrated fraction. In turn, FC had a significant energy advantage compared to the traditional concentration technologies, since FC uses only 14% of the total energy used in evaporative concentration [20]. Hence, FC has been positively applied in various liquid foods, presenting an increase in solutes and extraordinary retention of phenolic compounds and antioxidant activity [21–25]. However, to our best knowledge, there are a few studies available about the application of FC on valuable "superfruit" juices [26,27].

Thus, the objective of this study was to evaluate the use of block FC assisted by centrifugation technology on two wild endemic Chilean berry juices (murta and arrayan) in terms of physicochemical properties, process parameters, and quality characteristics.

## 2. Materials and Methods

### 2.1. Raw Materials

Fresh murta (*Ugni molinae*) and arrayán (*Luma apiculata*) were obtained from Valle Exploradores Ltda (46°37′27″ S, 72°40′36″ W, Puerto Río Tranquilo, XI Región de Aysén, Chile), with standard commercial maturity, i.e., uniform size and color, and without visual damage (injured) or being immature. Thus, the samples were transported in covered insulated boxes at Chillán (XVI Región del Ñuble, Chile) and were kept in a refrigerated chamber until their further analysis.

### 2.2. Fruit Juice Preparation

The fruits were washed with tap water to remove dust and dirt, and then subjected to manual pressure to obtain the juice, and then, the fresh juice was filtered using a nylon cloth (0.8 mm fine-mesh) to separate large fragments such as pulp, seeds, and peels from the liquid. Later, the fresh juice was stored at 4 °C until analysis or processing.

### 2.3. Freeze Crystallization (FC)

The FC process at three cycles was based on the protocol described by Orellana-Palma et al. [27] (Figure 1). First, the plastic centrifugal tubes with 45 mL of juice were isolated with foamed polystyrene. Then, the tubes with juice were frozen through axial freezing front propagation in a static freezer (model 280, M&S Consul, Sao Paulo, Brazil) overnight at −20 °C. Later, the cryoconcentrated fraction was separated from the ice fraction by centrifugation (Eppendorf 5430R, Hamburg, Germany) at 20 °C for 20 min at 4000 rpm, and thus, this procedure can be called the first cycle (C1). The cryoconcentrate juice obtained from C1 was collected. Subsequently, the C1 solution was used as the new feed solution for the second (C2) cycle, and thus, the C2 solution was used as new feed solution for the third (C3) cycle. All the cycles were performed under the same FC procedure (axial freezing front propagation at −20 °C and centrifugation conditions). Specifically, different quality properties such as physicochemical properties, total phenolic compound content, and antioxidant activity were determined in the cryoconcentrated fraction from C1, C2, and C3.

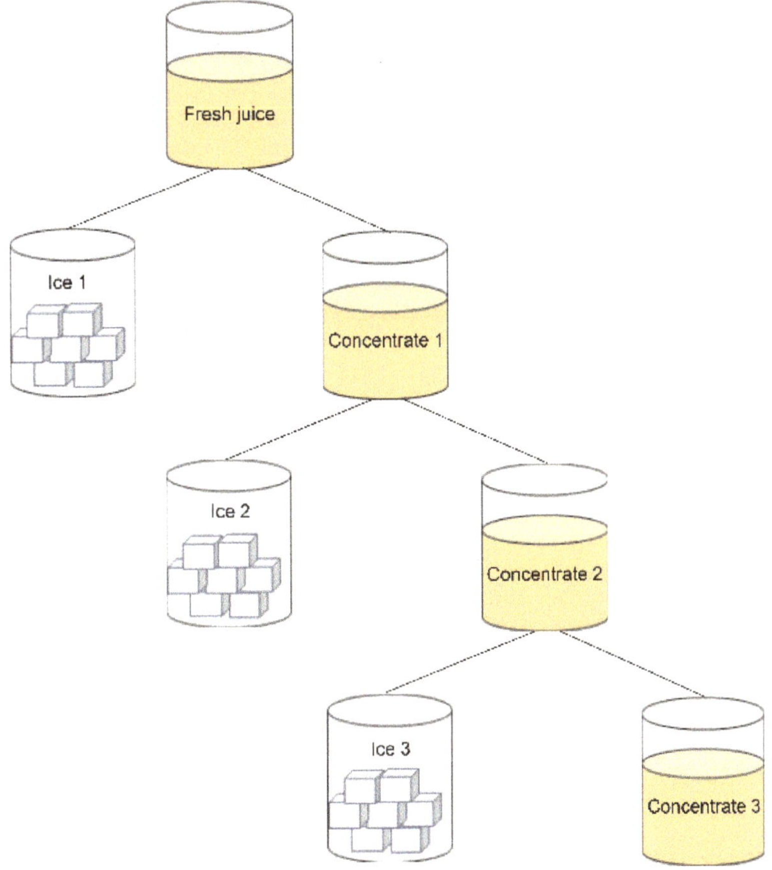


**Figure 1.** Freeze crystallization procedure at three centrifugation cycles.

### 2.4. Physicochemical Profile

The physicochemical profile in the samples was measured based on the method described by Orellana-Palma et al. [27]. Thus, the TSS content was measured using a digital refractometer (PAL-3, Atago Inc., Tokyo, Japan) with wide range (0–93%), and the results were expressed as Brix. The pH measurements were determined with a digital pH meter (HI-2221, Hanna Instruments, Woonsocket, RI, USA). The total titratable acidity (TTA) (grams of malic acid (MA) per liter of sample, g MA/L) was obtained by titrimetric method. The density (kg/m$^3$) was obtained using a pycnometer at 25 °C. The color properties were evaluated using a colorimeter (CM-5, Konica Minolta, Osaka, Japan), and the results were provided in accordance with CIELab system (L*: darkness-whiteness, a*; greenness-redness axis, and b*: blueness-yellowness axis). The total color difference (ΔE*) was calculated in accordance with Equation (1).

$$\Delta E^* = \sqrt{\left(L^* - L_0^*\right)^2 + \left(a^* - a_0^*\right)^2 + \left(b^* - b_0^*\right)^2}, \quad (1)$$

where $\Delta E^*$ is the change variation between fresh juices and cryoconcentrated samples. The subscript 0 corresponds to the initial CIELab values in the fresh juice, and L*, a*, and b* are the color properties in each cryoconcentrated sample.

### 2.5. Determination of Total Phenolic Compound Content (TPCC)

TPCC were determined by the total polyphenol content (TPC), total anthocyanin content (TAC), and total flavonoid content (TFC) assays.

TPC, TAC, and TFC assays were measured with a spectrophotometer UV/Vis (T70, Oasis Scientific Inc., Greenville, SC, USA) based on the method described by Sekizawa et al. [28], Lee et al. [29], and Zhishen et al. [30], where gallic acid (GA), cyanidin-3-glucoside (C3G), and quercetin (Q) were used for the standard curve, respectively, and the results were expressed in mg of GA equivalents (GAE) per grams (g) of dry matter (mg GAE/g d.m.), mg of C3G equivalents per grams (g) of dry matter (mg C3G/g d.m.), and mg of Q equivalents (QE) per grams (g) of dry matter (mg QE/g d.m.), respectively.

Additionally, the TPCC retention represents the TPCC retained from the fresh juice in the cryoconcentrated solution. The TPCC retention was determined by Equation (2) [31].

$$\text{TPCC retention } (\%) = \left(\frac{C_o}{C_c}\right) \times \left(\frac{TPCC_c}{TPCC_o}\right) \times 100, \quad (2)$$

where $C_o$ is the initial TSS; $C_c$ is the TSS at each cycle in the cryoconcentrated fraction; $TPCC_c$ is the value of TPC, TAC, and TFC at each cycle; and $TPCC_o$ is the initial value of TPC, TAC, and TFC.

### 2.6. Determination of Antioxidant Activity (AA)

Antioxidant activity was determined by means of the 2,2-diphenyl-1-picrylhydrazyl (DPPH), 2,2′-azino-bis(3-ethylbenzothiazoline-6-sulphonic acid) (ABTS), ferric reducing antioxidant power (FRAP), and oxygen radical absorbance capacity (ORAC) assays.

DPPH, ABTS, and FRAP assays were measured with a spectrophotometer UV/Vis (T70, Oasis Scientific Inc., Greenville, SC, USA) based on the method described by Zorzi et al. [32], Garzón et al. [33], and Chen et al. [34], respectively.

ORAC assay was evaluated based on the method described by Ou et al. [35] with a multimode plate reader (Victor X3, Perkin Elmer, Hamburg, Germany). The absorbance was measured at 485 nm ($\lambda_{excitation}$) and at 520 nm ($\lambda_{emission}$) every 1 min for 60 min.

Trolox (T) was used for the standard curve, and the AA assays were expressed as μM Trolox equivalents (TE) per gram (g) of dry matter (μM TE/g d.m.).

### 2.7. Process Parameters in FC

#### 2.7.1. Efficiency (η)

η (%) is defined as the solutes in the cryoconcentrated fraction relative to the solutes remaining in the ice fraction. The η (%) was determined by Equation (3).

$$\eta\ (\%) = \frac{C_c - C_I}{C_c} * 100\%, \tag{3}$$

where $C_c$ and $C_I$ are the TSS at each cycle in the cryoconcentrated and ice fractions, respectively.

#### 2.7.2. Solute Yield (Y)

Y (kg solutes/kg initial solutes, kg/kg) represents the relation between the recovered solute mass in the initial solution and cryoconcentrated solution. The Y (kg/kg) was determined by Equation (4).

$$Y\left(\frac{kg}{kg}\right) = \frac{m_c}{m_0}, \tag{4}$$

where $m_c$ is the solute mass in the cryoconcentrated fraction and $m_0$ is the initial solute mass.

#### 2.7.3. Percentage of Concentrate (PC)

PC (%) represents the weight in the initial sample relative to the weight remaining in the ice fraction. The PC was determined by Equation (5).

$$PC\ (\%) = \frac{W_0 - W_i}{W_0} \times 100\%, \tag{5}$$

where $W_0$ and $W_i$ are the initial and final weights in the ice fraction, respectively.

### 2.8. Statistical Analysis

The treatments were conducted in triplicate at ambient temperature (≈22 °C), and the results were presented as mean ± standard deviation. One-way analysis of variance (ANOVA) was used to the evaluation of statistical analysis and the treatment means were compared via Fisher's least significant difference (Fisher's LSD) test at a confidence level of 0.95 ($p \leq 0.05$). The data were analyzed through Statgraphics Centurion XVI software (v. 16.2.04, StatPoint Technologies Inc., Warrenton, VA, USA).

## 3. Results and Discussion

### 3.1. Physicochemical Characteristics

The physicochemical properties in the fruit juices and CBFC-treated juices are shown in Figure 2 and Table 1. Additionally, the TSS, pH, TTA, and color (L*, a*, b* and ΔE*) values presented significant differences when each BFC cycle was compared with their respective fresh juice.

The fresh murta juice presented lower physicochemical characteristics values than those indicated by Ah-Hen et al. [36]. However, our results were higher than previous results found in the same fruit juice by Ah-Hen et al. [37]. Moreover, the fresh arrayan juice values were higher than previous values reported by Fuentes et al. [13]. The differences can be connected to the fact that the studies were carried out with fruits from different areas of the Southern Chile, and these zones are characterized by various climatic conditions, equivalent to constant rains and low temperatures throughout the year [38]. Additionally, various factors such as type/time of harvesting, ripening process, and/or the interaction genotype-environment can change the properties of the fruit and juice [39].

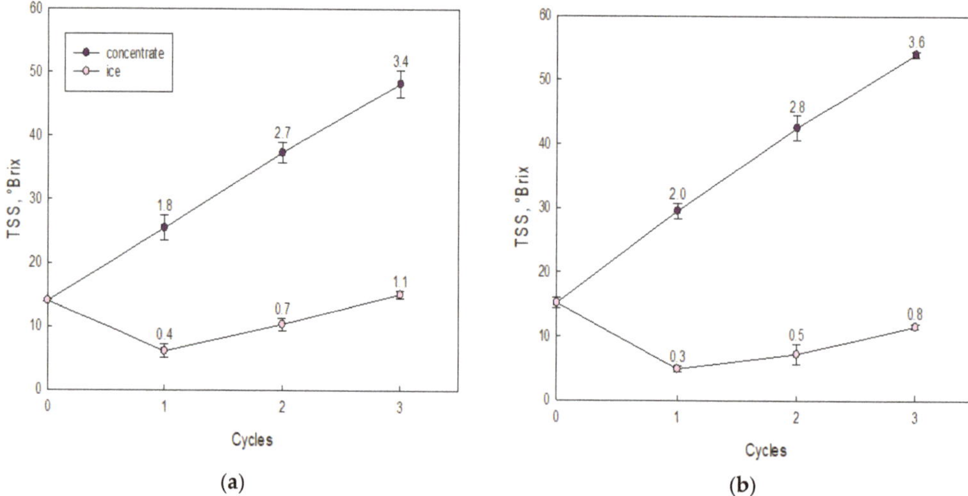


**Figure 2.** Total soluble solids (TSS) values at each BFC cycle: (**a**) murta juice and (**b**) arrayan juice. The bars represent the standard deviations. The number on the bar is the concentration index (CI), and CI is a dimensionless number that represents the increase in solutes at each BFC cycle (in both cryoconcentrated and ice fractions) with respect to the initial TSS value ($C_0$) (14.0 Brix for murta juice and 15.1 °Brix for arrayan juice), i.e., CI = $C_c/C_0$, where $C_c$ is the TSS value at each cycle.

First, independent of fresh juice, the TSS values showed an increasing trend with increasing BFC cycles (Figure 2). Hence, in the last cycle, for murta juice (Figure 2a), the solutes reached a TSS value close to 48.2 °Brix, and for arrayan juice (Figure 2b), the TSS value was approximately 54.0 °Brix, which is comparable to an increase in 3.4 and 3.6 times, in relation to the initial TSS values (14.0 and 15.1 °Brix), respectively. In comparison with previous studies in our laboratory, the results (third cycle) presented lower concentration index value than those obtained in orange juice (5.7) [40] and apple juice (3.9) [41], but higher than those achieved in calafate juice (3.0) [27], pineapple juice (3.3) [42], and blueberry juice (2.5) [43]. The variation in TSS values is due to the different and specific characteristics (pH, TTA, and density, among others) in the fresh juice, since each characteristic interacts under various forms in the cryoconcentration process [44]. It is worth to mention that, all the studies used the same freezing and centrifugation conditions.

The pH, TTA, and CIELab (L*, a*, and b*) values are given in Table 1.

In terms of pH, the values showed a significant decrease in both fresh murta juice (pH = 3.7) and fresh arrayan juice (pH = 5.1), since the values (third cycle) had a decrease in 27% (pH = 2.4) and 25% (pH = 3.8), in comparison to the correspondent pH value of the fresh juice, respectively. By contrast, the fresh murta juice and fresh arrayan juice increased the TTA values to 2.3 and 1.6 for C1, 3.1 and 2.1 for C2, and 3.8 and 2.6 for C3, indicating an increase in 224% and 217% (in the last cycle) in relation to the initial TTA values, respectively. The opposite behavior among pH and TTA values has been observed in various cryoconcentrated food liquids such as calafate juice [27], pineapple juice [42], and sapucaia nut cake milk [45]. The pH and TTA changes have been attributed to the continuous increases in TSS at each cycle, since it provokes an increase in the organic acids of the cryoconcentrates, causing the opposite phenomenon between pH and TTA values [41].

In terms of colorimetric CIELab parameters, the color measurement is considered an indicator of food quality and can also be used indirectly in the analysis of colored components contained in fruits, particularly in berries, providing an estimation of antioxidants and polyphenolic compounds [46,47].

**Table 1.** Physicochemical characteristics of the samples.

| Sample | pH | | TTA | | L* | | a* | | b* | | ΔE* | |
|---|---|---|---|---|---|---|---|---|---|---|---|---|
| | Murta | Arrayan | Murta | Arrayan | Murta | Arrayan | Murta | Arrayan | Murta | Arrayan | Murta | Arrayan |
| Fresh juice | $3.7 \pm 0.0^{a}$ | $5.1 \pm 0.1^{a}$ | $1.7 \pm 0.2^{a}$ | $1.2 \pm 0.0^{a}$ | $51.9 \pm 1.4^{a}$ | $12.2 \pm 0.1^{a}$ | $3.8 \pm 0.8^{a}$ | $8.7 \pm 0.8^{a}$ | $2.6 \pm 0.5^{a}$ | $1.1 \pm 0.2^{a}$ | - | - |
| Cycle 1 | $3.3 \pm 0.2^{b}$ | $4.7 \pm 0.2^{b}$ | $2.3 \pm 0.1^{b}$ | $1.6 \pm 0.2^{b}$ | $40.6 \pm 0.7^{b}$ | $7.5 \pm 0.5^{b}$ | $8.2 \pm 0.4^{b}$ | $22.0 \pm 0.2^{b}$ | $5.9 \pm 0.1^{b}$ | $14.0 \pm 0.2^{b}$ | $12.5 \pm 0.7^{a}$ | $19.1 \pm 0.5^{a}$ |
| Cycle 2 | $2.8 \pm 0.1^{c}$ | $4.2 \pm 0.2^{c}$ | $3.1 \pm 0.4^{c}$ | $2.1 \pm 0.1^{c}$ | $28.2 \pm 0.4^{c}$ | $4.1 \pm 0.2^{c}$ | $14.7 \pm 0.4^{c}$ | $31.0 \pm 1.4^{c}$ | $10.1 \pm 0.9^{c}$ | $22.6 \pm 1.3^{c}$ | $27.2 \pm 0.9^{b}$ | $32.0 \pm 1.1^{b}$ |
| Cycle 3 | $2.4 \pm 0.1^{d}$ | $3.8 \pm 0.0^{d}$ | $3.8 \pm 0.2^{d}$ | $2.6 \pm 0.3^{d}$ | $14.4 \pm 1.4^{d}$ | $0.4 \pm 0.5^{d}$ | $34.9 \pm 0.3^{d}$ | $38.9 \pm 1.2^{d}$ | $15.1 \pm 0.9^{d}$ | $33.4 \pm 0.4^{d}$ | $50.3 \pm 1.2^{c}$ | $45.7 \pm 0.7^{c}$ |

a–d: Different superscript letters in the same column indicate significant differences ($p \leq 0.05$) according to Fisher's LSD test. L*, a*, b*, and ΔE*, are the darkness-whiteness axis, greenness-redness axis, blueness-yellowness axis, and the total color difference, respectively.

Thus, all the samples presented an important change during the cycles, since each parameter had significant modifications, and thus, in C3, the L* values decrease from 52 to 14 CIELab units and from 12 to 0.4 CIELab units, for fresh murta and arrayan juices, which is equivalent to a decrease in 72% and 96%, respectively, signifying that the cryoconcentrated samples were darker than the original juice. However, as cycles advanced, a progressive increase in a* and b* values was observed, with values from 4 to 35 CIELab units and from 3 to 15 CIELab units, for murta juice, and from 9 to 39 CIELab units and from 1 to 33 CIELab units, for arrayan juice, respectively, indicating that the juices had a darkish red color (most noticeable in arrayan juice), and thus, the final cryoconcentrated samples presented a very dark visual appearance with red coloration (Figure 3). These results suggested that the changes in visual color of fresh juices treated by BFC can be linked to the notable increase in the TSS concentration and total phenolic compound content at each cycle [48]. Additionally, the ΔE* between the samples (fruit juices and CBFC-treated juices) specified that the highest ΔE* values were assessed for the final cycle, since the values ranged from 13 to 50 CIELab units and from 19 to 46 CIELab units, from the first to the third cycle, for murta juice and for arrayan juice, respectively. Specifically, Krapfenbauer et al. [49] defined that a $\Delta E^* \geq 3.5$ CIELab units denote visual differences by the consumers between food products. Therefore, our values indicate that the human eye can perceive visual differences between the original sample and each cryoconcentrated sample, since all the ΔE* values were higher than 13 CIELab units. All the data and visual color had similar trend to the results reported in various studies on FC applied to fresh juices [27,31,43].

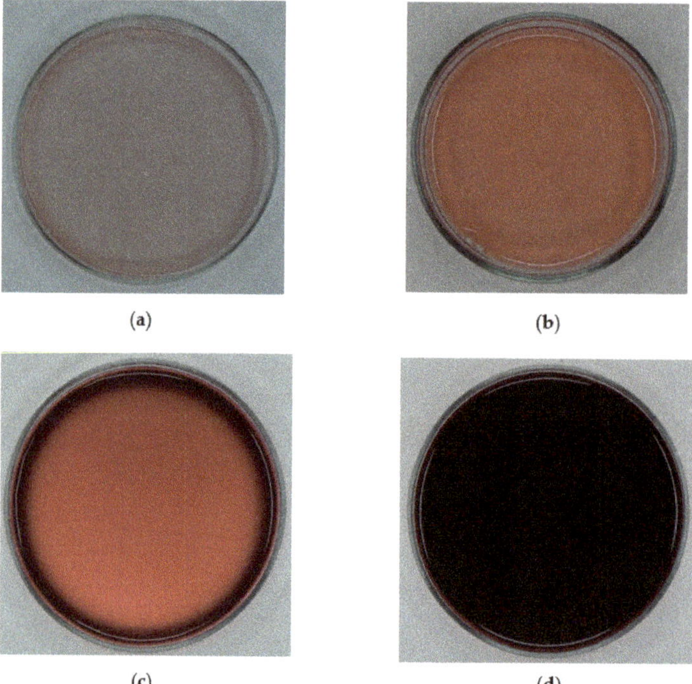


**Figure 3.** Effect on visual color of fresh murta juice and fresh arrayan juice under centrifugal block freeze crystallization (CBFC): (**a**) fresh murta juice, (**b**) cryoconcentrated murta juice, (**c**) fresh arrayan juice, and (**d**) cryoconcentrated arrayan juice.

### 3.2. *Total Phenolic Compound Content (TPCC)*

The levels of TPCC in the fruit juices and CBFC-treated juices are shown in Table 2.

**Table 2.** Total phenolic compound content of fresh juices and cryoconcentrated samples.

| Sample | TPC (mg GAE/g d.m.) | | TAC (mg C3G/g d.m.) | | TFC (mg QE/g d.m.) | |
|---|---|---|---|---|---|---|
| | Murta | Arrayan | Murta | Arrayan | Murta | Arrayan |
| Fresh juice | 6.4 ± 0.8 [a] | 20.5 ± 2.1 [a] | 4.2 ± 0.1 [a] | 8.2 ± 0.9 [a] | 3.2 ± 0.0 [a] | 6.0 ± 0.2 [a] |
| Cycle 1 | 8.1 ± 0.2 [b] | 28.7 ± 1.1 [b] | 5.1 ± 0.9 [b] | 10.5 ± 0.5 [b] | 3.6 ± 0.1 [b] | 6.9 ± 0.2 [b] |
| TPCC retention (%) | 70.0 ± 2.1 [A] | 76.9 ± 0.5 [A] | 66.6 ± 1.0 [A] | 70.8 ± 0.6 [A] | 62.1 ± 0.9 [A] | 63.2 ± 0.1 [A] |
| Cycle 2 | 13.4 ± 1.0 [c] | 47.0 ± 2.9 [c] | 8.3 ± 1.3 [c] | 17.0 ± 2.0 [c] | 5.8 ± 0.4 [c] | 11.4 ± 1.4 [c] |
| TPCC retention (%) | 78.6 ± 0.9 [B] | 85.9 ± 1.2 [B] | 73.1 ± 1.4 [B] | 77.9 ± 1.1 [B] | 67.8 ± 1.3 [B] | 71.1 ± 2.9 [B] |
| Cycle 3 | 19.9 ± 1.2 [d] | 65.6 ± 3.8 [d] | 12.5 ± 1.7 [d] | 24.8 ± 1.7 [d] | 8.6 ± 0.9 [d] | 17.0 ± 2.3 [d] |
| TPCC retention (%) | 91.1± 1.5 [C] | 93.1 ± 0.2 [C] | 85.6 ± 2.0 [C] | 88.0 ± 0.7 [C] | 78.1 ± 2.4 [C] | 82.5 ± 1.2 [C] |

a–d: Different small letters in the same column indicate significant differences ($p \leq 0.05$) between the fresh juice and their cycles, according to Fisher's LSD test. A–D: Different capital letters in the same column indicate significant differences ($p \leq 0.05$) between the TPCC retention at each cycle, according to Fisher's LSD test. TPCC, TPC, TAC, and TFC, are the total phenolic compound content, total polyphenol content, total anthocyanin content, and total flavonoid content, respectively.

First, fresh murta juice had TPCC values close to 6.4 mg GAE/g d.m., 4.2 mg C3G/g d.m., and 3.2 mg QE/g d.m., while fresh arrayan juice had TPCC values of approximately 20.5 mg GAE/g d.m., 8.2 mg C3G/g d.m., and 6.0 mg QE/g d.m., for TPC, TAC, and TAC, respectively. These results were lower than the ones reported by Ramirez [50], who reported the phenolic compounds content profile of six small berries from the VIII Region del Bio-Bío of Chile. However, the differences in TBCC values between the studies might be explained by the different harvest areas, since the VIII Region del Bio-Bío has dry temperate climates in summer and rainy temperate climates in winter, and thus, the temperature varies from 0 to 30 °C throughout the year, while the Southern Chile has constant rains and low temperatures throughout the year [38,51,52]. Hence, the climatic conditions and geographical characteristics contribute to various differences on fruit maturation, and in addition, other factors such as genetic and species variabilities, harvesting year, and growing season affect the phenolic compounds content in final fruits [53].

Concerning TPCC results, a similar behavior to the TSS values was observed, since a significant increase in TPC, TAC, and TFC values was detected as cycles advanced. Thus, in the last cycle, for murta juice, the values were close to 20.0 mg GAE/g d.m., 12.5 mg C3G/g d.m., and 8.6 mg QE/g d.m., and for arrayan juice, the values were approximately 65.6 mg GAE/g d.m., 24.8 mg C3G/g d.m., and 17.0 mg QE/g d.m., for the first, second, and third cycle, respectively, and thus, these values were over 2.7 times higher than the correspondent initial TPCC value. This upward trend in TPCC values post-FC has been observed in numerous food liquids such as green tea extract [21], fruit juices [22–24], broccoli extract [25], coffee extract [54], and wine [55]. Additionally, the retention specified a high amount of TPCC in the cryoconcentrated fraction, since the retention values (third cycle) were close to 91%, 86%, and 78%, for murta juice, and 93%, 88%, and 82%, for arrayan juice, for TPC, TAC, and TFC, respectively. The results were consistent with the values informed by Orellana-Palma et al. [41], Casas-Forero et al. [31], and Correa et al. [56], who reported TPCC retention close to 85%, 71–91%, and 90%, in apple juice, blueberry juice, and coffee extract, respectively. Therefore, the TPCC value and retention results allow corroborating that the FC technology can be visualized as a novel alternative due to the low temperatures used to concentrate, preserve, and retain an endless number of phenolic components such as polyphenols, anthocyanins, and flavonoids in the cryoconcentrated fraction [20].

### 3.3. Antioxidant Activity (AA)

Table 3 shows the AA values of the samples.

**Table 3.** Antioxidant activity (μM TE/g d.m.) of fresh juices and cryoconcentrated samples.

| Sample | DPPH | | ABTS | | FRAP | | ORAC | |
|---|---|---|---|---|---|---|---|---|
| | Murta | Arrayan | Murta | Arrayan | Murta | Arrayan | Murta | Arrayan |
| Fresh juice | $33.4 \pm 3.7^{a}$ | $62.0 \pm 7.4^{a}$ | $48.1 \pm 7.6^{a}$ | $84.8 \pm 9.9^{a}$ | $62.6 \pm 5.9^{a}$ | $92.6 \pm 3.1^{a}$ | $21.7 \pm 3.0^{a}$ | $43.4 \pm 4.4^{a}$ |
| Cycle 1 | $96.9 \pm 10.1^{b}$ | $167.3 \pm 11.6^{b}$ | $110.6 \pm 19.1^{b}$ | $229.0 \pm 10.4^{b}$ | $156.5 \pm 15.7^{b}$ | $287.0 \pm 17.2^{b}$ | $45.6 \pm 7.4^{b}$ | $99.8 \pm 11.7^{b}$ |
| Cycle 2 | $120.2 \pm 8.5^{c}$ | $266.4 \pm 20.3^{c}$ | $187.6 \pm 10.7^{c}$ | $373.1 \pm 20.5^{c}$ | $256.7 \pm 13.9^{c}$ | $416.6 \pm 20.7^{c}$ | $80.3 \pm 5.9^{c}$ | $204.0 \pm 17.0^{c}$ |
| Cycle 3 | $157.0 \pm 12.3^{d}$ | $353.1 \pm 14.0^{d}$ | $250.1 \pm 17.2^{d}$ | $407.0 \pm 7.6^{d}$ | $338.0 \pm 21.4^{d}$ | $509.1 \pm 23.1^{d}$ | $108.5 \pm 10.1^{d}$ | $221.3 \pm 20.5^{d}$ |

a–d: Different superscript letters in the same column indicate significant differences ($p \leq 0.05$) according to Fisher's LSD test.

First, in both murta and arrayan juices, the AA values (μM TE/g d.m.) were approximately 33.4, 48.1, 62.6, and 21.7, and 62.0, 84.8, 92.6, and 43.4, for DPPH, ABTS, FRAP, and ORAC, respectively. These values were lower than those found by Ramirez et al. [48], Augusto et al. [57], and Rodríguez et al. [58], who studied some quality properties of native berries species from different regions of Chile, explicating that the high variability in AA values can be justified by the interaction of factors such as edaphoclimatic conditions (climate, light, temperature, and other conditions), geographical zone, cultivar/genotype, harvesting period, cultivation practice, fruit maturation, type of storage, and even, the type of juice extraction from the fruit [59].

Once the fruit juices were subjected to three CBFC cycles, an important increase in the AA values (μM TE/g d.m.) was detected, with AA values from 96.9 to 157.0 and from 167.3 to 353.1, for DPPH, from 110.6 to 250.1 and from 229.0 to 407.0, for ABTS, from 156.5 to 338.0 and from 287.0 to 590.1, for FRAP, and from 45.6 to 108.5 and from 99.8 to 221.3, for ORAC, from the C1 to C3, for murta juice, and for arrayan juice, which is equivalent to an increase over 2.1, 2.5, 2.7, and 2.8 times, in comparison to the AA values from the fresh juices, respectively. In this way, Orellana-Palma et al. [27], Samsuri et al. [60], and Casas-Forero et al. [61] also observed higher AA values post-FC than the original sample, demonstrating the positive effects of the FC on TPCC, and in turn, these components allow concentrating and preserving the antioxidant activity from the fresh juice [21].

### 3.4. Process Parameters

The $\eta$, PC, and Y results at each cycle are presented on Figure 4.

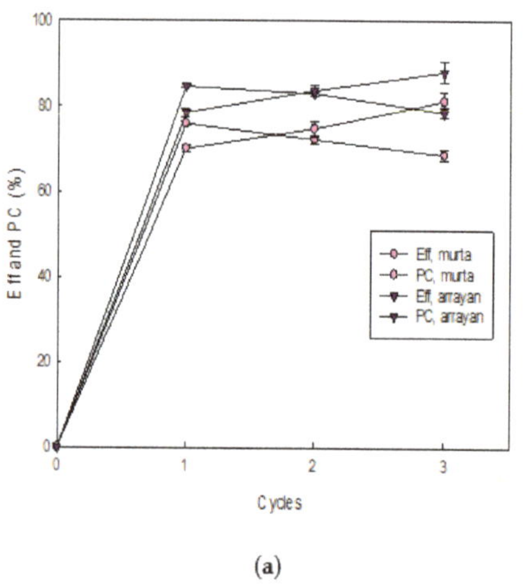


(a)

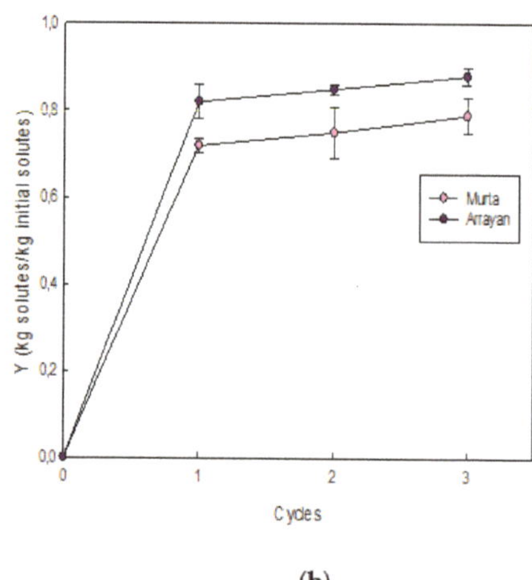


(b)

**Figure 4.** Process parameters at each CBFC cycle: (**a**) efficiency and percentage of concentrate and (**b**) solute yield.

The $\eta$ (Figure 4a) showed a significant decreasing trend from C1 to C3, with values close to 76%, 72%, and 69%, and 85%, 83%, and 79%, for murta juice and for arrayan juice, respectively. The results have similar ranges established by Bastías-Montes et al. [26], Orellana-Palma et al. [41], and Zielinski et al. [62], who studied the effect of FC in various liquid samples, specifying that the $\eta$ can be related to the TSS values. Specifically, the decrease in $\eta$ values can be related to the continuous increase in TSS values in the feed solution at each cycle, since the viscosity increases as the solutes increase, and thus, the flow of solutes was slowed in the third cycle than in the previous cycles, causing a difficulty in the separation process. In addition, the high viscosity reduces the ice purity postcentrifugation (increase in TSS values in the ice fractions cycle to cycle) [43].

The PC (Figure 4a) displayed an increasing tendency as the cycles passed, with values from 70% to 81%, for murta juice, and from 78% to 88%, for arrayan juice, from the C1 to C3, respectively. The values were similar to those reported in other samples [22,40,62], suggesting that the PC values can be connected to the components in the fresh juice, since each juice has multiple components that interact differently in the FC process, and later, in the separation process, these components can facilitate or complicate the extraction of cryoconcentrates from the ice fraction [63].

The Y parameter (Figure 4b) showed a significant increasing trend as the cycles advanced, with values of approximately 0.72, 0.75, and 0.79 (kg solutes/kg initial solutes), for murta juice, and 0.82, 0.85, and 0.88 (kg solutes/kg initial solutes), for arrayan juice, for the C1, C2, and C3, respectively. The results had comparable performance that prior studies [40,41,62], associating the Y values with the recovered mass from the original mass, i.e., for each applied cycle, a high quantity of mass can be extracted and collected as a cryoconcentrated sample [64].

## 4. Conclusions

The application of CBFC allowed obtaining higher solutes than the original sample, with an intensification of the natural color of the fresh juices, leading to increase in the phenolic compounds and antioxidant capacity in the cryoconcentrates samples, since the TPCC and AA results were over 2.7 and 2.1 times higher than the correspondent initial values, respectively. Moreover, the centrifugation improves the possibility to efficiently extract more cryoconcentrate from the ice fraction due to the high values in $\eta$, PC, and Y. These data suggest that FC can effectively improve the quality properties as well as visual appearance in any fruit juices. Thereby, the FC creates a unique opportunity for food industry, since this novel emerging non-thermal technology concentrates at low temperatures, and thus, various characteristics from the fresh juice can be retained in the final concentrated product. Therefore, the next challenge could be linked to FC in the production of concentrated juice rich in phenolic components and high antioxidant capacity from endemic fruits at pilot plant scale, and then at industrial scale, and in addition, the cryoconcentrated samples could be studied with a nutritional perspective.

**Author Contributions:** Conceptualization, M.G.-V., S.L.-P., P.O.-P., and G.P.; methodology, M.G.-V., S.L.-P., and P.O.-P.; software, M.G.-V., S.L.-P., and P.O.-P.; validation, M.G.-V. and S.L.-P.; formal analysis, M.G.-V., P.O.-P., and G.P.; investigation, M.G.-V., S.L.-P., and P.O.-P.; resources, P.O.-P. and G.P.; data curation, M.G.-V. and S.L.-P.; writing—original draft preparation, M.G.-V., S.L.-P., and P.O.-P.; writing—review and editing, P.O.-P. and G.P.; visualization, P.O.-P. and G.P.; supervision, P.O.-P. and G.P.; project administration, P.O.-P.; funding acquisition, P.O.-P. All authors have read and agreed to the published version of the manuscript.

**Funding:** This research was supported by ANID-Chile (Agencia Nacional de Investigación y Desarrollo de Chile) through the FONDECYT Postdoctoral Grant 2019 (Folio 3190420).

**Institutional Review Board Statement:** Not applicable.

**Informed Consent Statement:** Not applicable.

**Data Availability Statement:** Not applicable.

**Acknowledgments:** P.O.-P. acknowledges the financial support of ANID-Chile (Agencia Nacional de Investigación y Desarrollo de Chile) through the FONDECYT Postdoctoral Grant 2019 (Folio 3190420). M.G.-V. thanks at the Universidad del Bío-Bío for the doctoral scholarship (2018–2022) and for the "Beca de Investigación."

**Conflicts of Interest:** The authors declare no conflict of interest.

## References


1. Lavefve, L.; Howard, L.R.; Carbonero, F. Berry polyphenols metabolism and impact on human gut microbiota and health. *Food Funct.* **2020**, *11*, 45–65. [CrossRef] [PubMed]
2. Martini, D.; Marino, M.; Angelino, D.; del Bo', C.; del Rio, D.; Riso, P.; Porrini, M. Role of berries in vascular function: A systematic review of human intervention studies. *Nutr. Rev.* **2020**, *78*, 189–206. [CrossRef] [PubMed]
3. Vega-Galvez, A.; Rodríguez, A.; Stucken, K. Antioxidant, functional properties and health-promoting potential of native South American berries: A review. *J. Sci. Food Agric.* **2021**, *101*, 364–378. [CrossRef]
4. Schmeda-Hirschmann, G.; Jiménez-Aspee, F.; Theoduloz, C.; Ladio, A. Patagonian berries as native food and medicine. *J. Ethnopharmacol.* **2019**, *241*, 111979. [CrossRef]
5. Boeri, P.; Piñuel, L.; Dalzotto, D.; Monasterio, R.; Fontana, A.; Sharry, S.; Barrio, D.A.; Carrillo, W. Argentine Patagonia barberry chemical composition and evaluation of its antioxidant capacity. *J. Food Biochem.* **2020**, *44*, e13254. [CrossRef]
6. Chang, S.K.; Alasalvar, C.; Shahidi, F. Superfruits: Phytochemicals, antioxidant efficacies, and health effects–A comprehensive review. *Crit. Rev. Food Sci.* **2019**, *59*, 1580–1604. [CrossRef]
7. Fredes, C.; Parada, A.; Salinas, J.; Robert, P. Phytochemicals and traditional use of two southernmost Chilean berry fruits: Murta (*Ugni molinae* Turcz) and Calafate (*Berberis buxifolia* Lam.). *Foods* **2020**, *9*, 54. [CrossRef] [PubMed]
8. Carhuapoma, M.; Bonilla, P.; Suarez, S.; Villa, R.; López, S. Estudio de la composición química y actividad antioxidante del aceite esencial de *Luma chequen* (Molina) A. Gray "arrayán". *Ciencia e Investigación* **2005**, *8*, 73–79. [CrossRef]
9. López, J.; Vera, C.; Bustos, R.; Florez-Mendez, J. Native berries of Chile: A comprehensive review on nutritional aspects, functional properties, and potential health benefits. *J. Food Meas. Charact* **2020**, in press. [CrossRef]
10. Viktorová, J.; Kumar, R.; Řehořová, K.; Hoang, L.; Ruml, T.; Figueroa, C.R.; Valdenegro, M.; Fuentes, L. antimicrobial activity of extracts of two native fruits of Chile: Arrayan (*Luma apiculata*) and Peumo (*Cryptocarya alba*). *Antibiotics* **2020**, *9*, 444. [CrossRef]
11. Junqueira-Gonçalves, M.P.; Yáñez, L.; Morales, C.; Navarro, M.; Contreras, R.; Zúñiga, G.E. Isolation and characterization of phenolic compounds and anthocyanins from Murta (*Ugni molinae* Turcz.) fruits. Assessment of antioxidant and antibacterial activity. *Molecules* **2015**, *20*, 5698–5713. [CrossRef]
12. Jofré, I.; Pezoa, C.; Cuevas, M.; Scheuermann, E.; Freires, I.A.; Rosalen, P.L.; de Alencar, S.M.; Romero, F. Antioxidant and vasodilator activity of *Ugni molinae* Turcz. (murtilla) and its modulatory mechanism in hypotensive response. *Oxid. Med. Cell. Longev.* **2016**, *2016*, 6513416. [CrossRef]
13. Fuentes, L.; Valdenegro, M.; Gómez, M.G.; Ayala-Raso, A.; Quiroga, E.; Martínez, J.P.; Vinet, R.; Caballero, E.; Figueroa, C.R. Characterization of fruit development and potential health benefits of arrayan (*Luma apiculata*), a native berry of South America. *Food Chem.* **2016**, *196*, 1239–1247. [CrossRef] [PubMed]
14. Fuentes, L.; Figueroa, C.R.; Valdenegro, M.; Vinet, R. Patagonian berries: Healthy potential and the path to becoming functional foods. *Foods* **2019**, *8*, 289. [CrossRef]
15. Pires, T.C.; Caleja, C.; Santos-Buelga, C.; Barros, L.; Ferreira, I.C. *Vaccinium myrtillus* L. fruits as a novel source of phenolic compounds with health benefits and industrial applications-a review. *Curr. Pharm. Des.* **2020**, *26*, 1917–1928. [CrossRef] [PubMed]
16. Martins, I.B.A.; Rosenthal, A.; Ares, G.; Deliza, R. How do processing technology and formulation influence consumers' choice of fruit juice? *Int. J. Food Sci. Tech.* **2020**, *55*, 2660–2668. [CrossRef]
17. Meijer, G.W.; Lähteenmäki, L.; Stadler, R.H.; Weiss, J. Issues surrounding consumer trust and acceptance of existing and emerging food processing technologies. *Crit. Rev. Food Sci. Nutr.* **2021**, *61*, 97–115. [CrossRef] [PubMed]
18. Wen, L.; Zhang, Z.; Sun, D.W.; Sivagnanam, S.P.; Tiwari, B.K. Combination of emerging technologies for the extraction of bioactive compounds. *Crit. Rev. Food Sci. Nutr.* **2020**, *60*, 1826–1841. [CrossRef] [PubMed]
19. Siegrist, M.; Hartmann, C. Consumer acceptance of novel food technologies. *Nat. Food* **2020**, *1*, 343–350. [CrossRef]
20. Miyawaki, O.; Inakuma, T. Development of progressive freeze concentration and its application: A review. *Food Bioproc Tech.* **2021**, *14*, 39–51. [CrossRef]
21. Meneses, D.L.; Ruiz, Y.; Hernandez, E.; Moreno, F.L. Multi-stage block freeze-concentration of green tea (*Camellia sinensis*) extract. *J. Food Eng.* **2021**, *293*, 110381. [CrossRef]
22. Orellana-Palma, P.; Petzold, G.; Guerra-Valle, M.; Astudillo-Lagos, M. Impact of block cryoconcentration on polyphenol retention in blueberry juice. *Food Biosci.* **2017**, *20*, 149–158. [CrossRef]
23. Qin, F.G.; Ding, Z.; Peng, K.; Yuan, J.; Huang, S.; Jiang, R.; Shao, Y. Freeze concentration of apple juice followed by centrifugation of ice packed bed. *J. Food Eng.* **2021**, *291*, 110270. [CrossRef]
24. Das, D.; Das, D.; Gupta, A.K.; Mishra, P. Drying of *citrus grandis* (pomelo) fruit juice using block freeze concentration and spray drying. *Acta Aliment.* **2020**, *49*, 295–306. [CrossRef]
25. Azhar, A.N.; Panirselvam, M.; Amran, N.A.; Ruslan, M.S.; Samsuri, S. Retention of total phenolic content and antioxidant activity in the concentration of broccoli extract by progressive freeze concentration. *Int. J. Food Eng.* **2020**, *16*, 20190237. [CrossRef]
26. Bastías-Montes, J.M.; Martín, V.S.; Muñoz-Fariña, O.; Petzold-Maldonado, G.; Quevedo-León, R.; Wang, H.; Yang, Y.; Céspedes-Acuña, C.L. Cryoconcentration procedure for aqueous extracts of maqui fruits prepared by centrifugation and filtration from fruits harvested in different years from the same localities. *J. Berry Res.* **2019**, *9*, 377–394. [CrossRef]

27. Orellana-Palma, P.; Tobar-Bolaños, G.; Casas-Forero, N.; Zúñiga, R.N.; Petzold, G. Quality attributes of cryoconcentrated calafate (*Berberis microphylla*) juice during refrigerated storage. *Foods* **2020**, *9*, 1314. [CrossRef] [PubMed]
28. Sekizawa, H.; Ikuta, K.; Mizuta, K.; Takechi, S.; Suzutani, T. Relationship between polyphenol content and anti-influenza viral effects of berries. *J. Sci. Food Agric.* **2013**, *93*, 2239–2241. [CrossRef] [PubMed]
29. Lee, J.; Durst, R.W.; Wrolstad, R.E. Determination of total monomeric anthocyanin pigment content of fruit juices, beverages, natural colorants, and wines by the pH differential method: Collaborative study. *J. AOAC Int.* **2005**, *88*, 1269–1278. [CrossRef]
30. Zhishen, J.; Mengcheng, T.; Jianming, W. The determination of flavonoid contents in mulberry and their scavenging effects on superoxide radicals. *Food Chem.* **1999**, *64*, 555–559. [CrossRef]
31. Casas-Forero, N.; Orellana-Palma, P.; Petzold, G. Influence of block freeze concentration and evaporation on physicochemical properties, bioactive compounds and antioxidant activity in blueberry juice. *Food Sci. Technol.* **2020**, *40*, 387–394. [CrossRef]
32. Zorzi, M.; Gai, F.; Medana, C.; Aigotti, R.; Morello, S.; Peiretti, P.G. Bioactive compounds and antioxidant capacity of small berries. *Foods* **2020**, *9*, 623. [CrossRef]
33. Garzón, G.A.; Riedl, K.M.; Schwartz, S.J. Determination of anthocyanins, total phenolic content, and antioxidant activity in Andes berry (*Rubus glaucus* Benth). *J. Food Sci.* **2009**, *74*, C227–C232. [CrossRef] [PubMed]
34. Chen, L.; Xin, X.; Yuan, Q.; Su, D.; Liu, W. Phytochemical properties and antioxidant capacities of various colored berries. *J. Sci. Food Agric.* **2014**, *94*, 180–188. [CrossRef] [PubMed]
35. Ou, B.; Hampsch, M.; Prior, R.L. Development and validation of an improved oxygen radical absorbance capacity assay using fluorescein as the fluorescent probe. *J. Agric. Food Chem.* **2001**, *49*, 4619–4626. [CrossRef]
36. Ah-Hen, K.; Zambra, C.E.; Aguëro, J.E.; Vega-Gálvez, A.; Lemus-Mondaca, R. Moisture diffusivity coefficient and convective drying modelling of murta (*Ugni molinae* Turcz): Influence of temperature and vacuum on drying kinetics. *Food Bioprocess Technol.* **2013**, *6*, 919–930. [CrossRef]
37. Ah-Hen, K.S.; Mathias-Rettig, K.; Gómez-Pérez, L.S.; Riquelme-Asenjo, G.; Lemus-Mondaca, R.; Muñoz-Fariña, O. Bioaccessibility of bioactive compounds and antioxidant activity in murta (*Ugni molinae* T.) berries juices. *J. Food Meas. Charact.* **2018**, *12*, 602–615. [CrossRef]
38. Aceituno, P.; Boisier, J.P.; Garreaud, R.; Rondanelli, R.; Rutllant, J.A. Climate and Weather in Chile. In *Water Resources of Chile*, 2nd ed.; Fernández, B., Gironás, J., Eds.; Springer: Cham, Switzerland, 2020; pp. 7–29.
39. Cheng, G.; He, Y.N.; Yue, T.X.; Wang, J.; Zhang, Z.W. Effects of climatic conditions and soil properties on Cabernet Sauvignon berry growth and anthocyanin profiles. *Molecules* **2014**, *19*, 13683–13703. [CrossRef] [PubMed]
40. Orellana-Palma, P.; González, Y.; Petzold, G. Improvement of centrifugal cryoconcentration by ice recovery applied to orange juice. *Chem. Eng. Technol.* **2019**, *42*, 925–931. [CrossRef]
41. Orellana-Palma, P.; Lazo-Mercado, V.; Gianelli, M.P.; Hernández, E.; Zuñiga, R.; Petzold, G. Influence of cryoconcentration on quality attributes of apple juice (*Malus domestica cv. Red Fuji*). *Appl. Sci.* **2020**, *10*, 959. [CrossRef]
42. Orellana-Palma, P.; Zuñiga, R.N.; Takhar, P.S.; Gianelli, M.P.; Petzold, G. Effects of centrifugal block freeze crystallization on quality properties in pineapple juice. *Chem. Eng. Technol.* **2020**, *43*, 355–364. [CrossRef]
43. Petzold, G.; Moreno, J.; Lastra, P.; Rojas, K.; Orellana, P. Block freeze concentration assisted by centrifugation applied to blueberry and pineapple juices. *Innov. Food Sci. Emerg.* **2015**, *30*, 192–197. [CrossRef]
44. Ho, K.K.; Ferruzzi, M.G.; Wightman, J.D. Potential health benefits of (poly)phenols derived from fruit and 100% fruit juice. *Nutr. Rev.* **2020**, *78*, 145–174. [CrossRef] [PubMed]
45. Demoliner, F.; de Carvalho, L.T.; de Liz, G.R.; Prudêncio, E.S.; Ramos, J.C.; Bascuñan, V.L.A.F.; Vitali, L; Block, J.M. Improving the nutritional and phytochemical compounds of a plant-based milk of sapucaia nut cake using block freeze concentration. *Int. J. Food Sci. Technol.* **2020**, *55*, 3031–3042. [CrossRef]
46. Romero-Román, M.E.; Schoebitz, M.; Bastías, R.M.; Fernández, P.S.; García-Viguera, C.; López-Belchi, M.D. Native species facing climate changes: Response of calafate berries to low temperature and UV radiation. *Foods* **2021**, *10*, 196. [CrossRef]
47. Petzold, G.; Orellana-Palma, P.; Moreno, J.; Cerda, E.; Parra, P. Vacuum-assisted block freeze concentration applied to wine. *Innov. Food Sci. Emerg. Technol.* **2016**, *36*, 330–335. [CrossRef]
48. Casas-Forero, N.; Orellana-Palma, P.; Petzold, G. Comparative study of the structural properties, color, bioactive compounds content and antioxidant capacity of aerated gelatin gels enriched with cryoconcentrated blueberry juice during storage. *Polymers* **2020**, *12*, 2769. [CrossRef]
49. Krapfenbauer, G.; Kinner, M.; Gössinger, M.; Schönlechner, R.; Berghofer, E. Effect of thermal treatment on the quality of cloudy apple juice. *J. Agric. Food Chem.* **2006**, *54*, 5453–5460. [CrossRef] [PubMed]
50. Ramirez, J.E.; Zambrano, R.; Sepúlveda, B.; Kennelly, E.J.; Simirgiotis, M.J. Anthocyanins and antioxidant capacities of six Chilean berries by HPLC–HR-ESI-ToF-MS. *Food Chem.* **2015**, *176*, 106–114. [CrossRef]
51. Rubilar, R.; Hubbard, R.; Emhart, V.; Mardones, O.; Quiroga, J.J.; Medina, A.; Valenzuela, H.; Espinoza, J.; Burgos, Y.; Bozo, D. Climate and water availability impacts on early growth and growth efficiency of Eucalyptus genotypes: The importance of GxE interactions. *For. Ecol. Manag.* **2020**, *458*, 117763. [CrossRef]
52. Agurto, L.; Allacker, K.; Fissore, A.; Agurto, C.; De Troyer, F. Design and experimental study of a low-cost prefab Trombe wall to improve indoor temperatures in social housing in the Biobío region in Chile. *Sol. Energy* **2020**, *198*, 704–721. [CrossRef]
53. Ferreira, S.S.; Silva, P.; Silva, A.M.; Nunes, F.M. Effect of harvesting year and elderberry cultivar on the chemical composition and potential bioactivity: A three-year study. *Food Chem.* **2020**, *302*, 125366. [CrossRef] [PubMed]

54. Moreno, F.L.; Raventós, M.; Hernández, E.; Ruiz, Y. Block freeze-concentration of coffee extract: Effect of freezing and thawing stages on solute recovery and bioactive compounds. *J. Food Eng.* **2014**, *120*, 158–166. [CrossRef]
55. Wu, Y.Y.; Xing, K.; Zhang, X.X.; Wang, H.; Wang, Y.; Wang, F.; Li, J.M. Influence of freeze concentration technique on aromatic and phenolic compounds, color attributes, and sensory properties of Cabernet Sauvignon wine. *Molecules* **2017**, *22*, 899. [CrossRef] [PubMed]
56. Correa, L.J.; Ruiz, R.Y.; Moreno, F.L. Effect of falling-film freeze concentration on bioactive compounds in aqueous coffee extract. *J. Food Process Eng.* **2018**, *41*, e12606. [CrossRef]
57. Augusto, T.R.; Scheuermann-Salinas, E.S.; Alencar, S.M.; D'arce, M.A.B.R.; Costa de Camargo, A.; Vieira, T.M.F. Phenolic compounds and antioxidant activity of hydroalcoholic extracts of wild and cultivated murtilla (*Ugni molinae* Turcz.). *Food Sci. Technol.* **2014**, *34*, 667–673. [CrossRef]
58. Rodríguez, K.; Ah-Hen, K.; Vega-Gálvez, A.; López, J.; Quispe-Fuentes, I.; Lemus-Mondaca, R.; Gáñvez-Ranilla, L. Changes in bioactive compounds and antioxidant activity during convective drying of murta (*Ugni molinae* T.) berries. *Int. J. Food Sci. Tech.* **2014**, *49*, 990–1000. [CrossRef]
59. López, J.; Gálvez, A.V.; Rodríguez, A.; Uribe, E. Murta (*Ugni molinae* Turcz.): A review on chemical composition, functional components and biological activities of leaves and fruits. *Chil. J. Agric. Anim. Sci.* **2018**, *34*, 43–56. [CrossRef]
60. Samsuri, S.; Li, T.H.; Ruslan, M.S.H.; Amran, N.A. Antioxidant recovery from pomegranate peel waste by integrating maceration and freeze concentration technology. *Int. J. Food Eng.* **2020**, *16*, 20190232. [CrossRef]
61. Casas-Forero, N.; Moreno-Osorio, L.; Orellana-Palma, P.; Petzold, G. Effects of cryoconcentrate blueberry juice incorporation on gelatin gel: A rheological, textural and bioactive properties study. *LWT-Food Sci. Technol.* **2021**, *138*, 110674. [CrossRef]
62. Zielinski, A.A.; Zardo, D.M.; Alberti, A.; Bortolini, D.G.; Benvenutti, L.; Demiate, I.M.; Nogueira, A. Effect of cryoconcentration process on phenolic compounds and antioxidant activity in apple juice. *J. Sci. Food Agric.* **2019**, *99*, 2786–2792. [CrossRef] [PubMed]
63. Orellana-Palma, P.; Takhar, P.; Petzold, G. Increasing the separation of block cryoconcentration through a novel centrifugal filter-based method. *Sep. Sci. Technol.* **2019**, *54*, 786–794. [CrossRef]
64. Orellana-Palma, P.; Petzold, G.; Andana, I.; Torres, N.; Cuevas, C. Retention of ascorbic acid and solid concentration via centrifugal freeze concentration of orange juice. *J. Food Qual.* **2017**, *2017*, 5214909. [CrossRef]

 *foods*

*Article*

# Pre-Crystallization of Nougat by Seeding with Cocoa Butter Crystals Enhances the Bloom Stability of Nougat Pralines

**Birgit Böhme, Annika Bickhardt and Harald Rohm ***

Chair of Food Engineering, Institute of Natural Materials Technology, Technische Universität Dresden, 01062 Dresden, Germany; birgit.boehme@tu-dresden.de (B.B.); annika_sontje.bickhardt@tu-dresden.de (A.B.)
* Correspondence: harald.rohm@tu-dresden.de; Tel.: +49-3514-633-2420

**Abstract:** Fat bloom is an outstanding quality defect especially in filled chocolate, which usually comprises oils of different origins and with different physical properties. Dark chocolate pralines filled with nougat contain a significant amount of hazelnut oil in their center and have been reported as being notably susceptible to oil migration. The current study was designed to test the assumption that a targeted crystallization of nougat with cocoa butter seed crystals is an appropriate technological tool to reduce filling oil transfer to the outside of the praline and, hence, to counteract chocolate shell weakening and the development of fat bloom. For this purpose, the hardness of nougat/chocolate layer models and the thermal properties of chocolate on top of nougat were analyzed during storage at 23 °C for up to 84 days. Pronounced differences between layer models with seeded nougat and with control nougat that was traditionally tempered were observed. The facts that chocolate hardness increased rather than decreased during storage, that the cocoa butter melting peak was shifted towards a lower temperature, and that the hazelnut oil content in the chocolate was reduced can be taken as explicit indicators for the contribution of seeded nougat to the fat bloom stability of filled chocolate.

**Keywords:** praline; nougat; chocolate; fat bloom; DSC; hardness; crystal seeding; migration

**Citation:** Böhme, B.; Bickhardt, A.; Rohm, H. Pre-Crystallization of Nougat by Seeding with Cocoa Butter Crystals Enhances the Bloom Stability of Nougat Pralines. *Foods* **2021**, *10*, 1056. https://doi.org/10.3390/foods10051056

Academic Editors: Asgar Farahnaky, Mahsa Majzoobi and Mohsen Gavahian

Received: 30 March 2021
Accepted: 10 May 2021
Published: 11 May 2021



## 1. Introduction

One of the key quality attributes of dark chocolate is its appearance and, particularly, a smooth and glossy surface. Fat bloom is a prominent appearance defect that results in a loss of gloss and that devaluates the products by turning their surface dull and greyish. The main driving force behind fat bloom is the transformation of ßV cocoa butter crystals into the ßVI polymorph which, for instance, occurs when a small amount of cocoa butter is released from the chocolate matrix and recrystallizes at the surface [1–3]. In plain chocolate, the fat bloom is mainly triggered by (a) insufficient environmental conditions, for instance, pronounced temperature fluctuations or a storage temperature higher than the melting point of the ßV polymorph, or by (b) poor tempering during production so that the necessary crystal nuclei cannot be generated [4,5]. The situation becomes even more complicated when fats or oils other than cocoa butter are present in the system. These may either be comprised in the chocolate formulation or may be part of a second system that is in close contact with the chocolate.

The latter is true in the case of filled chocolate confectionery, for instance, pralines. A prominent and frequently consumed example is pralines filled with dark "Viennese" nougat. Similar to chocolate, nougat represents an oil-based multiphase suspension, made of different constituents. The main ingredients are hazelnuts and sugar which maybe, depending on the formulation, supplemented by a certain amount of cocoa butter and/or cocoa mass, milk solids including milk fat, and surface-active compounds. The fat content of nougat is typically in a range of 30–45 g/100 g and, to a large extent, made of hazelnut oil [6]. Nougat processing comprises the milling of roasted hazelnuts together with sugar and the other ingredients, traditionally in edge mills and, nowadays, mainly in five roller refiners or in agitated ball mills. Hazelnut oil differs largely from cocoa butter with respect to fatty acid and, hence, its triacylglycerol composition. As a consequence, the melting point and the solid fat content at a specific temperature also differ, and significant interactions between the fat phases of nougat and the surrounding chocolate can be expected.

The driving force behind fat bloom in confectionery products with different fats is migration [7]. The concentration difference at the interface between the chocolate shell and the filling induces a diffusive transport of the filling oil through the chocolate shell towards its outside, especially when the filling is rich in unsaturated fatty acids. Due to the higher mobility of the unsaturated oil, metastable cocoa butter crystals dissolve in the migrating oil and recrystallize at the surface as stable ßVI [8]. Apart from geometrical considerations, it has been reported that milk fat or some vegetable fats included in chocolate may help to decelerate fat bloom [9], or that specific barrier layers between chocolate and filling can be successfully introduced as a third component [10,11].

Seeding with cocoa butter crystals is an established alternative to the traditional stir/shear temper technology of chocolate [12]. Especially when containing cocoa butter, nougat also needs to be tempered during processing, regularly by applying the stir/shear technology. The current study aimed to evaluate whether a targeted pre-crystallization of dark nougat with cocoa butter crystals could be an innovative technology to enhance the long-term physical stability of filled confectionery products.

## 2. Materials and Methods

### 2.1. Materials

Dark chocolate mass, made of sugar, cocoa mass, cocoa butter, and soy lecithin, with a fat content of 32.4 g/100 g, was obtained from a German chocolate manufacturer. Four commercial nougat samples, encoded #A, #B, #C, and #D, and the respective formulations (Table 1) were obtained from two companies. SEED 100 cocoa butter crystals (ßV) were provided by Uelzena eG (Uelzen, Germany). The melting peak temperature measured by DSC was 34.0 ± 0.1 °C and the phase transition enthalpy was 132.5 ± 0.7 J/g, which is in line with literature data on the ßV polymorph [13,14] (Figure S1). It needs to be mentioned that other researchers (e.g., [12,15] associated such a melting temperature with polymorph ßVI, but this will not be discussed further. The seed crystal powder has a mean particle size of approx. 25 µm [16], which was confirmed by light microscopy (Figure S2). All chemicals used in the study were of analytical grade.

**Table 1.** Formulation of the nougat samples (g/100 g) used in this study.

| Ingredient | Nougat #A | Nougat #B | Nougat #C | Nougat #D |
|---|---|---|---|---|
| Hazelnuts | 44.2 | 47.0 | 38.0 | 30.0 |
| Sucrose | 39.3 | 42.0 | 49.0 | 49.0 |
| Cocoa mass | 5.0 | | 8.0 | 7.0 |
| Milk solids | 3.0 | | | 4.5 |
| Cocoa butter | 8.5 | 11.0 | 5.0 | 8.5 |
| Sunflower lecithin | | | <0.1 | <1.0 |
| Vanilla extract | | | | <0.1 |

### 2.2. Production and Storage of Nougat/Chocolate Layer Models

Layer models were produced to ensure controlled contact between nougat and chocolate. For this purpose, 16.0 ± 0.2 g liquid nougat was transferred into Petri dishes of 55 mm diameter, manually compacted by vibration, and solidified at 18 °C for 7 d in an environmental chamber. This nougat layer had a thickness of approx. 5 mm. The following samples were prepared: (a) Control plates were made at the nougat manufacturer sites after stir/shear tempering and, after solidification as specified above, sent to the laboratory. (b) For the experimental plates, nougat was melted at 50 °C overnight, mixed, and then cooled to 27 °C under continuous stirring. The amount of ßV cocoa butter crystals subsequently added was either 0.5%, 1.0%, or 2.0%, as related to the total fat content of the nougats. The average temper degree of the nougat immediately before weighing into the plates was determined using an E3 tempermeter (Sollich KG, Bad Salzuflen, Germany). The temper degree was 5.1 ± 0.2 when seeded with 0.5% crystals, 5.8 ± 0.3 when seeded with 1%, and 6.6 ± 0.3 when seeded with 2% cocoa butter crystals.

After solidification, the nougat plates were overlaid with 6.0 ± 0.2 g tempered chocolate, resulting in a layer of approx. 2 mm thickness. The chocolate used for this purpose was melted and conditioned to 50 °C overnight and tempered at 31 °C using a Minitemper® Turbo (Sollich KG, Bad Salzuflen, Germany). The final layer models were solidified at 18 °C for 24 h and subsequently stored at 23 ± 0.2 °C for up to 84 d in an environmental chamber. Individual samples of the layer models were analyzed after 1, 7, 21, 28, 63, and 84 d.

### 2.3. Methods for the Analysis of Raw Materials

Particle size distributions were analyzed using a HELOS KR laser diffractometer (Sympatec GmbH, Clausthal-Zellerfeld, DE). Prior to analysis, a 2 g sample conditioned to 50 °C was suspended in 10 g sunflower oil of the same temperature in duplicate [17]. Measurements were carried out in a 50 mL cuvette filled with sunflower oil to which the sample suspension was dropwise added until an optical density of approx. 30% was achieved. From the volume-specific density distribution the $x_{10}$, $x_{50}$, and $x_{90}$ diameters, representing the 10%, 50%, and 90% quantiles, respectively, were taken.

The amount of mobile fat was determined using a centrifugation method [18]. Fourteen g melted nougat was transferred into conical tubes and centrifuged at 35 °C for 15 min at 10,200 g (Heraeus Biofuge Stratos, Kendro Laboratory Products GmbH, Langenselbold, Germany). The amount of separated fat was weighed and represents the mobile fat, expressed as a gravimetric fraction of total fat.

Viscosity measurements were carried out with a MARS III rheometer equipped with a concentric CC25DIN geometry ($r_i$ = 12.54 mm, $r_o$ = 13.60 mm, h = 37.62 mm; Thermo Fisher GmbH, Karlsruhe, Germany). The temperature was kept constant at 40 °C using a Peltier device. After equilibration and pre-shearing for 300 s at 5/s, the shear rate was increased from 2/s to 50/s within 180 s, kept constant at 50/s for 60 s, and then reduced to 2/s in another 180 s [19]. Fifty data points per ramp were recorded in logarithmic spacing. The downward cycle was used for fitting the shear stress—shear rate data to the Casson model, and Casson yield stress and viscosity were taken as descriptors.

### 2.4. Methods for the Analysis of Raw Materials and Layer Models

The hardness of base chocolate, nougat, and chocolate/nougat layers was measured using a TA.XTplus texture analyzer (Stable Micro Systems Ltd., Surrey, UK) equipped with a 50 N force transducer [20]. By applying a crosshead velocity of 1 mm/s, samples were penetrated with a 2 mm diameter plunger. The maximum penetration depth was 2 mm. The maximum force was taken as a hardness indicator and calculated as an arithmetic mean of 8 replicate measurements (2 Petri dishes × 4 measurements per petri dish).

Thermal properties of chocolate were determined in duplicate with a DSC25 instrument connected to an RCS90 cooling unit (TA Instruments GmbH, Eschborn, Germany). Using an analytical balance, 5 ± 0.2 mg sample scratched from the surface of the layer models was transferred into a standard aluminum pan and closed. An empty pan served as a reference. After cooling to −50 °C and stabilization for 3 min, the samples were heated to 45 °C at 10 K/min under a continuous nitrogen flow of 50 mL/min. The thermograms were evaluated with respect to cocoa butter melting peak temperature and the enthalpy of the nut oil melting peak, which was calculated from the heat flow function.

### 2.5. Sensory Analysis

A duo-trio forced-choice test setup was used to evaluate whether stir/shear tempered and seeded nougat can be distinguished by tasting. For this purpose, 6 g aliquots of tempered or seeded nougat were dosed into paper cupcakes with an upper diameter of 35 mm and a height of 20 mm. The samples were then stored for 7–10 d at 18 °C.

A panel of 10 trained panelists took part in the study. Informed consent was obtained from all subjects involved in the study. Due to CoVid-19 restrictions, author A.B. who was in charge of the experiments delivered triads of the samples to the individual offices of the subjects. Samples were presented in Petri dishes encoded with 3-digit random numbers, and the panelists were asked to identify the odd specimen. The number of correct identifications was judged for significance using the tables of Roessler et al. [21].

### 2.6. Statistics

Statistical data evaluation was carried out using SAS® Studio 3.8 University Edition (SAS Institute Inc., Cary, NC, USA). Outliers were identified with the Dixon-Q test at $p < 0.05$. One-way analysis of variance was followed by Fisher's least significant difference post-hoc tests. All significance statements given in this study refer to an error probability level of $p < 0.05$.

## 3. Results and Discussion

### 3.1. Description of the Base Nougat

The total fat content of the base nougats ranged from 36.0–44.0 g/100 g (Table 2). The relative amount of cocoa butter in the oil fraction of the nougats was calculated based on an approximate fat content of the cocoa mass (53 g/100 g) and the amount of cocoa butter in the formulation, and ranged between 25.2% and 29.5% (nougat #A, #B, #C) but was 41.7% in case of nougat #D. The fraction of mobile fat was significantly higher for samples #A and #B than for nougat #D (approx. 26% vs. 18.5%, respectively). As for chocolate, the flow properties of liquid nougat depend on the volume fraction and surface properties of dispersed particles and are additionally affected by the presence of surface-active compounds [22,23]. Casson yield stress and viscosity were lowest for nougat #B which had the highest fat content and, hence, the lowest particle load, and highest for #D which showed the lowest fat content and, additionally, the lowest amount of mobile fat. Nougat hardness measured by penetration was approx. 1 N (samples #A, #B, #C) but significantly higher for nougat #D (3.76 ± 0.32 N).

**Table 2.** Fat distribution and physical properties of base nougat. Mean values with different superscripts in a column indicate statistical differences ($p < 0.05$).

| Nougat | Total Fat (g/100 g) [1] | Mobile Fat (%) [2] | Yield Stress (Pa) [2] | Viscosity (Pa.s) [2] | Hardness (N) [3] |
|---|---|---|---|---|---|
| #A | 37.6 | 27.2 ± 0.3 $^{a}$ | 10.78 ± 0.07 $^{b}$ | 1.75 ± 0.02 $^{b}$ | 0.95 ± 0.12 $^{b}$ |
| #B | 44.0 | 25.5 ± 0.8 $^{a}$ | 3.78 ± 0.05 $^{c}$ | 1.43 ± 0.04 $^{c}$ | 1.31 ± 0.16 $^{b}$ |
| #C | 37.0 | 23.1 ± 1.1 $^{ab}$ | 10.25 ± 0.17 $^{b}$ | 3.25 ± 0.04 $^{a}$ | 1.24 ± 0.24 $^{b}$ |
| #D | 36.0 | 18.5 ± 1.0 $^{b}$ | 12.90 ± 0.09 $^{a}$ | 3.87 ± 0.03 $^{a}$ | 3.76 ± 0.32 $^{a}$ |

[1] Data from manufacturer specifications; [2] Data is arithmetic mean ± half deviation range ($n = 2$); [3] Data is arithmetic mean ± standard deviation ($n = 8$).

Figure 1 displays the logarithmic volume-based particle size distributions, which were monomodal for all samples. The particle diameter median ($x_{50,3}$) was 6.1 µm for the chocolate, and slightly higher (7.4–8.5 µm) for the four nougats. The span between the $x_{10,3}$ and the $x_{90,3}$ quantiles was 22.4 µm and 21.8–33.6 µm for chocolate and nougat, respectively. The highest $x_{90,3}$ diameter (35.1 µm) was observed in the case of nougat #C, which is slightly above the critical particle size for chocolate concerning the sensory perception of graininess [24,25]. This value was, however, recently questioned by de Pelsmaeker et al. [26].

It is evident from the data that the nougat samples are representative of what is available on the market, starting with the high fat, low viscosity variant #B comprising of only hazelnuts, sugar, and cocoa butter. The fraction of dispersed particles for all other nougats were explicitly higher, with some also containing milk solids and/or cocoa mass and/or surface-active compounds such as lecithin at varying concentration.

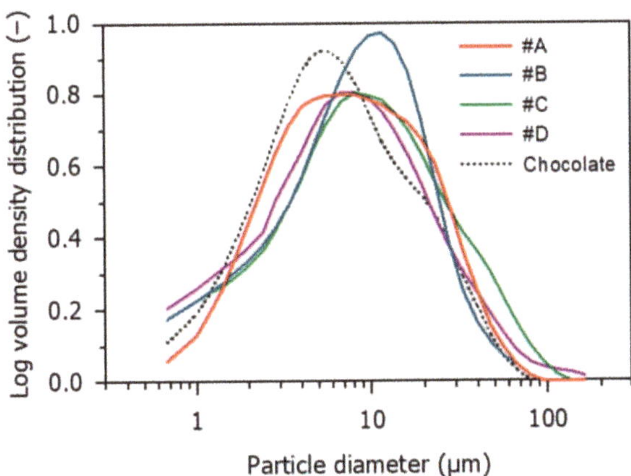


**Figure 1.** Particle size distributions of nougat samples #A–#D and of the chocolate used for the preparation of the layer models.

### 3.2. Storage Induced Changes and Nut Oil Migration

Figure 2 depicts the development of the hardness of the surface chocolate during storage as affected by nougat variety and pre-crystallization method and intensity. One day after production of the layer model, the hardness of the chocolate on top of the nougat was 13.9 ± 1.8 N for sample #A, 14.9 ± 1.8 N for #B, and 17.2 ± 2.1 N for #C. The mechanical response of nougat #D was 23.5 ± 2.3 N, thus reflecting the contribution of the stiffer bottom layer (see nougat hardness in Table 2) to chocolate hardness.

During storage at 23 °C, the hardness of the chocolate of the control sample decreased, especially within the first four weeks. After that, further changes in sample hardness were only minor. It is also evident that the relative deviation of the repeated measurements of individual samples increased with storage time. This can be attributed to local effects and increasing differences between local spots at the surface of the layer models. As indicated in the chart, the softening induced by 84 d storage was significant for chocolate on top of nougat #A, #B, and #D.

The situation is, however, different when cocoa butter crystals were used to induce pre-crystallization in nougat. In case an amount of 0.5% or 1% seed powder was used for seeding, initial chocolate hardness was approx. 10 N, with the highest values again for samples with nougat #D. In the first 20–30 d of storage, the systems slightly solidified, presumably because of an ongoing slow crystallization of cocoa butter and a resulting formation of a fat crystal network in the bottom nougat layer that presumably blocks the fat migration pathway due to microstructural changes [27,28]. At this particular time, chocolate hardness was almost similar to the hardness of chocolate on top of the industrially tempered control nougat. Using chocolate hardness as an indicator, it was not possible to detect further changes after that period of time. Except for nougat #A seeded with 1% cocoa butter crystals, analysis of variance with subsequent post-hoc tests revealed that the hardness of the chocolate layer at the end of storage was significantly higher than its initial hardness. The more pronounced hardness increase of the nougat #D sample can be attributed to the lower amount of nut oil that was present in this system. As regards nougat with 2% seed crystals, chocolate on top of it was even harder than the control immediately after production of the layer models. This system remained stable during approx. 4 weeks but, subsequently, showed a significant trend toward time-dependent softening. Although the investigated systems are not fully comparable, the mechanism behind a reduced nut oil migration is presumably similar to that observed with respect to oil separation in spreadable nougat creme [29,30], namely that the formation of some sort of fat crystal network in the nougat significantly lower filling oil mobility.

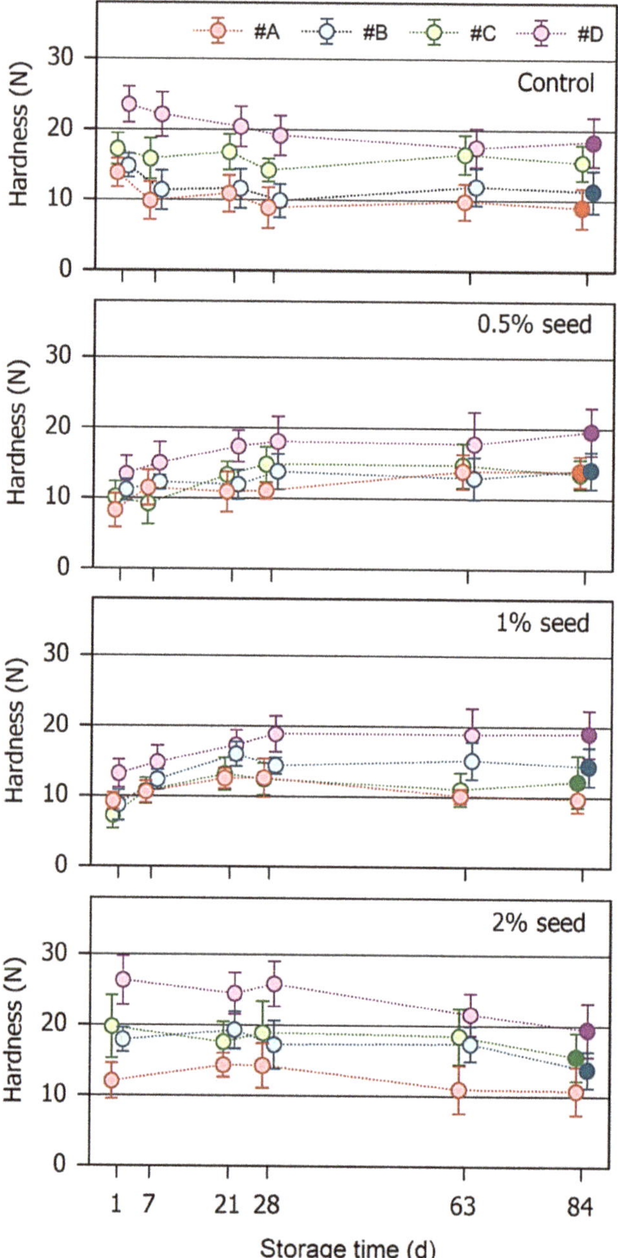


**Figure 2.** Development of the hardness of the nougat/chocolate layer models during storage at 23 °C for 84 days. Four different nougats (#A, #B, #C, #D) were either tempered using the stir/shear technique (control) or seeded with different amounts of cocoa butter crystals. For the sake of clarity, symbols are slightly shifted along the x-axis. Data are arithmetic mean ± standard deviation of 8-fold measurements. The hardness of samples at 84 days marked by a filled symbol is significantly ($p < 0.05$) different from the hardness of the fresh layer models.

Figure 3 depicts the effects of nougat crystallization on storage-induced changes of the cocoa butter melting peak temperature in chocolate taken from the surface of the samples. Immediately after production, the peak temperature ranged from 32.2–32.9 °C, which is typical for cocoa butter with ßV crystals [31,32]. In all control samples, the cocoa butter melting temperature increased continuously with ongoing storage and finally reached approx. 34.5 °C. This shift indicates that a significant fraction of ßV crystals transformed into the ßVI modification, which shows a typical melting temperature of approx. 35–36 °C. Their presence can be regarded as an indicator and expression of fat bloom [3,9,33].

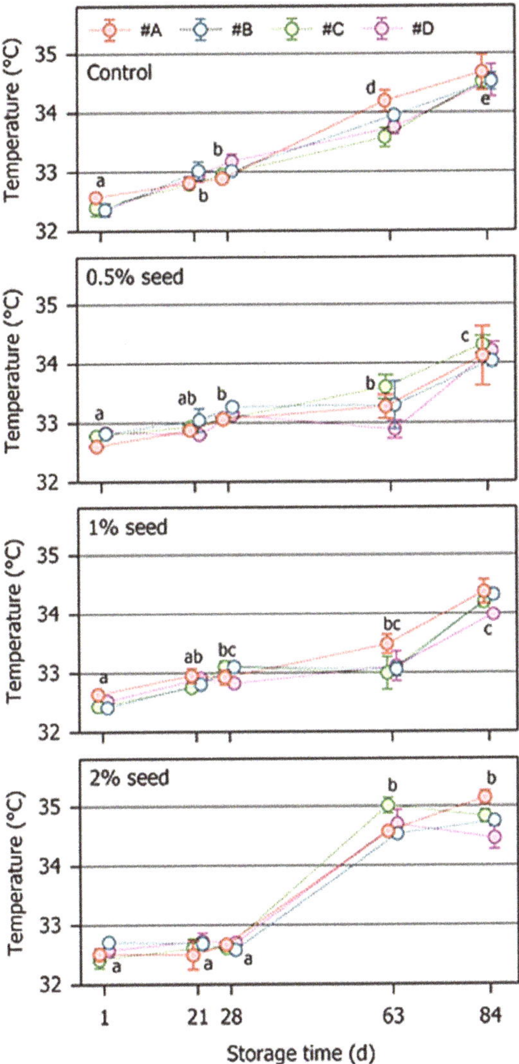


**Figure 3.** Development of the melting peak temperature of chocolate on top of the nougat/chocolate layer models during storage at 23 °C for 84 days. Four different nougats (#A, #B, #C, #D) were either tempered using the stir/shear technique (control) or seeded with different amounts of cocoa butter crystals. For the sake of clarity, symbols are slightly shifted along the x-axis. Individual data are arithmetic mean ± half deviation range of duplicate measurements. Mean values after different storage times labeled with different letters differ significantly ($p < 0.05$).

This increase in the temperature of the melting peak was significantly lower in chocolate placed on top of the nougat seeded with cocoa butter crystals than in the control sample. This effect was especially observed when 0.5% or 1% crystals were added, and the respective delay in the shift of the peak temperature was evident at least for a storage period of 63 d. When 2% seeding crystals were used, the melting temperature was stable for at least 28 d.

Figure 4 depicts temperature-resolved thermograms of the chocolate placed on top of control nougat #A and those of chocolate on nougat #A that was seeded with 1% cocoa butter crystals. On the day after producing the layer models, the progress of the heat flow in the samples was almost identical. After storage for 21 d the presence of a compound with a melting temperature of approx. −7 °C, obviously resulting from a small amount of hazelnut oil being present, started to become visible especially in the chocolate of the layer models with the control nougat. When plotted against time, the phase transition enthalpy corresponds to the area of the heat flow peak.

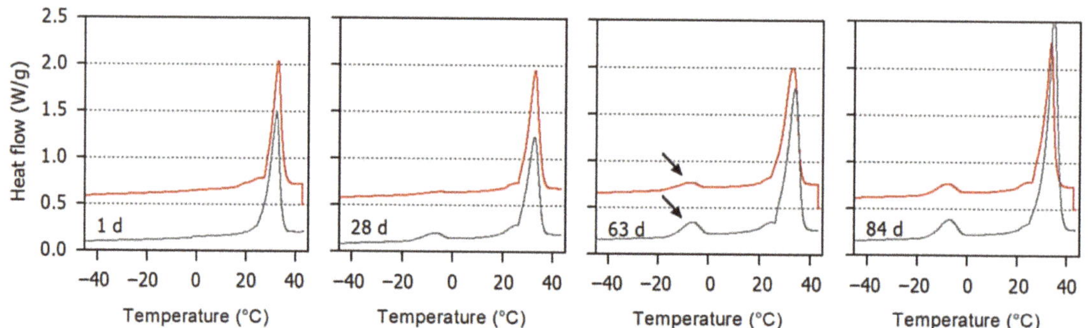


**Figure 4.** Average melting curves (endo up, $n = 2$) of chocolate on top of nougat stored for different periods of time (1, 28, 63, and 84 days). Grey lines, chocolate on top of control nougat; red lines, chocolate on top of nougat #A. The nut oil melting peak is marked by an arrow. For the sake of clarity, the red lines are shifted along the y-axis by 0.5 W/g.

For a storage time of 28 d and 84 d, the specific melting enthalpy of the chocolate surface as affected by the type of nougat and by the nougat crystallization method is outlined in Table 3. The relatively large deviation between some of the duplicate measurements can again be attributed to local effects at the surface where the samples were taken. However, it is, on average, evident from the data that, during the first period of storage, the seeding of nougat with cocoa butter crystals significantly reduced the migration of nut oil to the chocolate surface. This result is in good agreement with the melting peak data presented in Figure 3: a higher nut oil melting enthalpy in chocolate can be regarded to trigger the transformation of cocoa butter crystal morphology from ßV to ßVI with the accompanying shift in melting temperature. After long-term storage, the stabilizing effect of the seeding of nougat with cocoa butter crystals diminishes, and so do the significances of the differences.

In three sets of sensory discrimination tests, pure nougat seeded with 0.5% or 1% cocoa butter crystals was tested against the control prepared by the stir/shear technique and, in addition, against each other. In all cases, the number of correct identifications was seven at a maximum which means that the sensory panel could not discriminate the samples at an error probability level of $p < 0.10$.

Figure 5 finally shows the appearance of the nougat #A control layer models and, exemplary, the layer model with the same nougat seeded with 1% cocoa butter crystals. The photographs clearly show the differences in fat bloom intensity which is visible at the surface of the top layer chocolate and, especially, at the interface between the nougat layer and the chocolate.

**Table 3.** Effects of crystal seeding on specific enthalpies (J/g) of the nut oil melting peak in the chocolate on top of nougat layers after 28 or 84 days of storage at 23 °C. Mean values with different superscripts in a column indicate statistical differences ($p < 0.05$).

| Nougat Seeding | Nougat #A [1] | Nougat #B | Nougat #C | Nougat #D |
|---|---|---|---|---|
| Storage time: 28 days | | | | |
| Control | $1.83 \pm 0.25^{a}$ | $2.02 \pm 0.30^{a}$ | $1.64 \pm 0.27^{a}$ | $0.73 \pm 0.09^{a,b}$ |
| 0.5% seed crystals | $1.07 \pm 0.18^{b}$ | $1.56 \pm 0.14^{a}$ | $0.71 \pm 0.05^{b}$ | $1.09 \pm 0.32^{a}$ |
| 1% seed crystals | $1.09 \pm 0.02^{b}$ | $1.58 \pm 0.28^{a}$ | $0.76 \pm 0.12^{b}$ | $0.28 \pm 0.09^{b}$ |
| 2% seed crystals | $0.96 \pm 0.12^{b}$ | $0.42 \pm 0.04^{b}$ | $0.37 \pm 0.04^{b}$ | $0.32 \pm 0.12^{b}$ |
| Storage time: 84 days | | | | |
| Control | $9.21 \pm 0.73^{a}$ | $8.77 \pm 0.23^{a}$ | $7.34 \pm 0.77^{a}$ | $4.65 \pm 0.18^{a}$ |
| 0.5% seed crystals | $6.58 \pm 0.44^{b}$ | $4.45 \pm 0.36^{b}$ | $7.37 \pm 0.04^{a}$ | $4.73 \pm 0.52^{a}$ |
| 1% seed crystals | $6.05 \pm 0.46^{b}$ | $8.68 \pm 0.62^{a}$ | $7.38 \pm 0.58^{a}$ | $3.66 \pm 0.54^{a}$ |
| 2% seed crystals | $9.03 \pm 0.44^{a}$ | $7.78 \pm 0.89^{a}$ | $6.90 \pm 0.07^{a}$ | $4.51 \pm 0.24^{a}$ |

[1] Data is arithmetic mean ± half deviation range ($n = 2$). Mean values in a data block marked with different superscripts (a,b) differ significantly ($p < 0.05$).

**Figure 5.** Layer models after 28 days storage at 23 °C showing enhanced fat bloom at the interface between nougat and chocolate (**a**) and at the chocolate surface (**b**), and corresponding systems with nougat #A seeded with 1% cocoa butter crystals: interface between nougat and chocolate (**c**) and chocolate surface (**d**).

## 4. Conclusions

The current study shows that, under controlled conditions, the pre-crystallization of nougat by seeding with cocoa butter crystals may be considered appropriate for improving the physical storage stability of pralines and delaying the onset of fat bloom. In the layer models used in the experiments, the thickness of the chocolate layer was uniform across the tested surface. Such a uniform chocolate shell thickness can be achieved when molding techniques such as the frozen cone technology are used. It needs, however, to be tested in further studies whether the stability of filled systems can also be improved when shell thickness variations, resulting from standard molding techniques such as form inverting or spinning, are present.

**Supplementary Materials:** The following are available online at https://www.mdpi.com/article/10.3390/foods10051056/s1, Figure S1: Melting curves (duplicate measurement) of the ßV cocoa butter seed powder used in this study, Figure S2: Light microscopy images of the ßV cocoa butter seed powder used in this study.

**Author Contributions:** Conceptualization, B.B. and H.R.; data curation, investigation and data analysis, B.B. and A.B.; methodology, B.B. and H.R.; writing, B.B., A.B. and H.R. All authors have read and agreed to the published version of the manuscript.

**Funding:** This research received no external funding.

**Institutional Review Board Statement:** Not applicable.

**Informed Consent Statement:** Informed consent was obtained from all subjects involved in the study.

**Data Availability Statement:** The data presented in this study are available on request from the corresponding author.

**Acknowledgments:** Open Access Funding by the Publication Fund of the TU Dresden. We wish to thank Lübecker Marzipan Fabrik v. Minden & Bruhns GmbH & Co. KG, Stockelsdorf, Germany, and a second dark nougat producer who prefers to remain anonymous for providing the dark nougat samples.

**Conflicts of Interest:** The authors declare no conflict of interest.

## References

1. Sato, S.; Hondoh, H.; Ueno, S. Fat bloom caused by partial de-oiling on chocolate surfaces after high-temperature exposure. *J. Am. Oil Chem. Soc.* **2021**, *98*, 269–280. [CrossRef]
2. Lonchampt, P.; Hartel, R.W. Fat bloom in chocolate and compound coatings. *Eur. J. Lipid Sci. Technol.* **2004**, *106*, 241–274. [CrossRef]
3. Lonchampt, P.; Hartel, R.W. Surface bloom on improperly tempered chocolate. *Eur. J. Lipid Sci. Technol.* **2006**, *108*, 159–168. [CrossRef]
4. Bolliger, S.; Breitschuh, B.; Stranzinger, M.; Wagner, T.; Windhab, E.J. Comparison of precrystallization of chocolate. *J. Food Eng.* **1998**, *35*, 281–297. [CrossRef]
5. Miquel, M.E.; Stephen, D.; Couzens, P.J.; Wille, H.J.; Hall, L.D. Kinetics of the migration of lipids in composite chocolate measured by magnetic resonance imaging. *Food Res. Int.* **2001**, *34*, 773–781. [CrossRef]
6. Pajin, B.; Karlovic, D.; Omorjan, R.; Sovilj, V.; Antic, D. Influence of filling fat type on praline products with nougat filling. *Eur. J. Lipid Sci. Technol.* **2007**, *109*, 1203–1207. [CrossRef]
7. Choi, Y.J.; McCarthy, K.L.; McCarthy, M.J. Oil migration in a chocolate confectionary system evaluated by magnetic resonance imaging. *J. Food Sci.* **2005**, *70*, E312–E317. [CrossRef]
8. Smith, K.W.; Cain, F.W.; Talbot, G. Effect of nut oil migration on polymorphic transformation in a model system. *Food Chem.* **2007**, *102*, 656–663. [CrossRef]
9. Sonwai, S.; Rousseau, D. Controlling fat bloom formation in chocolate—Impact of milk fat on microstructure and fat phase crystallization. *Food Chem.* **2010**, *119*, 286–297. [CrossRef]
10. Nöbel, S.; Schneider, Y.; Böhme, B.; Rohm, H. Technofunctional barrier layers for preventing fat bloom in triple-shot pralines. *Food Res. Int.* **2009**, *42*, 69–75. [CrossRef]
11. Gray, M.P. Moulding, enrobing and cooling chocolate products. In *Beckett's Industrial Chocolate Manufacture and Use*, 5th ed.; Beckett, S.T., Fowler, M.S., Ziegler, G.R., Eds.; Wiley Blackwell: Oxford, UK, 2017; pp. 356–399.
12. Svanberg, L.; Ahrne, L.; Loren, N.; Windhab, E. Impact of pre-crystallization process on structure and product properties in dark chocolate. *J. Food Eng.* **2013**, *114*, 90–98. [CrossRef]
13. Wille, R.L.; Lutton, E.S. Polymorphism of cocoa butter. *J. Am. Oil Chem. Soc.* **1966**, *43*, 491–496. [CrossRef]
14. Bayés-García, L.; Aguilar-Jiménez, M.; Calvet, T.M.; Koyano, T.; Sato, K. Crystallization and melting behavior of cocoa butter in lipid bodies of fresh cocoa beans. *Cryst. Growth Des.* **2019**, *19*, 4127–4137. [CrossRef]
15. Huyghebaert, A.; Hendrickx, H. Polymorphism of cocoa butter, shown by differential scanning calorimetry. *Lebensmittel-Wissenschaft Technol.* **1971**, *4*, 59–63.
16. Fichtl, P.; Dietrich, H.; Schliehe-Dierks, R. Method for Producing Storage Stable Seed Crystals Made of Cocoa Butter or Chocolate Masses. European Patent EP2314170 B1, 23 October 2009.
17. Rohm, H.; Böhme, B.; Skorka, J. The impact of grinding intensity on particle properties and rheology of dark chocolate. *LWT Food Sci. Technol.* **2018**, *92*, 564–568. [CrossRef]
18. Ziegleder, G.; Amanitis, A.; Hornik, H. Thickening of molten white chocolates during storage. *Lebensm.-Wiss. Technol.* **2004**, *37*, 649–656. [CrossRef]
19. Servais, C.; Ranc, H.; Roberts, I.D. Determination of chocolate viscosity. *J. Texture Stud.* **2003**, *34*, 467–497. [CrossRef]

20. Böhme, B.; Schneider, R.; Harbs, T.; Rohm, H. Liberation of fat from milk powder particles during chocolate processing through moisture-induced lactose crystallisation. *LWT Food Sci. Technol.* **2020**, *126*, 109343. [CrossRef]
21. Rössler, E.B.; Pangborn, R.M.; Sidel, J.L.; Stone, H. Expanded statistical tables for estimating significance in paired-preference, paired-difference, duo-trio and triangle tests. *J. Food Sci.* **1978**, *43*, 940–947. [CrossRef]
22. Arnold, G.; Schade, E.; Schneider, Y.; Friedrichs, J.; Babick, F.; Werner, C.; Rohm, H. Influence of individual phospholipids on the physical properties of oil-based suspensions. *J. Am. Oil Chem. Soc.* **2014**, *91*, 71–77. [CrossRef]
23. Wolf, B. Chocolate flow properties. In *Beckett's Industrial Chocolate Manufacture and Use*, 5th ed.; Beckett, S.T., Fowler, M.S., Ziegler, G.R., Eds.; Wiley Blackwell: Oxford, UK, 2017; pp. 274–297.
24. Afoakwa, E.O.; Paterson, A.; Fowler, M. Factors influencing rheological and textural qualities in chocolate—A review. *Trends Food Sci. Technol.* **2007**, *18*, 290–298. [CrossRef]
25. Engelen, L.; Van Der Bilt, A.; Schipper, M.; Bosman, F. Oral size perception of particles: Effects of size, type, viscosity and method. *J. Texture Stud.* **2005**, *36*, 373–386. [CrossRef]
26. De Pelsmaeker, S.; Behra, J.S.; Gellynk, X.; Schouteten, J.J. Detection threshold of chocolate graininess: Machine vs. human. *Br. Food J.* **2020**, *122*, 2757–2767. [CrossRef]
27. Bricknell, J.; Hartel, R.W. Relation of fat bloom in chocolate to polymorphic transition of cocoa butter. *J. Am. Oil Chem. Soc.* **1998**, *75*, 1609–1615. [CrossRef]
28. Jin, J.; Hartel, R.W. Accelerated fat bloom in chocolate model systems: Replacement of cocoa powder with sugar particles and the effect sof lecithin. *J. Am. Oil Chem. Soc.* **2020**, *97*, 377–388. [CrossRef]
29. Hubbes, S.S.; Braun, A.; Foerst, P. Crystallization kinetics and mechanical properties of nougat creme model fats. *Food Biophys.* **2020**, *15*, 1–15. [CrossRef]
30. Hubbes, S.S.; Braun, A.; Foerst, P. Sugar particles and their role in crystallization kinetics and structural properties in fats used for nougat crème production. *J. Food Eng.* **2020**, *287*, 110130. [CrossRef]
31. Loisel, C.; Keller, G.; Lecq, G.; Bourgaux, C.; Ollivon, M. Phase transitions and polymorphism of cocoa butter. *J. Am. Oil Chem. Soc.* **1998**, *75*, 425–439. [CrossRef]
32. Windhab, E.J. Tempering. In *Beckett's Industrial Chocolate Manufacture and Use*, 5th ed.; Beckett, S.T., Fowler, M.S., Ziegler, G.R., Eds.; Wiley Blackwell: Oxford, UK, 2017; pp. 314–355.
33. Loisel, C.; Lecq, G.; Keller, G.; Ollivon, M. Fat bloom and chocolate structure studied by mercury porosimetry. *J. Food Sci.* **1997**, *62*, 781–788. [CrossRef]

 *foods*

*Article*

# Investigating the Use of Ultraviolet Light Emitting Diodes (UV-LEDs) for the Inactivation of Bacteria in Powdered Food Ingredients

**Laura Nyhan [1], Milosz Przyjalgowski [2], Liam Lewis [2], Máire Begley [1] and Michael Callanan [1,*]**

[1] Department of Biological Sciences, Munster Technological University, T12 P928 Cork, Ireland; l.nyhan@mycit.ie (L.N.); maire.begley@cit.ie (M.B.)
[2] Centre for Advanced Photonics and Process Analysis, Munster Technological University, T12 P928 Cork, Ireland; milosz.przyjalgowski@cit.ie (M.P.); liam.lewis@cit.ie (L.L.)
* Correspondence: michael.callanan@cit.ie; Tel.: +353-021-433-6127

**Abstract:** The addition of contaminated powdered spices and seasonings to finished products which do not undergo further processing represents a significant concern for food manufacturers. To reduce the incidence of bacterial contamination, seasoning ingredients should be subjected to a decontamination process. Ultraviolet light emitting diodes (UV-LEDs) have been suggested as an alternative to UV lamps for reducing the microbial load of foods, due to their increasing efficiency, robustness and decreasing cost. In this study, we investigated the efficacy of UV-LED devices for the inactivation of four bacteria (*Listeria monocytogenes, Escherichia coli, Bacillus subtilis and Salmonella* Typhimurium) on a plastic surface and in four powdered seasoning ingredients (onion powder, garlic powder, cheese and onion powder and chilli powder). Surface inactivation experiments with UV mercury lamps, UVC-LEDs and UVA-LEDs emitting at wavelengths of 254 nm, 270 nm and 365 nm, respectively, revealed that treatment with UVC-LEDs were comparable to, or better than those observed using the mercury lamp. Bacterial reductions in the seasoning powders with UVC-LEDs were less than in the surface inactivation experiments, but significant reductions of 0.75–3 $\log_{10}$ colony forming units (CFU) were obtained following longer (40 s) UVC-LED exposure times. Inactivation kinetics were generally nonlinear, and a comparison of the predictive models highlighted that microbial inactivation was dependent on the combination of powder and microorganism. This study is the first to report on the efficacy of UV-LEDs for the inactivation of several different bacterial species in a variety of powdered ingredients, highlighting the potential of the technology as an alternative to the traditional UV lamps used in the food industry.

**Keywords:** ultraviolet; LED; inactivation; bacteria; powder; foods

**Citation:** Nyhan, L.; Przyjalgowski, M.; Lewis, L.; Begley, M.; Callanan, M. Investigating the Use of Ultraviolet Light Emitting Diodes (UV-LEDs) for the Inactivation of Bacteria in Powdered Food Ingredients. *Foods* **2021**, *10*, 797. https://doi.org/10.3390/foods10040797

Academic Editors: Asgar Farahnaky, Mahsa Majzoobi and Mohsen Gavahian

Received: 9 March 2021
Accepted: 5 April 2021
Published: 8 April 2021



## 1. Introduction

The microbial contamination of powdered ingredients is not considered a major problem due to the limitation of growth by the low water activity ($a_w$) value. However, the addition of contaminated raw powder material to ready-to-eat (RTE) foods may result in the contaminants multiplying to high levels, thus posing a risk to public health. Van Doren et al. (2013) undertook a review of the Centers for Disease Control and Prevention's Foodborne Disease Outbreak Surveillance (CDCs FDOSS) System, finding that between 1973 and 2010, 14 reported foodborne outbreaks that were attributed to the consumption of contaminated spices such as red pepper and curry powder occurred across 10 countries and resulted in 1946 illnesses, 128 hospitalisations and 2 deaths [1]. *Salmonella enterica* and *Bacillus cereus* were identified as the main causative agents. Seventy per cent of illnesses were attributed to consumption of RTE foods prepared with spices which were applied after the food manufacturing pathogen reduction step, while in 75% of outbreaks, it was reported that no pathogen reduction step had been applied to the spice. Furthermore, the authors identified an additional seven spice-related foodborne outbreaks which lacked microbiological or epidemiological evidence and therefore did not meet the inclusion criteria of the study, highlighting the probability that the number of foodborne outbreaks associated with consumption of contaminated spices is under-reported [1]. Along with this, studies have found dried herbs and spices to be contaminated with various microorganisms including *Salmonella* spp., *Clostridium perfringens*, *Escherichia coli*, *Staphylococcus aureus* and *Enterobacter* spp. at the point of retail [2–4]. Thus, spices and seasonings should be subject to a decontamination process in order to protect the consumer and prevent foodborne disease.

Given the serious consequences of foodborne infection and the economic costs associated with destroying contaminated batches and product recalls, it is not surprising that food manufacturers are seeking novel approaches for the control of foodborne pathogens. Various methods have been used to reduce microbial levels in powdered seasonings including thermal processing and irradiation; however each of these methods has limitations. Studies have shown heat treatment to be less effective for microbial decontamination in low-moisture foods such as powders and spices [5,6], while high temperatures can alter the characteristics of the powders such as the flavour, colour and aroma [7,8]. Although irradiation is an effective method for reducing the microbial load of spices, studies have reported negative impacts on the sensory properties of powders, such as a change in appearance, difference in aroma and off-flavours [9,10]. Therefore, alternative decontamination processes are required, one of which could be UVC-LED radiation.

UV radiation is a well-established method for the reduction or elimination of pathogens from foods [11–14] and has also been tested for the decontamination of powdered ingredients such as flour powdered infant formula (PIF), black pepper and powdered red pepper [15–22]. The UV spectrum is subdivided into the UVA (315–400 nm), UVB (280–315 nm) and UVC (<280 nm) ranges, each with a specific effect on microorganisms. As maximum DNA absorption of UV light occurs at the peak wavelength of 260–265 nm, UVC radiation is usually the most effective for microbial inactivation [23]. Due to this, low-pressure (LP) or medium-pressure (MP) mercury lamps emitting UVC light at a wavelength of 254 nm have traditionally been used for decontamination; however, these lamps have several disadvantages, including a relatively large footprint and the requirement of a warm-up period prior to use [24]. In addition, they pose a health risk to consumers due to possible breakage of the lamp and lamp sleeve and, consequently, the potential release of mercury particles onto the food products undergoing treatment [25], while there is a risk of mercury waste accumulation if not disposed of correctly, which can have damaging effects on human health and the environment [26].

UV light-emitting diodes (UV-LEDs) have emerged as an alternative to traditional UV lamps in recent years. LEDs are composed of layers of semiconductor material which emit light when an electrical current is applied [27]. UV-LEDs are a more sustainable source of energy than UV lamps as they do not use mercury, they can reach maximum output power instantaneously and are becoming increasingly more efficient and economically viable as time goes on. Moreover, their compactness and robustness have highlighted the potential of UV-LEDs as a cost-effective inactivation technology within the food industry [28]. In recent years, UV-LED radiation has been applied to liquid beverages such as fruit juices [29–31] and solid food products such as cheese, lettuce, cabbage, tuna fillets and chicken [24,32–34] for the reduction of foodborne pathogens. Despite this, the efficacy of UV-LEDs for the inactivation of bacteria in powdered ingredients remains largely uninvestigated, with just two studies found in the literature which investigated the efficacy of UV-LEDs for the inactivation of bacteria in low $a_W$ foods, both of which focused on heat resistant strains of *Salmonella* spp. in wheat flour [35,36]. Therefore, the objective of this study was to firstly compare the performance of a UV lamp to UV-LEDs for the surface inactivation of four bacteria (*Listeria monocytogenes*, *E. coli*, *Bacillus subtilis* and *Salmonella* Typhimurium) and following this, investigate the efficacy of UVC-LEDs for inactivation of the bacterial strains in four different seasoning powders.

## 2. Materials and Methods

### 2.1. UV Devices

The UV and UV-LED devices used in this study were assembled by the Centre for Advanced Photonic and Process Analysis (CAPPA), MTU, Cork. Two types of UV-LED devices were used, emitting at both UVC and UVA wavelengths. The 270 nm UVC LED-based lamps were built around KL265-35R-SM-GD Crystal IS devices (Crystal IS, New York, NY, USA) and the LEDs were driven at a maximum current rating of 300 mA. The 365 nm UVA LED-based lamps were built around LED Engin LZ1 UV 365 nm Gen2 Emitter devices (Osram, Munich, Germany) and driven at a maximum current rating of 1 A. For comparison purposes, a traditional 254 nm mercury lamp was also included in the study, based around 100 mm low pressure Rexim MCCUV-CV-100 × 8 × 100 Hg bulbs (Rexim, Watertown, MA, USA) with 12VDC INV-1L-12V inverters. The emission spectra of each was measured using a UVPad E radiometer (Opstytec Dr. Groebel, Ettlingen, Germany) (Figure 1A). The distance between each of the emitters and the test samples in this study was set at 20 mm for both surface and powder inactivation experiments (Figure 1B).

(A)

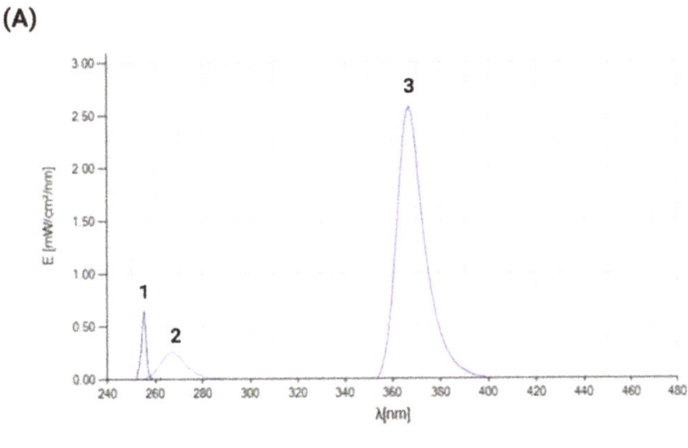


(B)

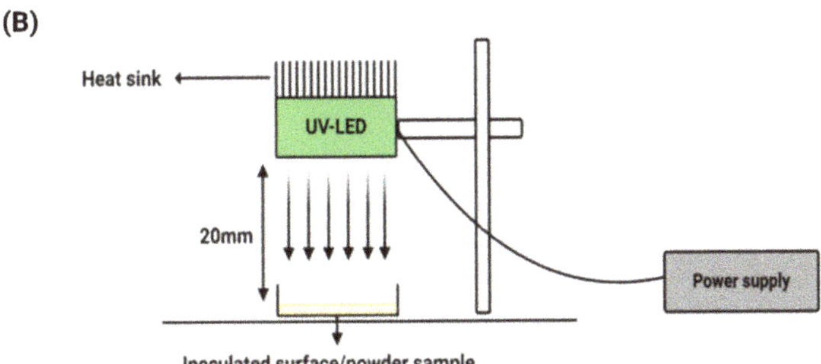


**Figure 1.** (**A**) Emission profiles of (1) mercury lamp at 254 nm; (2) UVC-LED at 270 nm; (3) UVA-LED at 365 nm. (1) and (2) were measured at 20 mm distance while (3) was measured at 100 mm distance. (**B**) Schematic diagram of the UV-LED experimental set-up. This image was created with BioRender.

### 2.2. Bacterial Strains and Growth Conditions

The strains used in this study were obtained from the MTU Cork culture collection. *L. monocytogenes* LO28, *Salmonella enterica* subsp. *enterica* serovar Typhimurium (*S.* Typhimurium) ATCC 49416 and *Bacillus subtilis* subsp. *Spizizenii* (*B. subtilis*) ATCC 6633 were grown aerobically in brain heart infusion (BHI) broth (LabM, Lancashire, UK) at 37 °C. *E. coli* DH5α*lux* contains a plasmid with a kanamycin resistance gene and was grown at 37 °C in Luria-Bertani (LB) broth (LabM, Lancashire, UK) supplemented with 100 µg/mL kanamycin (Merck, Darmstadt, Germany). Stocks of all strains were maintained at −20 °C and −80 °C in a final concentration of 40% glycerol (Merck, Darmstadt, Germany).

### 2.3. Sample Preparation

#### 2.3.1. Petri Dish Surfaces

A 1 mL sample from overnight cultures of each strain was centrifuged at 5000× *g* for 10 min, washed and resuspended in an equal volume of $\frac{1}{4}$ strength Ringers' solution (Merck, Darmstadt, Germany). Washed cultures were diluted to a starting concentration of ~$10^7$ CFU/mL. Sterile cotton swabs were soaked in the standardised cultures and were used to transfer the inoculum to a small circular area (diameter of 11 mm) in the centre of a sterile Petri dish. Inoculated Petri dishes were left to dry and cultures were re-applied two more consecutive times, resulting in an inoculation level of approx. $10^4$ CFU.

#### 2.3.2. Seasoning Powders

Four seasoning powders (garlic, onion, cheese and onion, and chilli) were sourced from a food manufacturing facility and used in the current study. Particle size of the powders were determined using a Horiba XploRA™ plus confocal Raman microscope. Powders were dispersed over a standard microscope slide. A mosaic image was created with a 50 × objective over a 2 × 2 mm area and processed using the ParticleFinder™ module in the Horiba LabSpec 6 software package. Briefly, this software analyses and counts particles in the image, measuring the particle diameter within defined limits and error margins. A statistical fit is then applied to the size distribution of the particles, which may show single or bimodal distribution. In the current study, approximately $10^2$–$10^3$ particles were detected per powder sample. The following imaging processing steps were applied prior to exporting the particle size data: remove edge particle, fill holes, erode filter: 3 × 3 and close filter: 3 × 3. The water activity ($a_w$) of the samples was measured at 25 °C using a LabMaster-aw neo water activity meter (Novasina AG, Lachen, Switzerland) (Table 1).

**Table 1.** Seasoning powder properties.

| Powder | Particle Size (μm) | $a_w$ |
|---|---|---|
| Garlic | 4.64 ± 2.58 | 0.358 |
| Onion | 24.14 ± 14.57 | 0.336 |
| Cheese and onion | 31.24 ± 14.69 | 0.371 |
| Chilli (large particles) [1] | 55.75 ± 55.28 | 0.430 |
| Chilli (fine particles) | 15.20 ± 20.99 | 0.430 |

[1] Two particle sizes were measured for chilli powder due to the presence of large and fine particles in the sample.

Prior to testing, the absence of the four target microorganisms was confirmed by diluting 1 g of powder in 9 mL of $\frac{1}{4}$ strength Ringer's solution and plating onto mannitol yolk polymyxin (MYP) agar (Oxoid), *Listeria* selective agar (LSA) (Merck, Darmstadt, Germany), xylose lysine deoxycholate (XLD) agar (LabM, Lancashire, UK) and eosin methylene blue (EMB) agar (LabM, Lancashire, UK) for selection of *B. subtilis*, *L. monocytogenes*, *S.* Typhimurium and *E. coli*, respectively. Samples were also plated onto BHI agar to assess background microbiota. Powders were inoculated following the method of Callanan et al. (2012) with some modifications [37]. Briefly, overnight cultures (40 mL) of each strain were centrifuged at 10,000× *g* for 10 min, the supernatant was discarded, and excess liquid removed from the pellet using a pipette. Pellets were air dried in a biological safety cabinet for 1.5 h. Powders were inoculated by mixing the dried pellet with 10 g aliquots of the powders using a sterile glass rod until a total of 50 g of powder was inoculated, resulting in an initial population of between 5 and 6 $\log_{10}$ CFU/g. To assess even distribution of the inoculum, 5 samples were taken from different areas of each inoculated powder samples and enumerated, with variations (standard deviation) of between 0.3 and 0.5 $\log_{10}$ CFU/g observed. Samples were stored in containers aerobically at room temperature for at least 48 h prior to use.

### 2.4. UV and UV-LED Inactivation

All UV inactivation experiments were performed in a Class II biological safety cabinet at room temperature. Samples were placed on a stainless-steel surface at a distance of 20 mm from the UV or UV-LED source. For surface decontamination experiments, the inoculated Petri dish surfaces were exposed to UV (mercury lamp), UVC-LED and UVA-LED sources emitting at 254 nm, 270 nm and 365 nm, respectively, with exposure times of 5 s, 10 s, 20 s and 40 s corresponding to the doses shown in Table 2. For treatment of powders, 1 g of inoculated powder was distributed in a thin layer onto the surface of a styrene container with an area of 36.5 $cm^2$ (73 mm × 50 mm × 11 mm), resulting in a powder layer thickness not exceeding 1.5 mm. Powders were exposed to UVC-LED radiation for treatment times of 5–40 s.

**Table 2.** Light intensity data for UV mercury lamp (254 nm), UVC-LED (270 nm) and UVA-LED (365 nm) measured at 20 mm at maximum current using a UVPad E radiometer.

| Time (s) | UV Radiation Dose (mJ/cm²) | | |
|---|---|---|---|
| | 254 nm | 270 nm | 365 nm |
| 5 | 20 | 16 | 1700 |
| 10 | 40 | 32 | 3400 |
| 20 | 80 | 64 | 6800 |
| 40 | 160 | 128 | 12,600 |

### 2.5. Bacterial Enumeration

Following UV and UV-LED exposure, surviving cells were recovered from the inoculated Petri dishes by soaking fresh sterile swabs in 1 mL of 1/4 strength Ringers' solution and swabbing the inoculated area three times. The recovered cells were serially diluted in 900 μL of 1/4 strength Ringers' solution and 100 μL aliquots of dilutions were spread plated onto BHI agar for *B. subtilis*, *L. monocytogenes* and *S.* Typhimurium or LB agar supplemented with 100 μg/mL kanamycin for *E. coli* DH5α*lux*. Following UVC-LED treatment of the powder samples, samples were serially diluted 1/4 strength Ringers' solution. As the powders provided were not sterile and contained low levels of background microbiota (<$10^2$ CFU/g), 100 μL aliquots of the sample dilutions were spread plated onto MYP agar, LSA, XLD agar or LB agar supplemented with 100 μg/mL kanamycin for selection of *B. subtilis*, *L. monocytogenes*, *S.* Typhimurium and *E. coli*, respectively. Plates were incubated at 37 °C for 48 h. Following incubation, colonies were counted and results were expressed as the mean $\log_{10}$ CFU/mL(g) ± standard deviation.

### 2.6. Modelling of Bacterial Inactivation Kinetics

#### 2.6.1. Log-Linear Model

The log-linear model is based on traditional first-order inactivation kinetics. It assumes that cells have equal susceptibility and that lethality occurs randomly over time during the inactivation treatment. The model equation is written as follows:

$$N_t = N_0 \exp(-k_{maxB} \cdot t) \tag{1}$$

where $N_t$ is the population at time $t$ (CFU/g), $N_0$ is the population at time 0, $k_{maxB}$ is the maximum specific inactivation rate ($s^{-1}$) and $t$ is the time (seconds) [38].

#### 2.6.2. Biphasic Model

Described by Cerf (1977), the biphasic model consists of two phases—an initial log-linear decrease (first-order kinetics) due to inactivation of a sensitive microbial population, followed by a second, slower rate of decrease of a more stress-resistant microbial population (tail):

$$log_{10} N_t = log_{10} N_0 + log_{10}(f \cdot \exp(-k_{max1} \cdot t) + (1-f) \cdot \exp(-kmax_2 \cdot t)) \tag{2}$$

where $N_t$, $N_0$ and $t$ are as defined above, $f$ and $(1 - f)$ are the UV-resistant and UV-sensitive population fractions, respectively. $K_{max1}$ and $K_{max2}$ ($s^{-1}$) are the maximum specific inactivation rates of the UV sensitive and the UV resistant populations, respectively [39].

2.6.3. Weibull Model

Due to heterogeneity in a sample, bacterial strains may not always follow first-order kinetics. Following UV light treatment, inactivation curves generally exhibit a sigmoidal shape and may display concavity or convexity behaviours, which can be described by the Weibull model:

$$Log_{10}\left(\frac{N_t}{N_0}\right) = -\left(\frac{t}{\delta}\right)^p \quad (3)$$

where $Nt$, $N_0$ and $t$ are as defined above, $\delta$ is the scale parameter and $p$ is the shape parameter. $p < 1$ represents upward concavity of the curve, indicating stress adaptation of surviving microorganisms following UV treatment. $p > 1$ represents downward concavity, indicating that the cells become increasingly damaged with increasing treatment time. $p = 1$ represents a linear curve [40].

2.6.4. Geeraerd shoulder–tail Model

Similar to the log-linear model, the Geeraerd model is based on first-order inactivation kinetics but includes additional parameters for shoulders and tailing. The model is described as follows:

$$N_t = (N_0 - N_{res})\cdot \exp(-k_{max}\cdot t)\,\frac{\exp(-k_{max}\cdot SL)}{1 + \exp((-k_{max}\cdot SL) - 1)\exp(-k_{max}\cdot t)} + N_{res} \quad (4)$$

where $Nt$, $N_0$ and $t$ are as defined above, $N_{res}$ is the UV resistant population, $k_{max}$ is the maximum specific inactivation rate ($s^{-1}$) and SL is the parameter representing the shoulder length (seconds). If log-linear kinetics with and without shoulder/tailing are observed, SL or $N_{res}$ can be equal to zero, resulting in reduced forms of the model, namely log-linear + shoulder and log-linear + tail [41,42].

Data fitting was performed using the GInaFiT application version 1.7 [42] in Microsoft Excel. Goodness of fit was determined using the root mean square error (RMSE) and the adjusted coefficient of determination ($R^2_{adj}$).

*2.7. Statistical Analysis*

All experiments were carried out using three biological replicates and data are expressed as the mean ± standard deviation (SD). CFU data were transformed to $\log_{10}$ prior to analysis. Statistical analysis was performed using R Studio software. Data were analysed using one-way analysis of variance (ANOVA) with post hoc comparison using Tukey's multiple comparisons test. Asterisks rating of *, ** or *** indicates statistically significant differences between groups ($p \leq 0.05$, $p \leq 0.005$ or $p \leq 0.001$, respectively).

**3. Results**

*3.1. UV-LED Surface Inactivation of Microorganisms*

Log reductions of each bacterial strain (log CFU) on a plastic surface following exposure to UV, UVC-LED and UVA-LED sources at wavelengths of 254 nm, 270 nm and 365 nm, respectively, are shown in Figure 2. The results showed that the cell numbers of all four strains were significantly reduced following 5 s exposure to the traditional 254 nm mercury lamp ($p < 0.05$), with *E. coli*, *B. subtilis* and S. Typhimurium reduced below the limit of detection (10 CFU) after 10 s exposure, and *L. monocytogenes* after 20 s exposure. Similar to the mercury lamp, a 5 s exposure to the UVC-LED also resulted in a significant decrease in cell numbers of all strains ($p < 0.05$). Reductions of 2.2 and 3.8 $\log_{10}$ CFU were observed for *L. monocytogenes* and *E. coli, respectively*, while *B. subtilis* (Figure 2C) and S. Typhimurium (Figure 2D) proved the most susceptible to UVC-LED inactivation, with cell numbers reduced to below the detection limit, a reduction of over 4 $\log_{10}$ CFU for each strain. A 10 s UV-LED exposure was sufficient for complete inactivation of *E. coli* cells, while *L. monocytogenes* again proved to be the most tolerant of the four strains, requiring a longer exposure time of 20 s for inactivation. In most cases, treatment of the bacteria with the UVC-LEDs resulted in significantly higher log reductions than those obtained following treatment with the mercury lamp. The only exception to this was the 5 s treatment of *L. monocytogenes*, whereby the cell numbers were reduced to similar levels ($p > 0.05$) following exposure to both the mercury and the UVC-LED lamps (Figure 2A). Of the three lamps tested, the 365 nm UVA-LED was the least effective for bacterial inactivation. Following 40 s exposure, reductions of approx. 1.07, 0.83 and 1.12 $\log_{10}$ CFU were observed for *L. monocytogenes*, *E. coli* and *S.* Typhimurium, respectively, while the lamp had little to no effect on *B. subtilis*, with $\leq$0.1 $\log_{10}$ reduction observed.

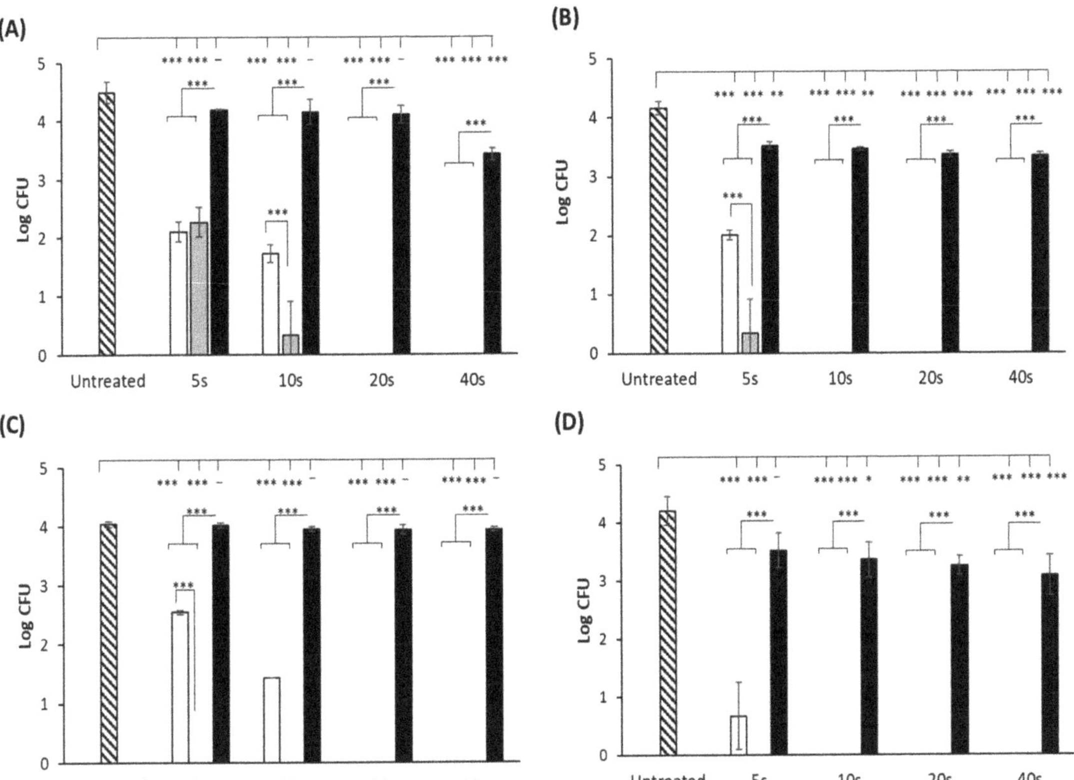


**Figure 2.** Mean log CFU values of (**A**) *L. monocytogenes;* (**B**) *E. coli;* (**C**) *B. subtilis;* (**D**) *S.* Typhimurium following exposure to UV light at wavelengths of 254 nm (white bars) and UV-LED light at 270 nm (grey bars) and 365 nm (black bars) on a plastic Petri dish surface. Error bars represent the standard deviation of replicate experiments. Absence of a bar indicates that bacteria were reduced to below detection level (10 CFU). *** denotes $p \leq 0.001$, ** denotes $p \leq 0.005$, * denotes $p \leq 0.05$ and – denotes no statistical significance.

### 3.2. UVC-LED Inactivation of Microorganisms in Powdered Ingredients

The surface inactivation experiments demonstrated that the performance of the 270 nm UVC-LED lamp was significantly superior to that of the 365 nm UVA-LED lamp, and resulted in bacterial reductions which were comparable to, or in most cases, better than those obtained using the mercury lamp. Therefore, only the UVC-LEDs were utilised in the powdered ingredient inactivation experiments. The inactivation curves of the four bacteria in each of the four seasoning powders are shown in Figure 3. The results showed powder type has an impact on bacterial inactivation. In particular, the cells were more susceptible to UVC-LED light in the cheese and onion powder rather than the onion or garlic powders. This is particularly evident in the case of *L. monocytogenes*, where a 20 s exposure time was required to reduce recoverable cell numbers in onion powder and garlic powder to the same levels obtained with a 10 s exposure in cheese and onion powder (Figure 3A). In chilli powder, despite a significant reduction in cell numbers for all four strains following 40 s treatment (between 0.75–1.3 $\log_{10}$ CFU/g), the overall levels of inactivation were lower than those observed in the other three powders. The physiochemical properties of the powder have been shown to influence UV inactivation efficacy in food matrices [11,33,43]. Therefore, particle size and water activity were determined (Table 1); however, in our study, smaller particle size and higher water activity were not associated with increased inactivation.

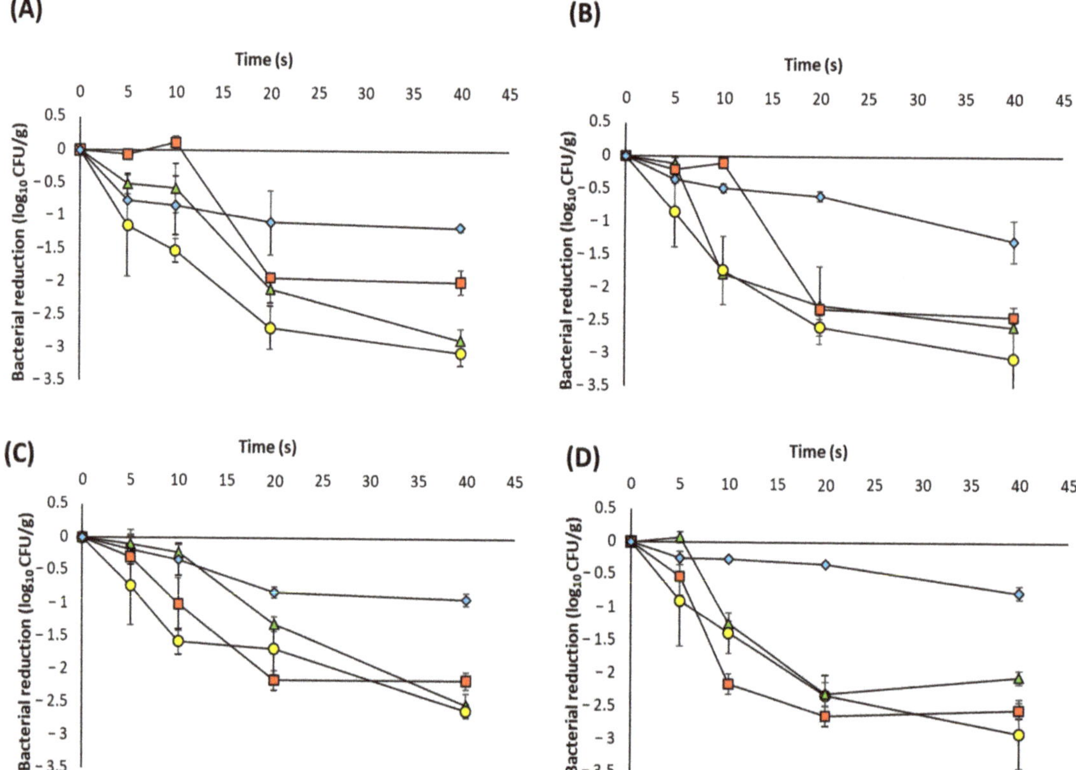


**Figure 3.** Inactivation curves of (**A**) *L. monocytogenes*; (**B**) *E. coli*; (**C**) *B. subtilis*; (**D**) *S.* Typhimurium following exposure to UVC-LED light at 270 nm in garlic powder (▲ green triangle), onion powder (■ red square), cheese and onion powder (• yellow circle) and chilli powder (♦ blue diamond). Data are represented as the mean ± standard deviation of three biological replicates.

### 3.3. Model Fitting

Mathematical models are used to quantify and compare the microbial inactivation efficacy of different technologies. Matrix effects such as shadowing and the shallow penetration of UV radiation means there is no consensus model for UV microbial inactivation kinetics, so we tested the ability of a number of commonly applied mathematical models (log linear, biphasic, Weibull and Geeraerd shoulder–tail) to accurately describe the observed inactivation kinetics. The UVC-LED inactivation profiles (Figure 3) differed depending on the powder and microbe, with some showing a shoulder phase, some a tailing phase and some with both shoulders and tails. In both onion and garlic powders, little or no decrease in cell numbers was observed within the first 5–10 s of treatment (shoulder), followed by a significant reduction after 20 s of UVC-LED treatment (between 1.3–2.6 $\log_{10}$ CFU/g), and a final gradual decrease of approx. 0.5–1 $\log_{10}$ CFU/g (tail). The data were fitted with the four different inactivation models and the goodness of fit of each model was assessed using the $R^2_{adj}$ and RMSE values (Table 3). Unsurprisingly, the log-linear model was a poor fit for the data in most of the conditions with the exception of *S.* Typhimurium in chilli powder ($R^2_{adj}$ = 0.884, RMSE = 0.088).

**Table 3.** Goodness of fit parameters of the log-linear, biphasic, Weibull and Geeraerd shoulder–tail models for inactivation of bacterial strains in garlic, onion, cheese and onion and chilli powders.

| Powder | Microorganism | Inactivation Model | | | | | | | |
|---|---|---|---|---|---|---|---|---|---|
| | | Log-Linear | | Biphasic | | Weibull | | Geeraerd Shoulder–Tail | |
| | | $R^2_{adj}$ | RMSE | $R^2_{adj}$ | RMSE | $R^2_{adj}$ | RMSE | $R^2_{adj}$ | RMSE |
| Garlic | *L. monocytogenes* | 0.902 | 0.357 | 0.928 | 0.305 | 0.906 | 0.349 | **0.947** [1] | **0.261** |
| | *E. coli* | 0.684 | 0.645 | 0.857 | 0.433 | 0.766 | 0.555 | **0.938** | **0.286** |
| | *B. subtilis* | 0.959 | 0.205 | - | - | 0.962 | 0.197 | **0.985** | **0.121** |
| | *S.* Typhimurium | 0.629 | 0.627 | 0.880 | 0.357 | 0.701 | 0.563 | **0.982** | **0.137** |
| Onion | *L. monocytogenes* | 0.720 | 0.533 | 0.756 | 0.497 | 0.701 | 0.550 | **0.986** | **0.121** |
| | *E. coli* | 0.761 | 0.567 | 0.789 | 0.531 | 0.752 | 0.577 | **0.987** | **0.130** |
| | *B. subtilis* | 0.738 | 0.487 | 0.908 | 0.289 | 0.812 | 0.413 | **0.956** | **0.200** |
| | *S.* Typhimurium | 0.575 | 0.742 | 0.929 | 0.303 | 0.771 | 0.545 | **0.979** | **0.164** |
| Cheese and onion | *L. monocytogenes* | 0.514 | 0.787 | **0.975** | **0.180** | 0.868 | 0.410 | - | - |
| | *E. coli* | 0.834 | 0.514 | 0.929 | 0.336 | **0.937** | **0.316** | - | - |
| | *B. subtilis* | 0.789 | 0.413 | **0.942** | **0.217** | 0.935 | 0.229 | - | - |
| | *S.* Typhimurium | 0.788 | 0.510 | 0.967 | 0.202 | **0.968** | **0.199** | - | - |
| Chilli | *L. monocytogenes* | 0.368 | 0.413 | 0.576 | 0.339 | **0.616** | **0.322** | - | - |
| | *E. coli* | 0.876 | 0.160 | **0.879** | **0.158** | 0.874 | 0.161 | - | - |
| | *B. subtilis* | 0.746 | 0.199 | **0.851** | **0.152** | 0.787 | 0.182 | 0.827 | 0.164 |
| | *S.* Typhimurium | **0.884** | **0.088** | 0.863 | 0.096 | 0.877 | 0.091 | 0.863 | 0.096 |

[1] Values highlighted in bold represent the best fitting model for each bacteria in each powder.

The biphasic model assumes the presence of two populations in the matrix, a major population which is more susceptible ($k_{max1}$) and a minor population which is more resistant ($k_{max2}$). Therefore, curves which show a sharp initial decline followed by a tailing phase are generally described well by the biphasic model and it was statistically the best fit for *L. monocytogenes* ($R^2_{adj}$ = 0.975, RMSE = 0.180) and *B. subtilis* ($R^2_{adj}$ = 0.942, RMSE = 0.217) in cheese and onion powder. In contrast, the behaviour of *E. coli* and *S.* Typhimurium in this powder were best described using the Weibull model which has been used in other studies to describe UV-C inactivation [43,44]. The Geeraerd-shoulder-tail model, which describes inactivation curves incorporating both shoulder and tailing phases, was the statistical best fit for all of the strain data in both garlic powder and onion powder. Interestingly, while this model showed the best goodness-of-fit for the both the garlic and onion powder data, the model could not be fitted to any of the data obtained in the cheese & onion powder, or to the *L. monocytogenes* and *E. coli* data in chilli powder. The correlation between experimentally observed bacterial log reductions and those predicted by the best-fitting model for each of the four powders is illustrated in Figure 4 and suggests that overall, the Geeraerd-shoulder-tail model was the most accurate.

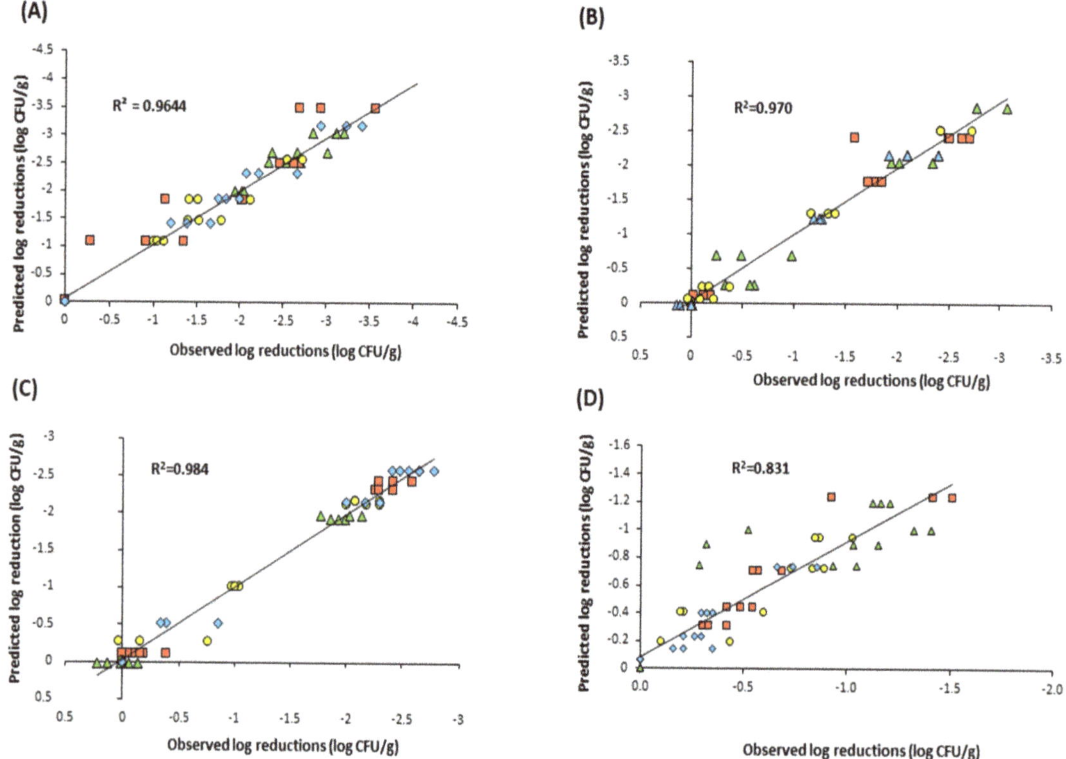


**Figure 4.** Correlation between experimentally observed log reductions and predicted log reductions of *L. monocytogenes* (▲ green triangle), *E. coli* (■ red square), *B. subtilis* (• yellow circle) and *S. Typhimurium* (♦ blue diamond) as predicted by (**A**) the biphasic model, in cheese and onion powder; (**B**) the Geeraerd shoulder–tail model, in garlic powder; (**C**) the Geeraerd-tail model, in onion powder and (**D**) the biphasic model, in chilli powder.

## 4. Discussion

The objective of this study was to investigate the use of UV-LED devices for inactivation of pathogenic microbes in powdered seasoning ingredients. Initially, the efficacy of both UVC-LED (270 nm) and UVA-LED (365 nm) devices for bacterial inactivation was investigated on a Petri dish surface, while a traditional mercury lamp was included for comparison purposes. The results showed that the UVC-LED was at least as effective or in some cases, more effective than the mercury lamp. Kim et al. (2016) reported similar findings, observing significantly higher reductions of *E. coli* O157:H7 in microbiological media with UV-LEDs emitting at 266 nm compared to a mercury lamp [24], while other authors have found similar success with UV-LEDs at wavelengths of 260–285 nm [29,45–49]. This highlights the advantage of using UV-LEDs which can be designed to produce specific wavelengths instead of mercury lamps which can only emit at a single wavelength of 254 nm and have a large environmental footprint.

Santos et al. (2014) investigated the effect of different wavelengths (UVA, UVB and UVC) on the inactivation of nine different bacterial isolates in media, reporting that UVC light was most effective for bacterial inactivation, with the highest survival rates observed following UVA exposure [50]. Likewise, other researchers have made similar observations regarding the inefficacy of UVA light for bacterial inactivation compared to UVC radiation [49,51]. Although some studies have reported significant bacterial reductions in liquids and on food surfaces following UVA-LED treatment [32,35,36,52,53], long exposure times (30–180 min) were required, highlighting the time impracticalities of UVA-LED inactivation. However, as the fluence emitted from the UVA-LED used in the current study was significantly higher than that of the mercury lamp or the UVC-LED source, it was included to investigate whether these increased energy levels would result in a significant reduction in bacterial cell numbers within a short treatment period. Similar to the results observed Hinds et al. (2019), the UVA-LED did not significantly reduce cell numbers of *B. subtilis* [49]; however, significant reductions (approximately 1-$\log_{10}$ CFU) in cell numbers of *L. monocytogenes*, *E. coli* and *S.* Typhimurium were obtained after 40 s treatment. Nevertheless, despite the increased energy doses, the performance of the UVA-LED source for bacterial inactivation was inferior to that of the mercury lamp or the UVC-LED, findings which are in agreement with the literature [49,50].

The Gram-positive pathogen *L. monocytogenes* appeared to be more difficult to eradicate on the plastic surface by UVC-LED radiation, requiring 64 mJ/cm$^2$ of UVC exposure for complete inactivation, compared to the Gram-negative *E. coli* and *S.* Typhimurium strains which required just 32 mJ/cm$^2$ and 16 mJ/cm$^2$, respectively. It has been reported that Gram-positive microorganisms tend to be more UV-resistant, requiring higher UV dosages than Gram-negative bacteria for inactivation. Gabriel et al. (2009) found that *L. monocytogenes* was significantly more resistant to UV treatment than *E. coli* and *S.* Enteritidis, with similar observations made by Shin et al. (2016) [47,54]. According to Gayán et al. (2014), differences between the UV-resistance of species can be attributed to several factors including cell wall thickness, cell size and DNA repair efficiency [55].

Bacterial strains were most and least susceptible to UVC-LED inactivation in cheese and onion powder and chilli powder, respectively. Other studies have also shown the efficacy of UV treatment to be dependent on the matrix [11,12,33,43,56], with differences commonly attributed to the ingredient composition and the chemical and physical properties of the food. The particle size of the food sample is one such characteristic which can impact UV-LED inactivation of microorganism. UV radiation penetration of powders is difficult due to the shadowing effect of food particles, which protects the bacterial cells from complete exposure, while bacterial cells can also protect each other from UV light [55], a phenomenon which has been observed in several studies [17,19–21]. As the particle sizes of the powders used in this study are generally larger than the typical cell size of each of the target microorganisms it is plausible that the crevices and cavities on the surface of the powders shielded a subpopulation of the microbial cells from the UVC light. Furthermore, the chilli powder used in the present study contained large spice particles which measured approx. 12-fold, 2.3-fold and 2-fold larger than the garlic, onion and cheese and onion particles, respectively; thus, it is likely that those particles physically shielded both the finer spice particles and the bacterial cells from the UVC-LED light, resulting in the lowest bacterial inactivation rates of all powders tested. Condón-Abanto et al. (2016) addressed the problem of shielding in flour by using sample sizes of 1 g and 0.1 g, resulting in reductions of approx. 2-$\log_{10}$ and 2.5-$\log_{10}$ CFU/g after 60 s treatment, respectively, most likely due to the reduction in the number of particles capable of producing a shadow effect [17]. Ha and Kang (2013, 2014) used a rotational mixer during UV radiation of *E. coli* and *S.* Typhimurium in red pepper powder [19] and *C. sakazakii* in PIF [20] with the aim of increasing the contact surface area of the UV light on the particles; however, the reduction in cell numbers obtained was more than 50 times less than those observed by Condón-Abanto et al. (2016) and in the current study, probably due to the large sample size of 250 g. In our study, we did not observe any significant differences in the log reductions obtained in samples which were stirred at intervals during UV-LED treatment (data not shown).

The fitting of inactivation models to the data confirmed that the bacteria did not show linear behaviour during inactivation, with both shoulders and/or tails observed in most curves and the Geeraerd shoulder–tail model being the most accurate of the models assessed. A number of studies have observed inactivation curves with shoulders following UV treatment of microbes in culture media [17,57,58] and liquid foods such as fruit and vegetable juices [43,59–61] However, as each study differs in the choice of UV device and treatment protocol, it is possible that the presence/absence of a shoulder may be dependent on the UV-LED dosage. For example, some studies report bacterial log reductions following a single UV treatment [16,22,56] while Liu et al. (2012) and Arroyo et al. (2017) did not observe shoulders following the inactivation of *C. sakazakii* in PIF; however, higher initial doses were used [15,21]. Similarly, Condón-Abanto et al. (2016) reported inactivation levels of >1.5 $log_{10}$ CFU/g of *S. Typhimurium* in flour during the first 60 s of UVC treatment [17]. Thus, it is possible that the absence of a shoulder in these studies may be due to the high initial UV dosage, leading to significant reductions in cell numbers during the early stages of the inactivation treatment. The tailing phenomenon observed in the inactivation curves can be attributed to resistant subpopulations but it is more likely due to subpopulations that are inaccessible or do not receive the same lethal dose [44] This is due to the previously described shadowing effect of the powder particles preventing complete UV penetration and, subsequently, the survival of a small population of cells. A similar scenario was observed by Gabriel et al. (2020), whereby total log reductions of *E. coli* and *S. aureus* on dried black peppercorns ranged from 1.92–3.60 $log_{10}$ CFU/g after 90 min of UVC treatment, with the majority of inactivation occurring within the first 20 min (0–500 $mJ/cm^2$) [62].

As with any food processing treatment, UV radiation can negatively impact the nutritional and organoleptic characteristics of food products. In particular, components which are capable of absorbing UV light, such as vitamin A, vitamin C, riboflavin and food colourings, are the most affected by photochemical reactions [63,64]. UV treatment has also been shown to induce increased lipid oxidation in foods [65,66]. However, few published studies are available which investigate effect of UV-LEDs on food quality characteristics, with most of the research focusing on microbial inactivation. Subedi et al. (2020) observed changes in the gluten structure and oxidation of gluten proteins and in wheat flour following UV-LED treatment at wavelengths ranging from 275–455 nm, with similar results reported by Du et al. (2020) [35,36]. Both Akgün and Unluturk (2017) and Ghate et al. (2016) noted colour changes in fruit juice following UV-LED treatment [29,67], while in contrast, Kim et al. (2017) reported no adverse effects on the nutritional or chemical properties of fresh-cut papaya or mango following UVA-LED exposure [68,69]. As the focus of this study was to investigate the efficacy of UV-LEDs for the microbial decontamination of powders, the impact of the treatment on the powders was not investigated and in general, studies in the literature describing the effect of UV-LED exposure on the nutritional and organoleptic properties of foods are scarce. Thus, it is evident that further work is required to address this area of UV-LED research.

## 5. Conclusions

In conclusion, we have demonstrated the potential of UVC-LEDs as an alternative to traditional UV lamps for bacterial inactivation. UVC-LED emission at 270 nm was just as effective or in some cases more effective than the 254 nm mercury lamp for surface decontamination of four bacterial strains. There are limited data available regarding the use of UV-LEDs and powdered foods and although previous studies have investigated the efficacy of UV-LEDs for inactivation of *Salmonella* spp. in wheat flour, to our knowledge the current study is the first to report the effect of UV-LEDs on four different bacteria in a range of different powdered ingredients. The results showed that bacterial numbers in each of the four powders were reduced significantly in just 40 s, highlighting the power and efficacy of UV-LEDs lamps in comparison to their mercury counterparts.

**Author Contributions:** Conceptualization, L.L. and M.P.; methodology, L.N., L.L., M.P., M.C.; investigation, L.N., M.P.; data curation, L.N.; writing—original draft preparation, L.N.; writing—review and editing, L.L., M.P., M.C., M.B.; supervision, L.L., M.B., M.C.; project administration, L.L.; funding acquisition, L.L., M.C. All authors have read and agreed to the published version of the manuscript.

**Funding:** This work was supported by the Department of Agriculture, Food and the Marine under the Food Institutional Research Measure, grant numbers 15F604 and 2019R495.

**Institutional Review Board Statement:** Not applicable.

**Informed Consent Statement:** Not applicable.

**Data Availability Statement:** The data presented in this study are available on request from the corresponding author.

**Conflicts of Interest:** The authors declare no conflict of interest.

## References

1. Van Doren, J.M.; Neil, K.P.; Parish, M.; Gieraltowski, L.; Gould, L.H.; Gombas, K.L. Foodborne illness outbreaks from microbial contaminants in spices, 1973–2010. *Food Microbiol.* **2013**, *36*, 456–464. [CrossRef] [PubMed]
2. Moreira, P.L.; Lourencao, T.B.; Pinto, J.P.A.N.; Rall, V.L.M. Microbiological quality of spices marketed in the city of Botucatu, Sao Paulo, Brazil. *J. Food Prot.* **2009**, *72*, 421–424. [CrossRef] [PubMed]
3. Sagoo, S.K.; Little, C.L.; Greenwood, M.; Mithani, V.; Grant, K.A.; McLauchlin, J.; de Pinna, E.; Threlfall, E.J. Assessment of the microbiological safety of dried spices and herbs from production and retail premises in the United Kingdom. *Food Microbiol.* **2009**, *26*, 39–43. [CrossRef] [PubMed]
4. Sospedra, I.; Soriano, J.M.; Mañes, J. Assessment of the microbiological safety of dried spices and herbs commercialized in Spain. *Plant. Foods Hum. Nutr.* **2010**, *65*, 364–368. [CrossRef]
5. Laroche, C.; Fine, F.; Gervais, P. Water activity affects heat resistance of microorganisms in food powders. *Int. J. Food Microbiol.* **2005**, *97*, 307–315. [CrossRef]
6. Santillana Farakos, S.M.; Frank, J.F.; Schaffner, D.W. Modeling the influence of temperature, water activity and water mobility on the persistence of *Salmonella* in low-moisture foods. *Int. J. Food Microbiol.* **2013**, *166*, 280–293. [CrossRef]
7. Rhim, J.W.; Hong, S.I. Effect of water activity and temperature on the color change of red pepper (*Capsicum annuum* L.) powder. *Food Sci. Biotechnol.* **2011**, *20*, 215–222. [CrossRef]
8. Baechler, R.; Clerc, M.-F.; Ulrich, S.; Benet, S. Physical changes in heat-treated whole milk powder. *Lait* **2005**, *85*, 305–314. [CrossRef]
9. Duncan, S.E.; Moberg, K.; Amin, K.N.; Wright, M.; Newkirk, J.J.; Ponder, M.A.; Acuff, G.R.; Dickson, J.S. Processes to preserve spice and herb quality and sensory integrity during pathogen inactivation. *J. Food Sci.* **2017**, *82*, 1208–1215. [CrossRef]
10. Jung, K.; Song, B.S.; Kim, M.J.; Moon, B.G.; Go, S.M.; Kim, J.K.; Lee, Y.J.; Park, J.H. Effect of X-ray, gamma ray, and electron beam irradiation on the hygienic and physicochemical qualities of red pepper powder. *LWT Food Sci. Technol.* **2015**, *63*, 846–851. [CrossRef]
11. Holck, A.; Liland, K.H.; Carlehög, M.; Heir, E. Reductions of *Listeria monocytogenes* on cold-smoked and raw salmon fillets by UV-C and pulsed UV light. *Innov. Food Sci. Emerg. Technol.* **2018**, *50*, 1–10. [CrossRef]
12. Keklik, N.; Demirci, A.; Puri, V.; Heinemann, P. Modeling the inactivation of *Salmonella* Typhimurium, *Listeria monocytogenes* and *Salmonella* Enteritidis on poultry products exposed to pulsed UV light. *J. Food Prot.* **2012**, *75*, 281–288. [CrossRef]
13. McLeod, A.; Hovde Liland, K.; Haugen, J.-E.; Sørheim, O.; Myhrer, K.S.; Holck, A.L. Chicken fillets subjected to UV-C and pulsed UV light: Reduction of pathogenic and spoilage bacteria, and changes in sensory quality. *J. Food Saf.* **2018**, *38*, e12421. [CrossRef]
14. Unluturk, S.; Atilgan, M.R.; Baysal, A.H.; Unluturk, M.S. Modeling inactivation kinetics of liquid egg white exposed to UV-C irradiation. *Int. J. Food Microbiol.* **2010**, *142*, 341–347. [CrossRef]
15. Arroyo, C.; Dorozko, A.; Gaston, E.; O'Sullivan, M.; Whyte, P.; Lyng, J.G. Light based technologies for microbial inactivation of liquids, bead surfaces and powdered infant formula. *Food Microbiol.* **2017**, *67*, 49–57. [CrossRef]
16. Cheon, H.L.; Shin, J.Y.; Park, K.H.; Chung, M.S.; Kang, D.H. Inactivation of foodborne pathogens in powdered red pepper (*Capsicum annuum* L.) using combined UV-C irradiation and mild heat treatment. *Food Control.* **2015**, *50*, 441–445. [CrossRef]
17. Condón-Abanto, S.; Condón, S.; Raso, J.; Lyng, J.G.; Álvarez, I. Inactivation of *Salmonella* Typhimurium and *Lactobacillus plantarum* by UV-C light in flour powder. *Innov. Food Sci. Emerg. Technol.* **2016**, *35*, 1–8. [CrossRef]
18. Fine, F.; Gervais, P. Efficiency of pulsed UV light for microbial decontamination of food powders. *J. Food Prot.* **2004**, *67*, 787–792. [CrossRef]
19. Ha, J.W.; Kang, D.H. Simultaneous near-infrared radiant heating and UV radiation for inactivating *Escherichia coli* O157: H7 and *Salmonella enterica* serovar Typhimurium in powdered red pepper (*Capsicum annuum* L.). *Appl. Environ. Microbiol.* **2013**, *79*, 6568–6575. [CrossRef]
20. Ha, J.W.; Kang, D.H. Synergistic bactericidal effect of simultaneous near-infrared radiant heating and UV radiation against *Cronobacter sakazakii* in powdered infant formula. *Appl. Environ. Microbiol.* **2014**, *80*, 1858–1863. [CrossRef]
21. Liu, Q.; Lu, X.; Swanson, B.G.; Rasco, B.A.; Kang, D.H. Monitoring ultraviolet (UV) radiation inactivation of *Cronobacter sakazakii* in dry infant formula using fourier transform infrared spectroscopy. *J. Food Sci.* **2012**, *77*, 86–93. [CrossRef] [PubMed]
22. Nicorescu, I.; Nguyen, B.; Moreau-Ferret, M.; Agoulon, A.; Chevalier, S.; Orange, N. Pulsed light inactivation of *Bacillus subtilis* vegetative cells in suspensions and spices. *Food Control.* **2013**, *31*, 151–157. [CrossRef]
23. Dai, T.; Vrahas, M.S.; Murray, C.K.; Hamblin, M.R. Ultraviolet C irradiation: An alternative antimicrobial approach to localized infections? *Expert Rev. Anti. Infect.* **2012**, *10*, 185–195. [CrossRef] [PubMed]
24. Kim, S.J.; Kim, D.K.; Kang, D.H. Using UVC light-emitting diodes at wavelengths of 266 to 279 nanometers to inactivate foodborne pathogens and pasteurize sliced cheese. *Appl. Environ. Microbiol.* **2016**, *82*, 11–17. [CrossRef] [PubMed]
25. Clarke, S. *Ultraviolet Light Disinfection in the Use of Individual Water Purification Devices*; Technical Information Paper #31-006-0211; U.S. Army Public Health Command: Aberdeen Proving Ground, MD, USA, 2006; pp. 1–12.

26. Hamamoto, A.; Mori, M.; Takahashi, A.; Nakano, M.; Wakikawa, N.; Akutagawa, M.; Ikehara, T.; Nakaya, Y.; Kinouchi, Y. New water disinfection system using UVA light-emitting diodes. *J. Appl. Microbiol.* **2007**, *103*, 2291–2298. [CrossRef] [PubMed]
27. Hinds, L.M.; O'Donnell, C.P.; Akhter, M.; Tiwari, B.K. Principles and mechanisms of ultraviolet light emitting diode technology for food industry applications. *Innov. Food Sci. Emerg. Technol.* **2019**, *56*, 102153. [CrossRef]
28. D'Souza, C.; Yuk, H.G.; Khoo, G.H.; Zhou, W. Application of light-emitting diodes in food production, postharvest preservation, and microbiological food safety. *Compr. Rev. Food Sci. Food Saf.* **2015**, *14*, 719–740. [CrossRef]
29. Akgün, M.P.; Ünlütürk, S. Effects of ultraviolet light emitting diodes (LEDs) on microbial and enzyme inactivation of apple juice. *Int. J. Food Microbiol.* **2017**, *260*, 65–74. [CrossRef]
30. Lian, X.; Tetsutani, K.; Katayama, M.; Nakano, M.; Mawatari, K.; Harada, N.; Hamamoto, A.; Yamato, M.; Akutagawa, M.; Kinouchi, Y.; et al. A new colored beverage disinfection system using UV-A light-emitting diodes. *Biocontrol Sci.* **2010**, *15*, 33–37. [CrossRef]
31. Xiang, Q.; Fan, L.; Zhang, R.; Ma, Y.; Liu, S.; Bai, Y. Effect of UVC light-emitting diodes on apple juice: Inactivation of *Zygosaccharomyces rouxii* and determination of quality. *Food Control.* **2020**, *111*, 107082. [CrossRef]
32. Aihara, M.; Lian, X.; Shimohata, T.; Uebanso, T.; Mawatari, K.; Harada, Y.; Akutagawa, M.; Kinouchi, Y.; Takahashi, A. Vegetable surface sterilization system using UVA light-emitting diodes. *J. Med. Investig.* **2014**, *61*, 286–290. [CrossRef]
33. Fan, L.; Liu, X.; Dong, X.; Dong, S.; Xiang, Q.; Bai, Y. Effects of UVC light-emitting diodes on microbial safety and quality attributes of raw tuna fillets. *LWT Food Sci. Technol.* **2020**, *139*, 110553. [CrossRef]
34. Haughton, P.N.; Grau, E.G.; Lyng, J.; Cronin, D.; Fanning, S.; Whyte, P. Susceptibility of *Campylobacter* to high intensity near ultraviolet/visible 395 ± 5 nm light and its effectiveness for the decontamination of raw chicken and contact surfaces. *Int. J. Food Microbiol.* **2012**, *159*, 267–273. [CrossRef]
35. Du, L.; Jaya Prasad, A.; Gänzle, M.; Roopesh, M.S. Inactivation of *Salmonella* spp. in wheat flour by 395 nm pulsed light emitting diode (LED) treatment and the related functional and structural changes of gluten. *Food Res. Int.* **2020**, *127*, 108716. [CrossRef]
36. Subedi, S.; Du, L.; Prasad, A.; Yadav, B.; Roopesh, M.S. Inactivation of *Salmonella* and quality changes in wheat flour after pulsed light-emitting diode (LED) treatments. *Food Bioprod. Process.* **2020**, *121*, 166–177. [CrossRef]
37. Callanan, M.; Paes, M.; Iversen, C.; Kleijn, R.; Bravo Almeida, C.; Peñaloza, W.; Johnson, N.; Vuataz, G.; Michel, M. Behavior of *Enterobacter pulveris* in amorphous and crystalline powder matrices treated with supercritical carbon dioxide. *J. Dairy Sci.* **2012**, *95*, 6300–6306. [CrossRef]
38. Bigelow, W.D.; Esty, J.R. The thermal death point in relation to time of typical thermophilic organisms. *J. Infect. Dis.* **1920**, *27*, 602–617. [CrossRef]
39. Cerf, O. Tailing of survival curves of bacterial spores. *J. Appl. Bacteriol.* **1977**, *42*, 1–19. [CrossRef]
40. Mafart, P.; Couvert, O.; Gaillard, S.; Leguerinel, I. On calculating sterility in thermal preservation methods: Application of the Weibull frequency distribution model. *Int. J. Food Microbiol.* **2002**, *72*, 107–113. [CrossRef]
41. Geeraerd, A.H.; Herremans, C.H.; Van Impe, J.F. Structural model requirements to describe microbial inactivation during a mild heat treatment. *Int. J. Food Microbiol.* **2000**, *59*, 185–209. [CrossRef]
42. Geeraerd, A.H.; Valdramidis, V.P.; Van Impe, J.F. GInaFiT, a freeware tool to assess non-log-linear microbial survivor curves. *Int. J. Food Microbiol.* **2005**, *102*, 95–105. [CrossRef]
43. Fenoglio, D.; Ferrario, M.; García Carrillo, M.; Schenk, M.; Guerrero, S. Characterization of microbial inactivation in clear and turbid juices processed by short-wave ultraviolet light. *J. Food Process. Preserv.* **2020**, 1–18. [CrossRef]
44. Mutz, Y.S.; Rosario, D.K.A.; Bernardes, P.C.; Paschoalin, V.M.F.; Conte-Junior, C.A. Modeling *Salmonella* Typhimurium inactivation in dry-fermented sausages: Previous habituation in the food matrix undermines UV-C decontamination efficacy. *Front. Microbiol.* **2020**, *11*. [CrossRef]
45. Cheng, Y.; Chen, H.; Sánchez Basurto, L.A.; Protasenko, V.V.; Bharadwaj, S.; Islam, M.; Moraru, C.I. Inactivation of *Listeria* and *E. coli* by Deep-UV LED: Effect of substrate conditions on inactivation kinetics. *Sci. Rep.* **2020**, *10*, 1–14. [CrossRef]
46. Sholtes, K.A.; Lowe, K.; Walters, G.W.; Sobsey, M.D.; Linden, K.G.; Casanova, L.M. Comparison of ultraviolet light-emitting diodes and low-pressure mercury-arc lamps for disinfection of water. *Environ. Technol.* **2016**, *37*, 2183–2188. [CrossRef]
47. Shin, J.Y.; Kim, S.J.; Kim, D.K.; Kang, D.H. Fundamental characteristics of deep-UV light-emitting diodes and their application to control foodborne pathogens. *Appl. Environ. Microbiol.* **2016**, *82*, 2–10. [CrossRef]
48. Kim, D.K.; Kim, S.J.; Kang, D.H. Bactericidal effect of 266 to 279 nm wavelength UVC-LEDs for inactivation of Gram positive and Gram negative foodborne pathogenic bacteria and yeasts. *Food Res. Int.* **2017**, *97*, 280–287. [CrossRef] [PubMed]
49. Hinds, L.M.; Charoux, C.M.G.; Akhter, M.; O'Donnell, C.P.; Tiwari, B.K. Effectiveness of a novel UV light emitting diode based technology for the microbial inactivation of *Bacillus subtilis* in model food systems. *Food Control.* **2019**, *114*, 106910. [CrossRef]
50. Santos, A.L.; Oliveira, V.; Baptista, I.; Henriques, I.; Gomes, N.C.M.; Almeida, A.; Correia, A.; Cunha, Â. Wavelength dependence of biological damage induced by UV radiation on bacteria. *Arch. Microbiol.* **2013**, *195*, 63–74. [CrossRef]
51. Wang, C.; Lu, S.; Zhang, Z. Inactivation of airborne bacteria using different UV sources: Performance modeling, energy utilization and endotoxin degradation. *Sci. Total Environ.* **2019**, *655*, 787–795. [CrossRef]
52. Lui, G.Y.; Roser, D.; Corkish, R.; Ashbolt, N.J.; Stuetz, R. Point-of-use water disinfection using ultraviolet and visible light-emitting diodes. *Sci. Total Environ.* **2016**, *553*, 626–635. [CrossRef] [PubMed]
53. Prasad, A.; Gänzle, M.; Roopesh, M.S. Inactivation of *Escherichia coli* and *Salmonella* using 365 and 395 nm high intensity pulsed light emitting diodes. *Foods* **2019**, *8*, 679. [CrossRef] [PubMed]

54. Gabriel, A.A.; Nakano, H. Inactivation of *Salmonella, E. coli* and *Listeria monocytogenes* in phosphate-buffered saline and apple juice by ultraviolet and heat treatments. *Food Control.* **2009**, *20*, 443–446. [CrossRef]
55. Gayán, E.; Condón, S.; Álvarez, I. Biological aspects in food preservation by ultraviolet light: A review. *Food Bioprocess. Technol.* **2014**, *7*, 1–20. [CrossRef]
56. Stoops, J.; Jansen, M.; Claes, J.; Van Campenhout, L. Decontamination of powdery and granular foods using continuous wave UV radiation in a dynamic process. *J. Food Eng.* **2013**, *119*, 254–259. [CrossRef]
57. Gayán, E.; Mañas, P.; Álvarez, I.; Condón, S. Mechanism of the synergistic inactivation of *Escherichia coli* by UV-C light at mild temperatures. *Appl. Environ. Microbiol.* **2013**, *79*, 4465–4473. [CrossRef]
58. Lasagabaster, A.; Martínez de Marañón, I. Impact of process parameters on *Listeria innocua* inactivation kinetics by pulsed light technology. *Food Bioprocess. Technol.* **2013**, *6*, 1828–1836. [CrossRef]
59. Arroyo, C.; Gayán, E.; Pagán, R.; Condón, S. UV-C inactivation of *Cronobacter sakazakii. Foodborne Pathog. Dis.* **2012**, *9*, 907–914. [CrossRef]
60. Gayán, E.; Serrano, M.J.; Raso, J.; Álvarez, I.; Condón, S. Inactivation of *Salmonella enterica* by UV-C light alone and in combination with mild temperatures. *Appl. Environ. Microbiol.* **2012**, *78*, 8353–8361. [CrossRef]
61. Gouma, M.; Gayán, E.; Raso, J.; Condón, S.; Álvarez, I. Inactivation of spoilage yeasts in apple juice by UV-C light and in combination with mild heat. *Innov. Food Sci. Emerg. Technol.* **2015**, *32*, 146–155. [CrossRef]
62. Gabriel, A.A.; David, M.M.C.; Elpa, M.S.C.; Michelena, J.C.D. Decontamination of dried whole black peppercorns using ultraviolet-c irradiation. *Food Microbiol.* **2020**, *88*, 103401. [CrossRef] [PubMed]
63. Islam, M.S.; Patras, A.; Pokharel, B.; Vergne, M.J.; Sasges, M.; Begum, A.; Rakariyatham, K.; Pan, C.; Xiao, H. Effect of UV irradiation on the nutritional quality and cytotoxicity of apple juice. *J. Agric. Food Chem.* **2016**, *64*, 7812–7822. [CrossRef] [PubMed]
64. Guneser, O.; Karagul Yuceer, Y. Effect of ultraviolet light on water- and fat-soluble vitamins in cow and goat milk. *J. Dairy Sci.* **2012**, *95*, 6230–6241. [CrossRef]
65. Molina, B.; Sáez, M.I.; Martínez, T.F.; Guil-Guerrero, J.L.; Suárez, M.D. Effect of ultraviolet light treatment on microbial contamination, some textural and organoleptic parameters of cultured sea bass fillets (*Dicentrarchus labrax*). *Innov. Food Sci. Emerg. Technol.* **2014**, *26*, 205–213. [CrossRef]
66. Matak, K.E.; Sumner, S.S.; Duncan, S.E.; Hovingh, E.; Worobo, R.W.; Hackney, C.R.; Pierson, M.D. Effects of ultraviolet irradiation on chemical and sensory properties of goat milk. *J. Dairy Sci.* **2007**, *90*, 3178–3186. [CrossRef]
67. Ghate, V.; Kumar, A.; Zhou, W.; Yuk, H.G. Irradiance and temperature influence the bactericidal effect of 460-nanometer light-emitting diodes on *Salmonella* in orange juice. *J. Food Prot.* **2016**, *79*, 553–560. [CrossRef]
68. Kim, M.J.; Bang, W.S.; Yuk, H.G. 405 ± 5 nm light emitting diode illumination causes photodynamic inactivation of *Salmonella* spp. on fresh-cut papaya without deterioration. *Food Microbiol.* **2017**, *62*, 124–132. [CrossRef]
69. Kim, M.J.; Tang, C.H.; Bang, W.S.; Yuk, H.G. Antibacterial effect of 405 ± 5 nm light emitting diode illumination against *Escherichia coli* O157:H7, *Listeria monocytogenes*, and *Salmonella* on the surface of fresh-cut mango and its influence on fruit quality. *Int. J. Food Microbiol.* **2017**, *244*, 82–89. [CrossRef]

MDPI
St. Alban-Anlage 66
4052 Basel
Switzerland
www.mdpi.com

*Foods* Editorial Office
E-mail: foods@mdpi.com
www.mdpi.com/journal/foods

www.ingramcontent.com/pod-product-compliance
Lightning Source LLC
LaVergne TN
LVHW070907170726
843515LV00005B/720

* 9 7 8 3 0 3 6 5 8 9 3 4 3 *